Intertidal Ecology

Intertidal Ecology

David Raffaelli

Senior Lecturer
Department of Zoology
University of Aberdeen
Aberdeenshire, UK

and

Stephen Hawkins

Professor of Environmental Biology and
Director, Centre for Environmental Sciences
University of Southampton
Southampton
Hampshire, UK

Formerly at Port Erin Marine Laboratory, University of Liverpool,
Isle of Man, UK

 CHAPMAN & HALL
London • Weinheim • New York • Tokyo • Melbourne • Madras

Published by
Chapman & Hall, 2–6 Boundary Row, London SE1 8HN

Chapman & Hall, 2–6 Boundary Row, London SE1 8HN, UK

Chapman & Hall GmbH, Pappelallee 3, 69469 Weinheim, Germany

Chapman & Hall USA, 115 Fifth Avenue, New York, NY 10003, USA

Chapman & Hall Japan, ITP-Japan, Kyowa Building, 3F, 2-2-1 Hirakawacho, Chiyoda-ku, Tokyo 102, Japan

Chapman & Hall Australia, 102 Dodds Street, South Melbourne, Victoria 3205, Australia

Chapman & Hall India, R. Seshadri, 32 Second Main Road, CIT East, Madras 600 035, India

First edition 1996
Reprinted 1997

Typeset in 10/12 pt Palatino by Saxon Graphics Ltd, Derby
Printed in Great Britain by the Alden Press, Osney Mead, Oxford

ISBN 0 412 29950 X (HB)   0 412 29960 7 (PB)

A catalogue record for this book is available from the British Library

Library of Congress Catalog Card Number: 96-83049

♾ Printed on permanent acid-free text paper, manufactured in accordance with ANSI/NISO Z39.48-1992 and ANSI/NISO Z39.48-1984 (Permanence of Paper).

Contents

Preface

The seashore has long been the subject of fascination and study – the Ancient Greek scholar Aristotle made observations and wrote about Mediterranean sea urchins. The considerable knowledge of what to eat and where it could be found has been passed down since prehistoric times by oral tradition in many societies – in Britain it is still unwise to eat shellfish in months without an 'r' in them.

Over the last three hundred years or so we have seen the formalization of science and this of course has touched intertidal ecology. Linnaeus classified specimens collected from the seashore and many common species (*Patella vulgata* L., *Mytilus edulis* L., *Littorina littorea* (L.)) bear his imprint because he formally described, named and catalogued them. Early natural historians described zonation patterns in the first part of the 19th century (Audouin and Milne-Edwards, 1832), and the Victorians became avid admirers and collectors of shore animals and plants with the advent of the new fashion of seaside holidays (Gosse, 1856; Kingsley, 1856). As science became professionalized towards the end of the century, marine biologists took advantage of low tides to gain easy access to marine life for taxonomic work and classical studies of functional morphology. The first serious studies of the ecology of the shore were made at this time (e.g. Walton, 1915) at the new marine biological stations established throughout the world, at first using a qualitative approach but increasingly using quantitative methods. The shore has always been a training ground for young scientists – both for taught courses and for the first steps in research. For example, Russell, who in the 1930s was instrumental in introducing an analytical approach in fisheries science, began his career working on limpets (Russell, 1907).

At the turn of the century, ecologists began to appreciate the tremendous opportunities the shore provided to test ideas developing in their very young science. Shore organisms were taken into the laboratory and experiments carried out (e.g. Herdman, 1890). Just a few years later, Baker (1909) summarized her culture experiments on intertidal brown algae in what Connell (1972) called a prophetic passage: 'On the whole it seems as though the greatest competition has been called into play in the lowest zones, the dry and uncongenial regions of the upper shore being left to the most tolerant forms, which, if left to themselves, are able to grow anywhere on the shore'. This neatly summarized the end product of another 70 years of often contradictory endeavour!

In the 1920s one of the first British textbooks on shore ecology appeared (Flatteley and Walton, 1926), with descriptions of zonation patterns and hints on how to survey shores. Several books aimed at the general reader (Wilson, 1937; Yonge, 1949), but also jolly useful for students, were produced. Seaside field courses were well established by this time and identification guides were beginning to be produced (for example, in the UK, Eales, 1950; Barratt and Yonge, 1958).

The period between the two world wars saw much research interest in shore ecology. Classic broadscale studies of rocky shores were undertaken in this period (e.g., Stephenson, 1936; Ricketts and Calvin, 1939; Stephenson and Stephenson, 1949). In Europe these set the scene for more detailed quantitative studies of zonation (Colman, 1933) and of individual species including distribution, population structure, reproductive cycles and sex change (Orton, 1929; Moore, 1934; Moore and Kitching, 1939). Similar studies were under way in the United States (Hewatt, 1935, 1937) and South Africa (Broekhuysen, 1941). The first experiments and detailed long-term descriptions were made in the 1930s by pioneering French workers (Fischer-Piette, 1932; Hatton, 1938) and South African workers (Bokenham, 1938). Fischer-Piette also undertook broadscale biogeographic surveys on both sides of La Manche (English Channel) and in France, Spain, Portugal and North Africa.

After World War II the experimental approach gathered steam but did not attract much attention (Jones, 1948; Lodge, 1948; Southward, 1956; Connell – but not published until the 1960s; Kitching and Ebling, 1967). Quantitative surveys of zonation were made (Southward and Orton, 1954) and biogeographic studies undertaken (Southward and Crisp, 1954; Crisp and Southward, 1958; Muus, 1967). Experimental studies of the ecophysiology and behaviour of shore animals were pursued (Knight-Jones, 1953; Crisp and Barnes, 1954; Barnes and Barnes, 1957) and the Stephensons inspired broadscale descriptions of zonation throughout the world (Dakin, 1953; Morton and Miller, 1968; Stephenson and Stephenson, 1972; also Dahl, 1952 on soft shores). Much of this work on rocky shores is summarized in the excellent reviews of

Southward (1958) and Lewis (1964), and that for the soft shores in
Eltringham (1971). Lewis (1964) and Stephenson and Stephenson (1972)
mark the zenith of the broadscale qualitative descriptive approach.
Lewis in particular also asks many questions in his book which were the
subject of study by his group over the next 30 years. Crisp, Southward
and Lewis all anticipated the current interest in 'supply-side ecology' –
but in a jargon-free way.

The 1960s saw renewed interest in ecology sparked by the work of
Connell (1961a,b) and Paine (1966, 1969). They and their kindred spirits
in North America were essentially ecologists using the shore to
experimentally test ecological ideas and theories (e.g. Paine and Vadas,
1969; Dayton, 1971, 1972; Menge, 1976; Peterson, 1977, 1979). This
approach spread rapidly to Australia (Underwood, 1975, 1976a,b, 1978),
New Zealand (Luckens, 1976), Germany (Reise, 1985) and South Africa
(Branch, 1976; Griffiths, 1981).

Both the authors (D.R. and S.J.H.) were research students in the 1970s
when interest in shore ecology was re-awakened in Britain (e.g. Lewis,
1976; Hartnoll and Wright, 1977; Hughes, 1980a). This renewed interest
was because shores were amenable to studies of energetics (Wright and
Hartnoll, 1981), of ecophysiology (Foster, 1971a), of animal behaviour
and behavioural ecology (Hughes, 1980b). There was also considerable
debate about the use of shore communities for monitoring pollution in
coastal waters (Lewis, 1976) and in measuring human impacts on shores
(Southward and Southward, 1978), especially estuaries (Knox, 1986). On
soft shores much work was prompted because of their importance as
nursery grounds for fish (McIntyre, 1970; McIntyre and Murison, 1973)
and feeding grounds for birds (Baird *et al.*, 1985). Low-shore and shallow
subtidal kelp beds were also being studied because of their considerable
contribution to nearshore productivity (e.g. Mann, 1973, Field *et al.*,
1980a).

The use of shores for developing and testing ecological theory has
proceeded apace in the 1980s and 1990s (Reise, 1985; Paine, 1994); their
potential as an ecological laboratory seems limitless. Emphasis remains
on understanding dynamic processes, with ecologists now addressing
interactions between physical oceanography, larval supply and
community dynamics (Gaines *et al.*, 1985), as well as relationships
between patterns and processes at different spatial and temporal scales
(Giller *et al.*, 1994). These are exciting developments.

Our book is an attempt at interpreting studies on both rocky and
depositing shores, in particular their usefulness in demonstrating and
exploring general ecological principles. However, we do feel that the
coastline and nearshore waters often need to be considered as a single

system. The low water is a convenient and arbitrary dividing line at which we have stopped (with occasional straying) to keep things manageable and within the experience of most students. Moreover, most students taking courses in biology do a field course involving shore work at some time. We hope that pressures on budgets do not further reduce this admirable tradition.

Chapter 1 outlines the major environmental gradients on the shore and Chapter 2 describes distribution patterns along these gradients of tidal height, wave action, sediment and salinity, as in estuaries. Salt marshes and mangroves are also briefly considered to widen the view. The rest of the book focuses mainly on rocky and depositing sandy and muddy shores. Chapter 3 discusses the causes of the distribution patterns along these gradients, emphasizing the importance of manipulative field experiments. Chapter 4 explores the factors involved in structuring shore communities, again emphasizing experimental approaches. Chapter 5 considers how shore organisms cope with the physical environment and respond to biological interaction, and Chapter 6 is concerned with how shores function as ecosystems. Chapter 7 then widens the scope of the book with an account of human impacts on the shore. Chapter 8 contains some advice on studying shores. We hope some of our readers will be inspired to put their wellies on, get out on the shore and do some experiments!

David Raffaelli and Stephen Hawkins

Acknowledgements

Many people have contributed to and maintained our interest in intertidal systems over the years, but we are especially grateful to early guidance from Roger Hughes, Richard Hartnoll, Harry Milne and George Russell. This interest has been sustained by our own research students and fellows, all of whom have played a vital role in expanding our horizons and keeping us on our toes. We thank them all. The atmosphere and friendships at Culterty and Port Erin Marine Laboratory (where a lot of the book was written) are much appreciated.

Many friends and colleagues read chapters and provided much helpful criticism and literature: Dan Baird, Pat Boaden, Debora Cha, Bruce Coull, John Crothers, Paul Dayton, Bob Elner, Teresa Fernandes, Steve Hall, Terry Holt, Roger Hughes, Stuart Jenkins, Sara Lawrence, Sarah Lawrie, Anton McLachlan, Alasdair McIntyre, Roger Mitchell, Geoff Moore, Derek Murison, Trevor Norton, Bob Paine, Keith Probert, Sarah Proud, Karsten Reise, John Taylor, Richard Thompson, Simon Thrush, Tony Underwood, Richard Warwick, Gray Williams and Bev Wilson. Thanks.

During the last frantic weeks of assembly, the efforts of Sara Lawrence, Debbie Jones, Mark Williams, Elspeth Jack and Sue Way saw the book home. Thanks are due to their patience, tolerance and rigorous checking. Bruce MacGregor produced most of the illustrations. Various editors at Chapman & Hall who initiated the project, nagged us along the way and eventually bullied a final manuscript deserve thanks, especially Clem Earle and Bob Carling. Chuck Hollingworth copy-edited the final manuscript, and Martin Tribe saw it through production.

Finally, we wish to thank Professor A.J. Southward for his help with the preface, and to acknowledge the inspiration he has provided over the last 45 years by his endeavours in both rocky and sediment shore ecology, amongst many other research interests.

The shore environment: major gradients

1

At first sight, wave-beaten cliffs and sheltered estuarine mudflats do not seem to have much in common. Indeed, the physical nature of these shores is so different that they have to be investigated using quite different techniques – thereby encouraging researchers to become specialist rocky shore or sandy beach or mudflat ecologists. Yet these habitats have common environmental and ecological features that enable us to group them together as 'the shore'. In this chapter we argue that rocky cliffs and estuarine mudflats are only extremes of a spectrum of habitat and community types. If we look hard enough, we can find the intermediates. To put it more formally, each of these shore types is a point within a field of continuously varying and intersecting environmental gradients. This view of shores is best appreciated by considering the nature of these environmental gradients. A large number of ecological gradients may be associated with the shore, but it is possible to understand the physical and biological properties of most shores by focusing on a few major ones and the interactions between them. Once this framework is established, it is then possible to consider how both small-scale (e.g. topographic) and large-scale (e.g. regional, geographic) factors modify shore patterns and processes.

1.1 FOUR MAJOR ENVIRONMENTAL GRADIENTS

1.1.1 GRADIENT 1: THE VERTICAL GRADIENT, FROM SEA TO LAND

Unidirectional stress gradient

Most shore plants and animals are closely related to fully marine species. For them, the environmental gradient from low water to the limit of the influence of the sea in the splash zone is one of increasingly harsh physical conditions. The sea itself is a relatively constant environment, with annual temperature fluctuations of less than 10 °C, and plants and animals are bathed in a benign nutrient broth of more or less constant concentration. At increasingly higher levels on the shore, marine organisms will be stressed by experiencing increasingly longer periods of time spent in air (**emersion**). Air temperatures are much more variable than those in the sea (changes of 10–20 °C are common over a 24 hour period); relative humidity can also vary considerably; precipitation and evaporation can affect the salinity of rock pools and water on the rock surface (see below). The most important stress, however, in temperate and tropical areas is greater desiccation due to the combined effects of heat and low relative humidity. In polar and boreal regions, extreme cold is a major problem at high shore levels: sea temperatures never get below about –0.2 °C, but intertidal air temperatures can plummet to –40 °C! The time available for respiration and feeding also decreases towards higher shore levels. Filter feeders, like barnacles and mussels, can only feed if surrounded by water since these, and most other intertidal animals, use gill-like structures to breath. Many mobile animals forage more efficiently in water than in air. Similarly, seaweeds take up their nutrients via their fronds or thalli and nutrient uptake can be limited if too much time is spent out of water. Whether it is desiccation or cold temperatures or lack of nutrients or food, the shore represents an essentially unidirectional vertical stress gradient for most marine organisms. This is often simply referred to as the **intertidal gradient** or **vertical gradient**. The only stress that may increase at lower shore levels is light availability for plants. This is unlikely on most shores although it may occur in areas with large tidal ranges and murky water as found in some large estuaries (e.g. the Mersey and Severn in the UK, which have tidal ranges of 8–10 m).

In the minds of many readers this gradient will be intimately associated with the rise and fall of the tides, but in many parts of the world tidal range is insignificant. Yet the biological response to this gradient between the land and the sea – **zonation** – can still be recognized on these shores (Chapter 2). It is important to realize that what has come to be called the 'intertidal' gradient is not created by the tides as such, but by the interface between air and water. In fact, the term 'sea–land gradient' would be more appropriate. The action of tides is to amplify this existing gradient by moving the water's edge up and down the land in a predictable and regular fashion, thereby increasing the amount of living space for shore organisms (Fig. 1.1). Species are ordered along this gradient according to their different abilities to cope with physical factors (e.g. desiccation) and variation in their responses to biological processes (e.g. competition and predation). This ordering is reflected in the familiar zonation patterns most obvious along the intertidal gradient of rocky shores. Zonation is a feature of all environmental gradients, not just marine ones, being seen in terrestrial ecosystems such as up mountain sides.

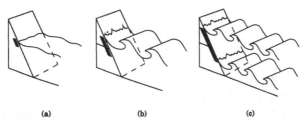

(a) (b) (c)

Fig. 1.1 (a) The gradient of wetness (solid black) set up by the interface between air and water, is (b) extended upwards by wave action, and (c) may be greatly amplified by tidal rise and fall.

A few intertidal organisms, such as lichens, flowering plants, trees and some arthropods, such as insects, centipedes and pseudoscorpions, are essentially terrestrial groups that can partly tolerate exposure to the marine environment and are restricted mainly to the upper shore. These species respond not to an emersion gradient, but to a gradient of increased immersion and salt inundation that runs in the opposite direction to the stress gradient described above for marine organisms. The best examples which are restricted to higher shore levels include saltmarshes and mangroves (Fig. 1.2).

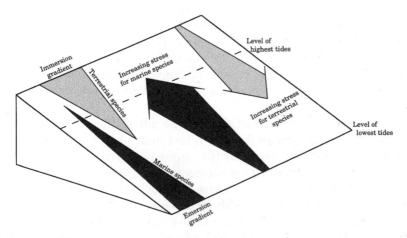

Fig. 1.2 The number of species of terrestrial organisms (light shading) which decrease with stress due to immersion, and marine species (dark shading), which decrease due to stress caused by emersion.

The biological features that develop in response to the vertical gradient add their own variation. For example, habitat structure and complexity, biological productivity, species diversity and the intensity of biological interactions all increase towards the lower shore. The presence of dense stands of seaweeds, beds of mussels and tunicates, or reefs built by tube worms, all modify the shore environment. In many cases the modification is positive, preventing desiccation or providing shelter from water movement. Shading by canopies, however, will reduce the amount of light available to young plants of the same species and to understorey species. Seaweed turfs also trap significant quantities of silt which can smother other plants and animals, but can also provide a sedimentary environment for essentially burrowing species (Fig. 1.3 summarizes the factors changing on the vertical stress gradient).

Tides

Although tides are not the primary cause of zonation, they are the dominant physical feature of many shores. Tides are produced by a combination of forces caused by the spin of the Earth and gravitational forces between the Earth and the Moon and the Sun. The centrifugal force of the Earth has the same magnitude and direction at all points on the Earth's surface (Fig. 1.4(a)), but the gravitational force exerted by the Moon varies in magnitude and direction as the Moon changes its position relative to that of the Earth (Fig. 1.4(b)). The resultant of these centrifugal and gravitational forces is the tide-producing force, which causes two 'bulges' in the oceans in line with the Moon. In the simplest situation, these two bulges encircle the Earth once in every 24.5 hours (not 24 because the Moon also moves around the Earth as the latter rotates). Thus, at least in theory, shores should experience two high and two low tides during this period as they move through this standing wave.

To this Earth–Moon interaction must be added the pull of the Sun. The tide-producing force of the Sun is just under half that of the Moon because it is about 360 times further from the Earth. The interaction between solar and lunar tides is shown in Fig. 1.4(c). At full moon and new moon both Sun and Moon cause a bulge in the ocean in the same direction and at these times the shore experiences the highest or spring tides. Conversely, when the Moon is in its first and third quarters the bulges produced by the Sun and Moon do not coincide and the difference in height between high and low waters is small. These are called neap tides (Fig. 1.4 (c)).

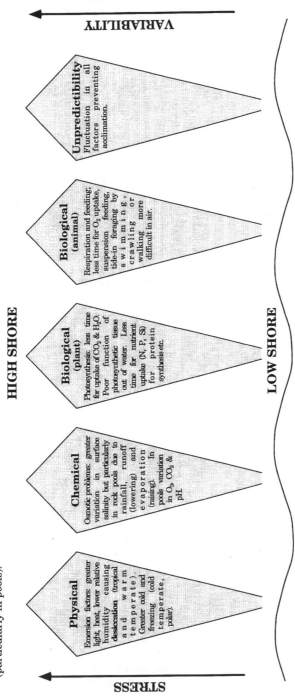

LAND

Highly variable terrestrial environment: variation in heat and light leading to extremes of hot and cold; winds interacting with sunlight to create low relative humidity and hence dessication stress; variable salinity due to rainfall and evaporation (particularly in pools).

HIGH SHORE

VARIABILITY

STRESS

Physical
Emersion factors: greater light, heat, lower relative humidity causing desiccation (tropical and warm temperate). Greater cold and freezing (cold temperate, polar).

Chemical
Osmotic problems: greater variation in surface salinity but particularly in rock pools due to rainfall, runoff (lowering) and evaporation (raising). In pools variation in O_2, CO_2 & pH

Biological (plant)
Photosynthesis: less time for uptake of CO_2 & H_2O. Poor function of photosynthetic tissue out of water. Less time for nutrient uptake (N, P, Si) for protein synthesis etc.

Biological (animal)
Respiration and feeding; less time for O_2 uptake, suspension feeding, tide-in foraging by swimming, crawling or walking more difficult in air.

Unpredictibility
Fluctuation in all factors preventing acclimation.

LOW SHORE

SEA

Highly stable marine environment: high and constant salinity; little temperature and pH change; plentiful raw materials for photosynthesis (CO_2, H_2O), nutrients (N, P, Si), food for suspension feeders; supportive medium for swimming or crawling.

Fig. 1.3 Stress on the vertical gradient associated with tidal height. The arrows point to greater stress.

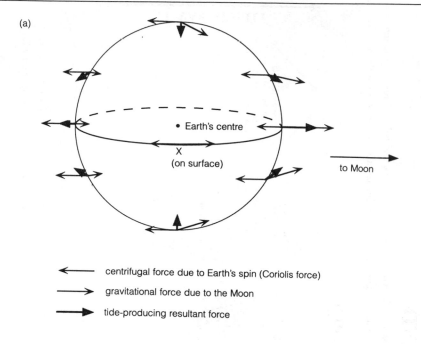

(a)

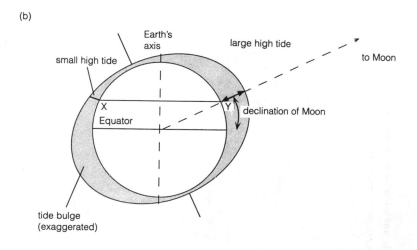

(b)

Fig. 1.4 (a) Tide-producing forces generated by the Earth and the Moon. (b) The standing wave created by the gravitational pull of the Moon as it orbits the Earth.

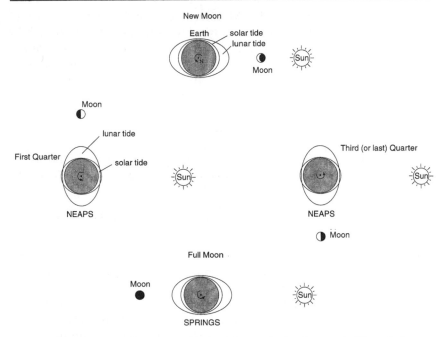

Fig. 1.4 (c) Interaction of solar and lunar tide-producing forces at different phases of the Moon. (Adapted with permission from OUCT, 1989.)

This very simple picture is complicated by various factors. The presence of land masses can obstruct the circumnavigation of the globe by this bulge of water. The shapes of the ocean basins constrain the direction of tidal flow. The **Coriolis force**, generated by the spinning of the Earth, deflects water movement to the right in the Northern Hemisphere and to the left in the Southern Hemisphere. Finally, the rapid rotation of the Earth causes a time lag in the generation of the bulge and prevents the standing wave reaching any kind of equilibrium. These factors largely account for the tremendous variation in the behaviour of the tides in different parts of the world and their interactions can be such that some locations have zero tidal range. These locations are known as **amphidromic points**, around which the crest of the high-water tidal wave rotates. The further a shore is from one of these points, the greater will be the tidal rise and fall. On a more local scale, the configuration of the coastline can affect tidal range. In areas with shallow continental shelves, and on coastlines with bays and indentations, the tidal range may be amplified. In rapidly narrowing estuaries, like the Severn in the UK, this can produce a tidal-wave or bore.

In some places such as the Mediterranean and the west coast of Sweden, tidal range is very small, only a few centimetres at most; whilst in the Bay of Fundy, Canada, and the Bristol Channel, UK, the range is more than 10 metres. Figure 1.5 illustrates several variants on the basic pattern of two similar tides per day. In the British example, there are two high tides per day of more or less equal height, whereas in the Philippines there is sometimes only one high tide per day and successive high waters are of very different height. At Do-San in Vietnam, there is always only one high and one low tide per day (Fig. 1.5(d)).

A much fuller and more formal treatment of tides can be found in the very readable UK Open University text on waves and tides (OUCT, 1989).

1.1.2 GRADIENT 2: THE HORIZONTAL GRADIENT OF EXPOSURE TO WAVE ACTION

Exposure means how much wave action a shore experiences. The term is potentially confusing, because it could also imply that the shore is uncovered by the tide, that is, exposed to the air. Intertidal ecologists rarely use exposure in this sense, using instead the term **emersed**. Whereas the intertidal gradient runs vertically from the lower to the upper shore, wave exposure is a horizontal gradient running from sheltered bays out to exposed headlands. This gradient of water movement is determined by the aspect a shore presents to the prevailing winds, which generate waves, coupled with the **fetch** or distance over which those winds blow. The stronger the wind, the larger the waves. If fetch is sufficiently large, shores will experience heavy wave action. Thus shores on remote oceanic islands or facing prevailing oceanic winds are usually pounded by breakers, because the wind can generate waves over many hundreds of kilometres. Even on windless days such shores can experience large waves in the form of swell arriving as waves from some distant storm. In contrast, in enclosed seas, such as the Baltic or the Irish Sea, waves will rapidly dampen down after the wind responsible stops blowing.

All waves can be characterized using a few simple parameters. **Wave length** is the horizontal distance between successive crests. **Wave period** is the time between successive crests. **Wave height** is the vertical distance between the crests and the troughs. Finally, **wave amplitude** is defined as half wave height (Fig. 1.6(a)). The waves that mostly concern intertidal ecologists are known as **gravity waves** and have periods ranging from seconds to minutes. These waves are generated by wind blowing over the sea surface, the friction between the air and the sea stretching the sea surface. As the surface is restored, partly by the surface springing back through surface tension and partly by the collapse of the crest through gravity, undulations or waves are created.

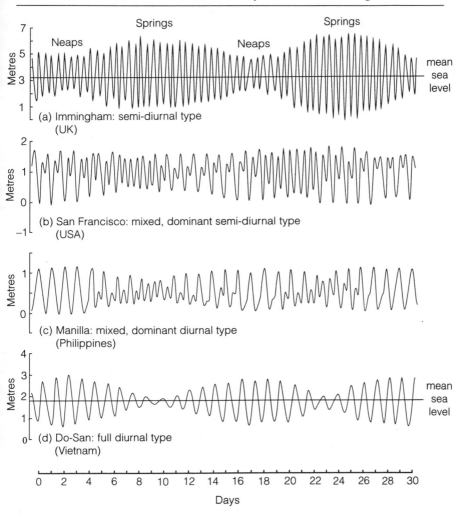

Fig. 1.5 Tidal curves from different parts of the world (adapted with permission from OUCT, 1989). (a) Semi-diurnal (two tides per day); (b) mixed, dominant semi-diurnal (some double high waters once a day, but dominated by two tides per day); (c) mixed, dominant diurnal type (single tide per day but some largely asymmetrical double tides per day); (d) full diurnal type (only one tide per day).

It is important to realize that the transfer of energy in waves involves no transfer of water – only the wave form. Thus the water that crashes onto the shore has not been blown from further offshore by the wind. The movement of water particles themselves is restricted to rotating orbits (Fig. 1.6(b)) which at the surface are almost circular but which become squashed and elliptical near the seabed. As waves approach the shore, and the distance between the seabed and the sea surface decreases rapidly, the orbits become so distorted that they fail to close. At this point the entire mass of water within the wave moves forward and dissipates its energy as a breaker onto the shore.

Waves often approach shorelines obliquely: that is, the wave crest may not be parallel to the shore. When this happens, the water depth at the inshore end of the wave crest will be much less than that at the offshore end. The speed at which inshore waves move depends on water depth, so that wave speeds will differ along the crest, leading to wave refraction and the concentration of wave energy at particular points along the shore, usually on headlands (Fig. 1.6(c)).

Wave action has a profound effect on the biological characteristics of a shore. In the most exposed situations, such as steep cliffs facing huge breakers, only those species able to maintain a firm hold on rocky surfaces are found (Chapter 5). The most sheltered shores are where fetch is insignificant, as in sealochs, fjords, rias and estuaries, or where offshore topographic features, such as barriers and reefs, dissipate wave action. Water movement may be so gentle that grasses and trees can root in the sediment. In extreme shelter, seaweeds may be unattached, merely rising and falling vertically with the tide. Strong tidal currents will have similar effects to wave action. Wave action is difficult to measure quantitatively (Chapter 8), but extreme values of measurements of water movement are impressive (Riedl, 1971): maximum current speeds in shallow areas of 5 ms^{-1}, particle speeds in rocky surf of 15 ms^{-1}, wave depths down to 500 m, wave lengths of 45 m and surf pressures of 100 tm^{-2}.

In contrast to the vertical gradient, which is a unidirectional stress gradient (Fig. 1.3), stress along the exposure gradient is more difficult to define (Fig. 1.7). Exposed conditions are good for suspension feeders like mussels and goose-barnacles, or sessile predators such as sea anemones, because water movement brings plenty of food. For other species, attachment to the shore or foraging are difficult in exposure and sheltered conditions seem more favourable. Examples would include seaweeds with air flotation devices and predators like crabs. On the other hand, for species like limpets, siltation can be a problem in sheltered conditions. Slow or non-turbulent water movement might also restrict the supply of oxygen for both plants and animals and dissolved nutrients for seaweeds. Some species may just not reach sheltered areas because of problems of larval supply. Clearly, the assumption often made that conditions are more stressful on exposed shores need not always apply – for species living high on the shore in the splash zone, nothing could be worse than sheltered conditions, particularly in the tropics or at high latitudes where organisms can be exposed to extreme heat or extreme cold respectively.

(a)

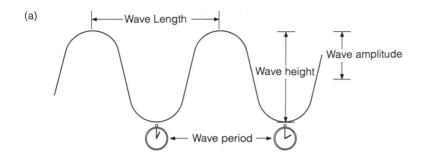

(b)

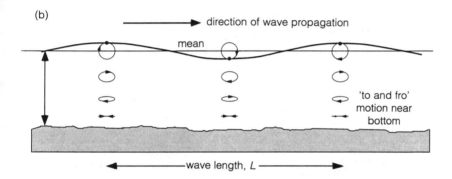

(c)

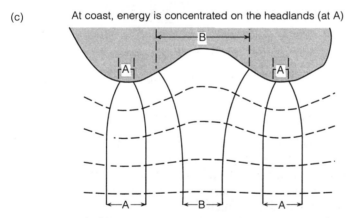

Fig. 1.6 (a) Terms used to characterize waves; (b) particle motion in waves; (c) wave refraction on an indented coastline (adapted with permission from MAFF, 1993. *Coastal Defence and the Environment*. Ministry of Agriculture, Fisheries and Food. © Crown Copyright).

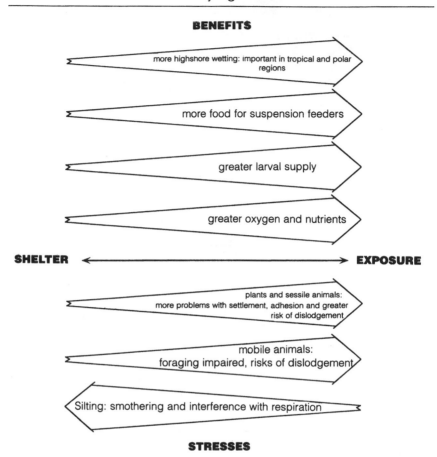

Fig. 1.7 Beneficial and stressful changes on the horizontal wave action gradient.

1.1.3 GRADIENT 3: THE PARTICLE SIZE GRADIENT

Shores, including rocky shores, differ in their average particle size. Thus, particle size may only be a few micrometres on mudflats, several hundred micrometres on sandy beaches, a few centimetres in gravel and shingle habitats, tens of centimetres on boulder shores and tens of metres for massive boulders, which are functionally no different from cliffs (Fig. 1.8). The average size of the particles constituting the shore reflects in part its exposure to water movement due to wave action and currents, and in part the geological history of the area which determines the availability of sediment sizes. Thus fine silt only accumulates under the most sheltered conditions, but cliff-type habitats will occur at both ends of the exposure gradient. The interaction between particle size and exposure is discussed below.

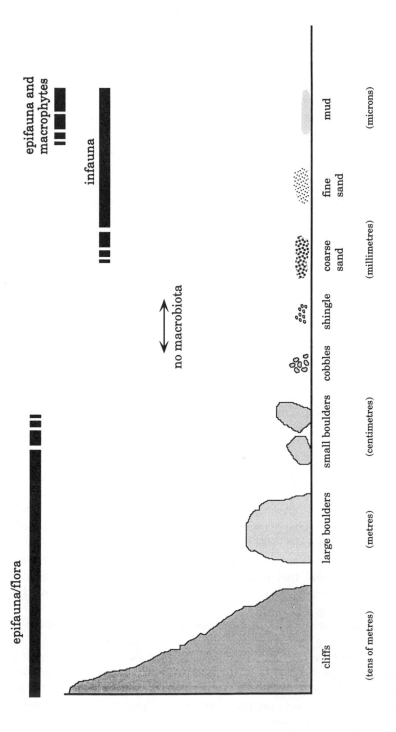

Fig. 1.8 The particle size gradient. Surface-dwelling species are present at both ends of this gradient, whilst infauna are restricted to smaller particle sizes. Macrobiota are absent from the middle section of the gradient.

1.1.4 GRADIENT 4: THE MARINE–FRESHWATER GRADIENT OF SALINITY

The salinity of oceanic water is about 35‰, but in coastal areas this is usually reduced slightly by freshwater run-off. Where the freshwater influence is significant, a salinity gradient is generated running from marine to completely freshwater conditions. This is most often seen in estuaries but the effect can occur on a massive scale, as in the Baltic Sea. Only those marine species that can tolerate lower and often fluctuating salinities can exploit the less saline section of this gradient. Species richness is low, but populations attain very high densities and biomasses. Few freshwater species are capable of occupying the more saline sections of the gradient (Chapter 2), and they penetrate only to a limited extent into the upper, slightly brackish section of estuaries.

Fluctuations in salinity also occur all along the vertical gradient: wherever pools of water are trapped in rock depressions and hollows, where there is a film of water covering organisms attached to rocky shores, or in the interstitial water trapped by finer sediments. Throughout most of the shore these fluctuations are cyclical because they are driven by the tides and hence they are more or less predictable; but the higher the shore level, the less predictable these changes become. For instance, as the tide ebbs, a midshore rock pool will experience increasing salinity as water evaporates or decreasing salinity if rainfall is sufficiently heavy. When the tide returns the pool will be flushed out with sea water. The higher the pool, the more pronounced the fluctuations. Salinities can vary dramatically in pools at the top of the shore which are not reached by every tide (see later in this chapter and Fig. 1.18).

On coarse sandy beaches, drainage of water is sufficiently rapid that there is usually little opportunity for standing water to evaporate or be diluted by rainwater. On mudflats there is often a great deal of standing water such that heavy rainfall or evaporation at low tide might considerably change its salinity. However, these shores are most often found in estuaries where the plants and animals are well able to cope with the stress that this imposes. Furthermore, water trapped in the sediment also acts to damp down salinity changes in any overlying water column (Fig. 1.9).

In a typical estuary the fresh water flowing down a river meets the sea. The resulting salinity gradient from this riverine input has both horizontal and vertical components. Where river flow is great and completely dominates the circulation system, the denser sea water intrudes up the estuary as a wedge-shaped bottom layer with fresh water flowing outwards on the surface. This results in water which is composed of different vertical layers of salinity – in other words **salinity**

stratification (Fig. 1.10(a)). Where tidal flow is great relative to the freshwater discharge, the water column becomes well mixed through current-induced turbulent mixing and the vertical salinity gradient may disappear altogether (Fig. 1.10(c)). Between these two extremes, a range of partially mixed conditions are found (Fig. 1.10(b)). The location of a particular estuary along this continuum will of course change according to rainfall, sea-state and tidal cycle. Most estuaries can, however, be broadly categorized for much of their time as a particular type such as salt-wedge or partially-mixed. These circulation processes of salt and fresh water within a large estuary are also affected by the spinning of the Earth – the Coriolis phenomenon. In large estuaries the inflowing tidal water and the outflowing river water are swung to the right in the Northern and to the left in the Southern Hemisphere. Consequently, in the Northern Hemisphere, sea water flows up-estuary on the left-hand side (when looking downstream).

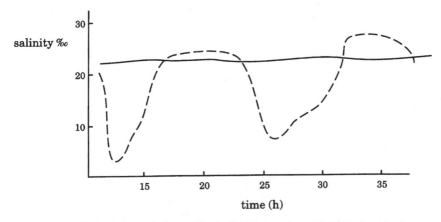

Fig. 1.9 Salinity fluctuations in interstitial (solid line) and overlying (dashed line) water at one station in the Pocasset estuary, USA (after Barnes, 1980 with permission).

Unlike the normal estuarine situation, salinity gradients in lagoons tend to be stable and do not fluctuate with the tidal ebb and flow. Because of their enclosed nature, salinity often varies seasonally due to evaporation during hot, dry periods and heavy rainfall at other times. The salinity of tropical lagoons may vary from virtually zero to five times normal seawater concentration, and in the Caimanero lagoon, Mexico, evaporation has been shown to increase salinity by 2‰ per day up to 300‰ (Schramm, 1991). The fauna and flora of lagoons have to cope therefore with dramatic fluctuations in salinity, often from almost fresh water to over 60‰. Several species are more characteristic of these lagoonal habitats than of estuarine environments (Barnes, 1980).

a) salt-wedge

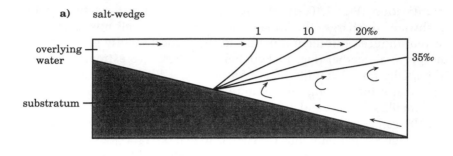

b) partially mixed

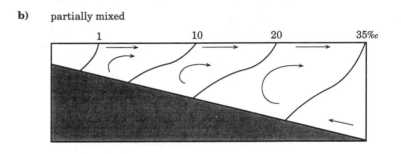

c) well-mixed

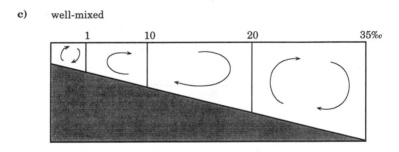

Fig. 1.10 Section through different types of estuary showing lines (isohalines) of different salinity. (a) Salt wedge; (b) partially mixed; (c) well mixed.

1.2 INTERACTIONS BETWEEN GRADIENTS

Species of animals and plants are distributed or zoned along each of the four major gradients described above according to their abilities to cope with the changes in physical and biological factors associated with each gradient. The assemblage of species that we find on a shore is therefore largely a result of the way these gradients combine or interact (as indicated in Fig. 1.11). Chance events may also be responsible for some of the biological features of a shore, as discussed below. Nonetheless, it is possible to understand much of what we see on the shore in terms of the major gradients described above or their interactions.

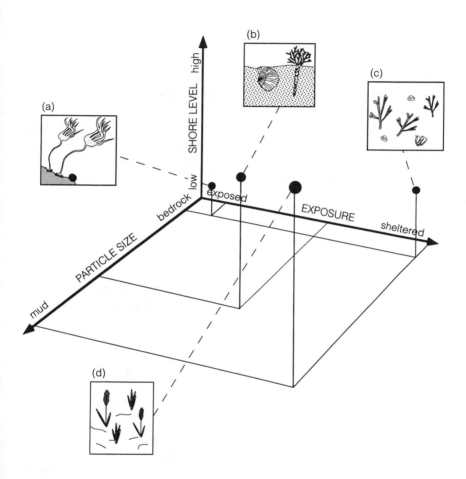

Fig. 1.11 The continuous nature of the major environmental gradients on an open-coast, fully marine shore. Salinity gradients have not been shown for the sake of clarity. Shore assemblages are points within this multidimensional space. (a) Kelps occur at low shore levels on exposed bedrock cliffs; (b) cockles and tube worms are found at mid-tide levels on a moderately sheltered muddy sand beach; (c) fucoids, algal turfs and occasional limpets occur low on a sheltered bedrock shore; (d) pioneer salt marsh grasses occur towards higher shore levels on fine-particle, sheltered shores. The examples given are typical of British shores.

1.2.1 INTERACTION 1: PARTICLE SIZE AND EXPOSURE

The interaction between wave exposure and particle size is possibly the most important process shaping the biological characteristics of the shore. For the very-large-particle habitats (rocky shores), organisms are sufficiently small, relative to the size of particle, that they can maintain themselves on the rock surface. Rocky shore organisms occur throughout the exposure gradient (Fig. 1.11). The actual assemblages found on rocks or boulders will vary (Chapter 2) because different species can cope with different degrees of wave action (Chapter 5) and because intensity of biological interaction will vary along this gradient (Chapters 3, 4).

At the other end of the particle size gradient, on muddy and sandy shores, most species are larger than the average particle size. They can live below the surface by pushing the particles aside (burrowing) or by remaining more or less permanently within the sediment in a tube. These are called the **infauna**. Only in extreme shelter, where water movement is too weak to significantly disturb the sediment, do large surface-living fauna and flora occur (Fig. 1.12). Some metazoans (multicellular organisms) are small enough to inhabit the spaces or interstices between small particles and these are termed the **interstitial meiofauna**.

Shores of small stones and shingle are found midway between the two extreme habitats of exposed rocky shores and sheltered mudflats (Fig. 1.8). Shingle and gravel beaches are common at high latitudes where the stones that originate in relict glacial deposits are smooth and of similar size. Very coarse particle beaches are also derived from erosion of cliffs and here the particles are more mixed and angular in shape. These coarse particle beaches are usually found in conditions where they are continually disturbed by plunging breakers. These locations are often exposed to moderate wave action, so that the beach is physically unstable and the moving particles crush and grind against one another. The constant grinding never allows successful colonization by organisms to last very long so that the surfaces of the particles are usually devoid of any life other than microscopic species. On the ebbing tide, water drains freely because the gaps between the stones are large and the capillary forces which hold onto the water are weak. This is particularly true for shingle, gravel and very coarse sand beaches. These habitats are therefore inhospitable to marine life and are intertidal deserts.

The relationships between factors such as beach slope, particle size and wave action are complex and are used by coastal geomorphologists to classify beaches into various **morphodynamic states**. Put simply, a continuum of these states ranges from **reflective** to **dissipative** (defined in Table 1.1). When a breaking wave surges up the beach, the **swash** carries sediment upshore and some of it is carried back downshore in the **backwash**. The balance between these two processes depends on the particle size and will in turn determine the steepness of the beach (Fig. 1.13). On coarse particle beaches, the water in the retreating wave will quickly percolate down through the sediment and little material will be carried back downshore in the backwash. The net effect of this is to create a shore with a steep profile. Where particle size is smaller, the beach is more waterlogged so that the backwash component will be correspondingly greater and the shore profile flatter.

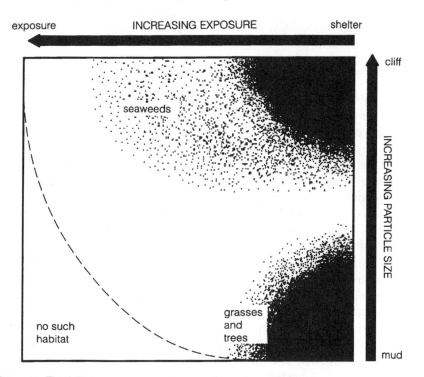

Fig. 1.12 The interaction between exposure and particle size on the distribution of shore macrophytes.

It should be noted that the terms 'reflective' and 'dissipative' tend to be used mainly by coastal geomorphologists, not biologists, and they are not interchangeable with 'exposed' and 'sheltered'. Being able to classify a beach as belonging to one of the six recognized morphodynamic states (described in Carter, 1988) permits more rigorous ecological comparisons between regions. Subjective terms, such as exposed or sheltered, conjure up quite different impressions to different shore ecologists – an exposed European beach might be regarded as moderately sheltered by South African and Australian workers (see also Chapter 8).

Table 1.1 Characteristics of beaches of differing morphodynamic state. Intermediate states have intermediate characteristics*

Reflective	Dissipative
Coarse sand	Fine sand
Small waves	Large waves
Steep slope	Shallow slope
Small tidal range	High tidal range
Short wave period	Long wave period
Harsh swash conditions	Benign swash conditions
Impoverished fauna	Rich fauna

* A. McLachlan (pers. comm.)

The interactions between exposure, shore profile and particle size have profound effects on living conditions within the sediment. The finer the particles making up the beach and the shallower the shore profile, the more water will be retained at low tide because of slower run-off and increased capillarity. Whilst this will clearly affect the amount of desiccation experienced by the beach fauna, there are also consequences for other, perhaps more important, physico-chemical characteristics of the beach (Figs. 1.14 and 1.15). Micro-organisms, especially bacteria, are small enough to attach to the surfaces of individual sand grains, although bacteria tend not to be present on particles smaller than 10 μm (Watling, 1991). A small particle will have a much larger surface area relative to its volume than a larger particle, so that a given volume of mud will provide a much greater total surface for bacterial attachment than an equivalent volume of sediment from a sandy beach (Fig. 1.15(a,b)). The relationship between surface area and volume of particles is a simple exponential one and shores that differ by only a small amount in their average particle sizes may have enormous differences in their bacterial biomasses (Fig. 1.15(c)). Muds, in particular, support very high numbers of bacteria. For instance, one gram of sediment of particle size 10–50 μm has a total surface area of 3–8 m² and may contain 10^9 bacteria (Watling, 1991). Of course, the supply of organic material to a beach will also affect its bacterial biomass (Fig. 1.15(d)) (Chapter 7). It must also be remembered that organic material often binds sediment together in a matrix body to form larger particles.

Bacteria and other micro-organisms provide a food resource for deposit feeders but they also have a major influence on sediment chemistry. Bacteria decompose organic material imported into the beach by the flooding tide, ranging in size from an entire kelp plant to pieces of material only a few microns (micrometres) across. The micro-organisms and this organic material are together termed **detritus** (Chapter 6). The trophic activities of the bacteria create a demand for oxygen in the beach, particularly in the interstitial water. As the sediment becomes progressively less oxygenated, anaerobic bacteria predominate. Forms like *Desulphovibrio* use the rich supply of sulphates found in marine sediments as a chemical substrate, reducing them to sulphides and releasing the toxic gas hydrogen sulphide. Iron is one of the more abundant elements in marine sediments and this is reduced under anaerobic conditions to black ferrous sulphide, or, in aerobic sediments, is oxidized to orange ferric oxide (Fig. 1.14).

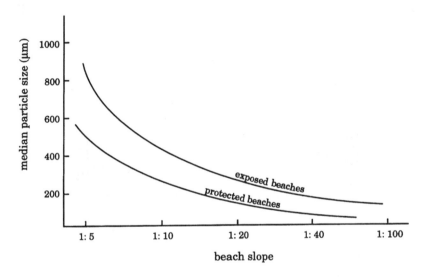

Fig. 1.13 The relationship between shore profile (slope) and particle size (median size) on exposed and sheltered beaches (after Brown and McLachlan, 1990, with permission).

Exposed, well-drained, coarse sandy beaches, with their reduced organic matter and hence limited bacterial activity, are therefore well oxygenated and have an orange-golden appearance. Conversely, the high organic loading and large bacterial biomass of sheltered and waterlogged mudflats is responsible for their dark, almost black appearance as well as a strong smell of hydrogen sulphide. Anaerobic chemistry predominates in the deeper sediment layers because of the restricted oxygen supply from the surface waters, so that the switch from aerobic to anaerobic conditions occurs near to the sediment surface on mudflats but may be many tens of centimetres, even metres, deep on sandy shores (Fig. 1.14).

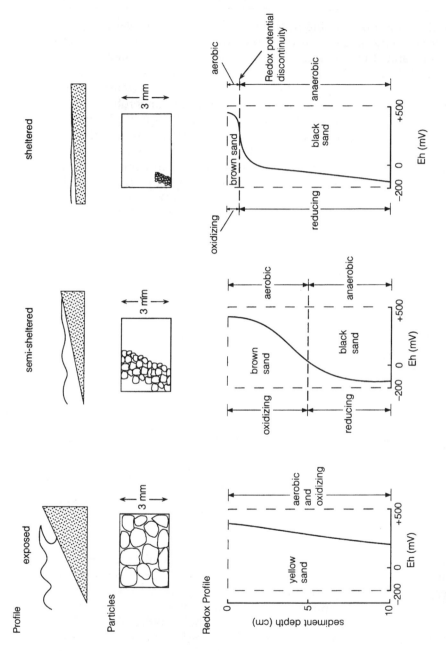

Fig. 1.14 Effects of exposure and particle size on sediment chemistry. Top, section through beach showing profile; middle, impression of size and arrangement of particles in beach; lower, redox profile through sediment column.

This change with depth from aerobic to anaerobic conditions is reflected in the **sediment redox potential (Eh)**, which can be easily measured using a platinum electrode (Pearson and Stanley, 1979). The technique can provide a rough and ready way to characterize the sediment environment within and between shores. But positive redox values occurring above the discontinuity layer do not necessarily reflect an environment rich in free oxygen, merely the presence of oxidized compounds (Parks and Buckingham, 1986). Free oxygen only occurs in the top millimetre or so of sheltered mudflats, but can extend down to a metre or more on the most exposed and coarse sandy beaches (Fig. 1.15).

The interactions between exposure and particle size, and their consequences for sediment chemistry, have important effects on the distribution and abundance of sediment organisms. Species will occupy different sections of the exposure/particle size gradients according to their relative abilities to cope with low oxygen conditions, on the one hand, and on the other to maintain themselves in a turbulent environment (see also Chapter 5). On exposed sandy beaches the interactions between physical processes may be even more complex. For instance, on flat (dissipative) sandy beaches with offshore surf zones of different widths, the frequency of swash waves will differ considerably. On these beaches the fauna utilize the swash to move around the beach and for suspension feeding, but cannot effectively do so with short swash periods of less than about 5 to 10 seconds (McLachlan, 1990). The longer the period between swashes, the better the conditions for the fauna on these beaches, and McLachlan *et al.* (1993) have suggested that the **swash climate** directly controls the composition and structure of these sandy beach communities.

1.2.2 INTERACTION 2: THE VERTICAL INTERTIDAL GRADIENT AND EXPOSURE TO WAVE ACTION

The extent of the vertical gradient – how much the influence of the marine environment is spread up the shore – is also determined by exposure. Where wave action is strong, there is much wave splash and spray. Parts of the shore, theoretically well above the predicted tidal level of extreme high water springs, will in fact be quite wet and will accommodate species found at a lower tidal level on more sheltered shores. In other words, the zones are uplifted on exposed shores (Fig. 1.16) and it is not uncommon to find high-shore species many tens of metres above the theoretical tidal limit on very exposed cliffs.

This raises an interesting question: should we in fact use the term 'intertidal' for describing the shore? Not only are shore species found above the highest tidal levels, but many extend well below the level of the lowest low tide. The term **littoral** is therefore probably more appropriate, but intertidal is in such common usage that it would now be difficult to replace.

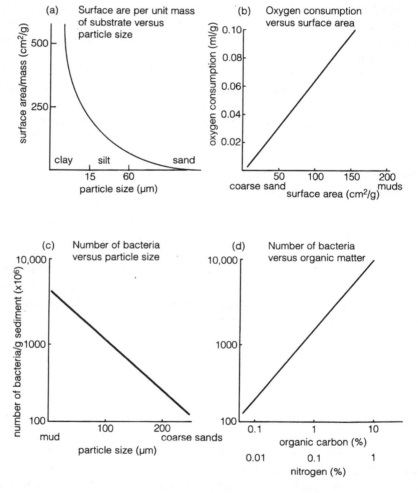

Fig. 1.15 Factors affecting abundance of bacteria in sediments composed of different particle sizes. (a) Beaches with a smaller average particle size have a larger surface area per unit mass. (b) Microbial activity in terms of oxygen consumption, and (c) number of bacteria increase with smaller particle size. (d) Microbial biomass also increases with higher organic carbon and nitrogen, which occur in finer sediments.

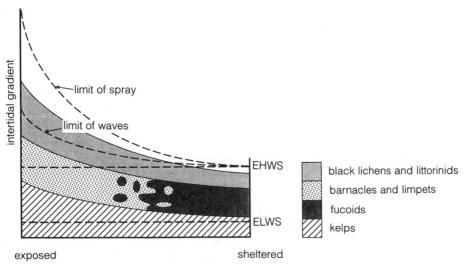

Fig. 1.16 The interaction between exposure and intertidal gradients on British rocky shores. Note the uplift of zones towards exposure and the shift in balance of plant and animal domination. ELWS, extreme low water of spring tides; EHWS, extreme high water of spring tides. (Modified with permission from Lewis, 1964.)

The effects of exposure on the vertical extent of the shore are more complex on beaches and mudflats. Here, the overall physical structure of the shore is affected by wave action (Figs. 1.12 and 1.13) and it is probably not sensible to talk about an uplift of zones on such shores. What is clear is that physical and biological conditions change more rapidly with shore level on exposed sandy beaches than on sheltered tidal flats.

Within a shore, and where the tidal amplitude is significant, the time for which organisms are subjected to wave action will vary along the vertical intertidal gradient. On average, lower parts of the shore experience more wave action than the highest, just because they are covered by water for longer periods. Similarly, where alongshore tidal currents are strong, the lower parts of the shore may experience quite high energetic conditions at low tide. On sediment beaches this often leads to coarser particles dominating towards the bottom of the shore. The shore profile itself can also have a marked effect on the amount of energy that waves can dissipate on the shore; a gentle slope at lower shore levels will reduce much of the energy in a wave breaking higher up.

1.2.3 INTERACTION 3: SALINITY AND PARTICLE SIZE

Where fresh water and sea water meet, as in estuaries, there will be marked effects on the particle size distribution of sediments. The sediment properties of the outer reaches and the more open parts of estuaries are determined by the same hydrodynamic processes that operate on open coasts. Thus, regions of high wave energy or strong currents tend to have coarse particles. Away from the mouth of the estuary, tidal currents and non-tidal estuarine circulation have the greatest effect on sediment distributions. Tidal currents on the flood tide transport sand up-estuary, and as current strength declines, part of the load of suspended material is deposited to form shoals in the lower reaches. Large particles entering the estuary in fresh water are deposited at the tip of the saline intrusion. Thus, theoretically, sediments should become finer away from the mouth to the upper reaches. However, because of the complex behaviour of tidal currents, the presence of separate ebb and flood channels and the irregular shoreline of estuaries, such a clear pattern is almost never seen.

The distribution of the fine particles that characterize estuarine mudflats is due in large part to the phenomenon of **tidal asymmetry**. This occurs where the slack water period at high tide lasts considerably longer than the slack water period at low tide. As a consequence there is a longer time available for particles to settle at higher shore levels and these levels accumulate finer sediments. There is also a time lag between the tidal flood (and ebb) rhythm and the time of maximum concentration of suspended material in the water. Suspended particles will settle out of the water column once water velocities become sufficiently low for them to do so. Because very fine particles take so long to settle, they will be deposited on the flood tide much further landward than the point at which their critical current settling velocity was reached. On the ebb tide there is a corresponding time lag for scouring and resuspending particles from the sediment. The net result of these processes is that fine particles tend to accumulate at higher shore levels. Detailed accounts of these processes can be found in Kennish (1986).

Sediment distributions in estuaries are also determined by internal circulation processes. These differ in salt-wedge, partially mixed and well-mixed estuaries and generate different sediment distributions. Where river flow dominates, as in a salt-wedge estuary, most of the suspended material is transported by the fresh water to the coast to form a zone of active sedimentation. In contrast, a partially mixed estuary with strong tidal forces is characterized by re-suspended sediments and a high turbidity in the middle reaches. When tiny particles of clay, silt and organic material carried down in fresh water meet higher-salinity water, **flocculation** of the material occurs. Flocculation is caused by molecular-scale attraction (van der Waals forces) which, although weak, becomes important when particles are brought close together. Flocculation does not occur in the freshwater inflow because all the clay particles carry a net negative charge and repel each other. When these particles come into contact with free positive ions in sea water, the negative charges are sufficiently neutralized to allow the van der Waals forces to operate and flocculation takes place. The particles become heavier and start to sink through the water column, travelling seaward all the while until they meet marine sediment travelling upstream at the tip of the salt intrusion. Here, vigorous mixing concentrates particles to form a **turbidity maximum** and the middle and upper reaches become a sediment trap, commonly concentrating particles in the size range 5–10 μm. It is these particles which, in combination with tidal asymmetry, form extensive areas of mudflats and tidal marshes.

In a large, well-mixed estuary, the effects of the Coriolis force also come into play. In the Northern Hemisphere, this force generates a net seaward flow of low-salinity water on the right-hand side of the estuary (when looking downstream) and a net landward flow of higher-salinity water on the left. Fine sediment will then move landward along the left side to the tip of the saline intrusion.

The lower flats of estuarine shores are usually exposed to considerable water movement through strong tidal currents and scouring by the river itself. The sediments at these shore levels therefore tend to be sandy and the higher flats muddy. However, on many European sheltered sandy beaches with a large (> 4 m) tidal range, this situation can be reversed. Here the shore has a concave profile so that the low shore is dissipative (with finer sediment) and the high shore is reflective (with coarser sands). Local topographic and other physical and biological conditions can modify these general schemes greatly and many exceptions can occur (OUCT, 1989).

1.3 MODIFYING FACTORS

1.3.1 GEOGRAPHY

Geographic location can be considered a fifth important and all-embracing environmental gradient. Biological responses along this gradient have been documented in detail for some parts of the world where there is a sharp change from one biogeographic region to another, as in the British Isles (Lewis, 1972) and southern Africa (Stephenson and Stephenson, 1972; Branch and Branch, 1981). Many species reach their geographic limits around the British coastline and there are marked differences between the **oceanic** influenced west coasts and the more neritic regions of the North Sea and the eastern

English Channel (see maps in Lewis, 1964 and Hawkins and Jones, 1992). In southern Africa, differences between the west coast with its strong upwelling of cold water and essentially cold temperate biota, and the southern (warm temperate) and eastern (subtropical) coasts have also received much attention (see Chapter 6 for a fuller account of these aspects).

The stresses associated with the vertical gradient of increasing time spent emerged will change in intensity with latitude. At middle latitudes, stress will be less extreme high on the shore compared with low latitudes, where extremes of heat occur, and with high latitudes, where freezing is common. Similarly, wave action will depend on the weather patterns and the fetch of a particular ocean: shores bordering enclosed seas will experience less severe wave action. **Upwelling** will also have a profound influence on the biota of the shore, bringing cold, nutrient-rich water and advecting high primary production onto the shore (Chapter 6). Upwelling has a profound influence on various shorelines, including California, northern Chile and Peru, the west coast of South Africa, Oman, northern Spain and Western Australia.

As will be shown in Chapter 2, entire groups of characteristic taxa are present only within particular temperature ranges and this has allowed the recognition of different major biogeographic groups (Fig. 1.17). Thus **boreal and cold temperate species** of the midshore survive temperatures below the freezing point of sea water; **tropical sublittoral species** show only limited tolerance below 10 °C with upper lethal limits of 33–35 °C; **cold temperate and polar sublittoral species** do not survive freezing and, like mid-intertidal forms, may suffer lethal damage around –20 °C (Lüning and Asmus, 1991).

1.3.2 GEOLOGY, TOPOGRAPHY AND MICROHABITAT

Rocky shores are rarely smooth slabs of rock. Dependent on their geology they will be crossed with cracks, crevices, gullies and pools. Hard rocks, like granite, provide a more secure anchorage for large plants and animals than soft rocks, such as sandstone or chalk. Drainage will be slower on porous rock and some species will burrow in soft rock. Boulder fields provide a mosaic of physical conditions, with conditions under stones approximating to sediment shores and the tops of the boulders akin to rocky shores. Depositing shores are also uneven, with ridges and banks and ripple marks; this results in considerable small-scale variation in physical conditions, some areas being damper, more shaded or better protected from wave action.

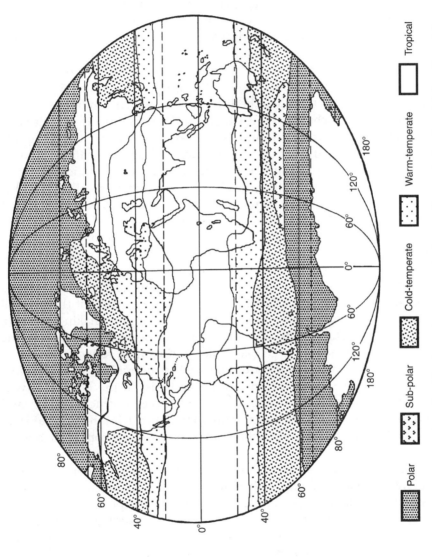

Fig. 1.17 Biogeographical regions defined by thermal tolerances (redrawn with permission from Lüning, K. and Asmus, R. (1991) *Ecosystems of the World*, Elsevier Science, pp 10–11).

Polar   Sub-polar   Cold-temperate   Warm-temperate   Tropical

Small-scale variation in environmental conditions, **spatial heterogeneity**, provides refuges from physical extremes as well as biological factors. On rocky shores these refuges can also provide a base from which predators and grazers (Chapter 4) emerge to forage.

Rock pools and deep crevices provide special habitats with their own sets of advantages and problems. Low and midshore pools are generally more beneficial habitats than nearby open drained rock. They usually support plants and animals more typical of lower on the shore or even the subtidal. The fluctuations in a rock pool will vary with tidal height: there will be little time in which conditions in low-shore pools can change; whilst high on the shore, pools can often go for several days without inundation (Naylor and Slinn, 1958). Slightly lower down, the area around high water of neap tides can be exposed to the air for up to 11 hours or so. High-shore pools will therefore vary considerably in temperature, getting much warmer or colder than the sea. They also show considerable variation in salinity, becoming saltier due to evaporation or diluted due to rainfall or freshwater run-off. During the day the plants in the pool can produce oxygen and the pool will become supersaturated. At night net production of carbon dioxide will occur and the oxygen in the pools can decrease, which will stress both plants and animals. The amount of dissolved carbon dioxide or oxygen will affect the pH of the pool. At night the pH will go down as more carbon dioxide goes into solution, making the pool more acidic. During the day the pool will be more alkaline. If the volume of the pool is large, particularly in relation to its surface area, then changes in physical conditions will be minimized over the tidal cycle. Changes also occur on a seasonal basis. In the Firth of Clyde, greater variations in temperature and oxygen concentration and salinity occur in the late spring and winter (Fig. 1.18). Because of such dramatic fluctuations, high-shore rock pools often contain a limited suite of species.

Crevices collect sediment and can trap air. They often support a highly specialized fauna with many air-breathing arthropods such as centipedes, millipedes, beetles and pseudoscorpions of terrestrial origin. These take refuge in pockets of air and emerge from crevices to forage when the tide is out. Crevice fauna also include the primitive onchidellid pulmonates which have been confused with opisthobranchs in the past. These have a lung-like structure and can be found foraging at low tide. Sediment-dwelling forms typical of muddy or sandy shores such as terebellid worms and sipunculids are often abundant, deposit feeding on organic material which collects in the crevices.

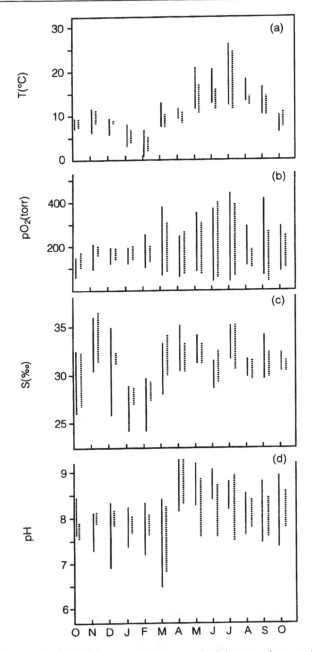

Fig. 1.18 Seasonal variation in the maximum and minimum values and ranges recorded for (a) temperature, (b) oxygen, (c) salinity and (d) pH in a high-shore pool (solid lines) and a low-shore pool (dotted lines) (adapted with permission from Morris and Taylor, 1983).

1.3.3 TIME OF LOW WATER

Finally, the timing of low water will influence the degree of stress experienced when the tide recedes. Low-shore organisms are particularly vulnerable in this respect. If spring tides coincide with the middle of the day, as in many temperate areas, then desiccation stress will be high. In polar regions, if spring tides occur during the night then freezing stress will be greater.

1.4 COMPARISONS WITH OTHER ECOSYSTEMS

By adopting a gradient perspective towards understanding shores, it becomes clear that many parallels can be drawn between the ecological responses of the intertidal fauna and flora and assemblages along other, non-marine gradients. Whittaker (1974) recognized these similarities many years ago and illustrated the principles by reviewing four major terrestrial environmental gradients (Fig. 1.19), which he termed **ecoclines**. These are:

1. a temperature gradient associated with altitude up the sides of hills and mountains;
2. a temperature gradient associated with latitude running from the poles to the Equator;
3. a climatic moisture gradient in a temperate region, for example running westward from the southern Appalachians, which receive heavy rainfall, to the deserts of New Mexico;
4. a similar climatic moisture gradient in tropical regions, for example South America.

Each of these gradients is most obviously reflected in its vegetation zonation, with lush, massive vegetation at what Whittaker terms the 'favourable' end of the gradient and sparse vegetation at the 'unfavourable' end (Fig. 1.19). These gradients also share several interesting ecological features (Table 1.2). Biomass, primary production, amount of area covered by vegetation, habitat complexity and species richness are all high at the 'favourable' end and low at the 'unfavourable' end.

Such trends are also apparent down the intertidal gradient of most rocky shores (Fig. 1.20) and this raises the interesting possibility that the rocky shore might make a useful model for terrestrial gradients as well as the other marine gradients such as salinity and exposure. In other words, any generalizations that can be made about the nature and workings of gradient-dependent processes on rocky shores might well apply to other ecoclines. Additionally, there are many advantages in studying such processes on rocky shores (Chapter 8). The vertical gradient is usually compressed within a short distance. Even the horizontal gradient of

ALTITUDINAL

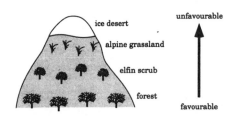

CONTINENTAL MOISTURE GRADIENT - TEMPERATE

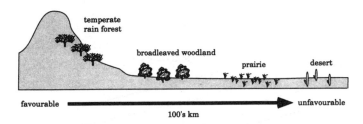

CONTINENTAL MOISTURE GRADIENT - TROPICAL

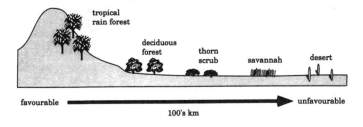

LATITUDINAL

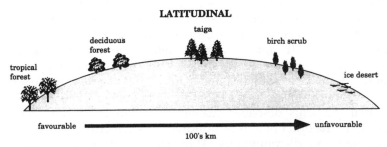

Fig. 1.19 Zonation of vegetation along major terrestrial gradients (after Whittaker, 1974, with permission).

Table 1.2 Ecological features of major terrestrial environmental gradients*

	Favourable	Unfavourable
Biomass (kgm^{-2})	> 40	< 1
Vegetation height (m)	~ 40	< 1
Ground cover (%)	> 100[†]	< 10
Primary production	High	Low
Habitat complexity	High	Low
Species richness	High	Low

*After Whittaker (1974).
[†]By convention, a value > 100% signifies canopy overlap.

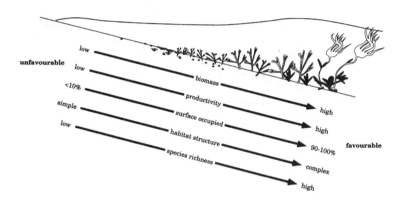

Fig. 1.20 Ecological features of a rocky shore which show strong parallels with terrestrial gradients (compare with Table 1.2).

exposure to wave action can be compressed into tens or hundreds of metres. The communities on rocky shores are relatively simple and are composed of easily identifiable species the abundance of which can be readily assessed. Patterns of distribution can be easily described. Most organisms compete for a clearly defined resource, space. Most importantly, shores are relatively easy to manipulate experimentally because of their two-dimensional nature and the fast turnover (and hence response times) of the dominant components. In contrast, the terrestrial gradients described above are on an enormous scale, with distances in the order of tens or hundreds of kilometres. They are usually fragmented and discontinuous. They have a species richness several orders of magnitude higher than that of intertidal systems. Their complex and often massive three-dimensional structure makes detailed description of their biological components a nightmare. Experimental manipulation of these systems is logistically difficult and the life spans, turnover and response times of the dominant structuring species can be depressingly long – hundreds of years in the case of trees. Similarly, other environmental gradients in the intertidal zone are often fragmented and have to be tackled over a much larger scale than the compressed vertical gradient between the sea and the land.

Attractive as the idea is, one needs to be cautious in using the rocky shore as such a model. Extrapolating processes between apparently identical shores, let alone ecosystems, can be dubious (Chapters 3 and 4) and there are some real and substantial differences between rocky shore and other gradient systems. For instance, there is no terrestrial equivalent of the regular short-term immigration of highly mobile consumers onto shores with the tide. On rocky shores, there is little comparable to the build-up of soils and nutrients on land. Nevertheless, we offer this idea as an intriguing possibility for the reader to bear in mind as he or she studies the shore – it is more convenient than climbing a mountain!

OVERVIEW

We believe that an appreciation of the major environmental gradients and the interactions between them provides a sensible framework for organizing an otherwise complex array of shore types. Also, it will become apparent in the following chapters that almost all of the physico-chemical limiting factors and dynamic complex biological processes that shape the ecology of the shore vary in their intensity along these gradients. These include desiccation stress, problems of osmoregulation, oxygen availability, nutrients, disturbance, living and detrital food stocks, competition, grazing and predation. In other words, all these factors and processes are gradient dependent. The vertical gradient is essentially a unidirectional stress gradient for marine organisms. To some extent the estuarine salinity gradient is as well. The other gradients are more complex, with different types of organisms being positively or negatively affected at different points along them.

Very few studies have looked at more than two gradients in combination. Most have focused on the combined effects of shore level and wave action on patterns of distribution on rocky shores or the interactions between particle size, shore level and wave action on depositing shores. Most studies are undertaken at a limited set of points on one, two or rarely three of the gradients. In the next chapter we explore how these gradients influence distribution patterns.

Patterns of distribution 2

Most biology or environmental sciences students will at some time or other visit the shore as part of their education, many taking a formal course in intertidal ecology. One of the reasons why the intertidal area is such a valuable teaching aid is that the distribution patterns of the major groups of organisms change rapidly on the sharp gradients between sea and land and these are clearly modified by wave action. Moreover, these patterns can be easily seen and described – especially the horizontal bands or zones of conspicuous organisms like barnacles and seaweeds on rocky shores.

Zonation patterns on rocky shores have long been studied (e.g. Audouin and Milne-Edwards, 1832; Vaillant, 1891; Baker, 1909; Walton, 1915; Colman, 1933). Because the main biological features of rocky shores can be so easily recorded with a sketch book or camera, these early intertidal ecologists were able to adopt a broad-brush qualitative approach. Such studies have revealed some consistent distribution patterns worldwide. In this chapter, we first describe the so-called universal features of zonation on the vertical (intertidal) gradient of rocky shores (Stephenson and Stephenson, 1949). We then describe how these major patterns change along the exposure and particle size gradients described in Chapter 1. We then assess the evidence for universal zonation patterns along the intertidal gradient of sandy beaches and mudflats. The specialized interstitial meiofauna is briefly described before considering macrophytes on sheltered shores. Finally zonation along the salinity gradient of estuaries is briefly examined.

2.1 UNIVERSAL ZONATION SCHEMES FOR ROCKY SHORES

Alan and Anne Stephenson started their rocky shore work in South Africa in the 1930s and subsequently travelled to many parts of the world, recording the biological patterns they observed. This catalogue is presented in a beautifully illustrated book (Stephenson and Stephenson, 1972) which all intertidal ecologists should dip into to get a feel for other shores worldwide. It soon became evident to the Stephensons that rocky shores from different geographical areas had certain biological features in common. The same general zones could be recognized on shores as far apart as South Africa and Scotland. In 1949, they proposed their **universal** classification scheme of zonation for all rocky shores (Fig. 2.1), in which the shore is divided into three major zones:

- a high-shore area, called by the Stephensons the **supralittoral** fringe, characterized by encrusting lichens, Cyanobacteria (bluegreen bacteria) and small snails, the periwinkles;
- a broad midshore zone, called by the Stephensons the **midlittoral** zone, dominated by suspension-feeding barnacles, mussels or sometimes oysters;
- a narrower low-shore zone, called by the Stephensons the **infralittoral** fringe, dominated by red algae including pink, calcareous, encrusting forms, often with large kelps (brown seaweeds), or in some places in the Southern Hemisphere, large filter-feeding tunicates (sea squirts).

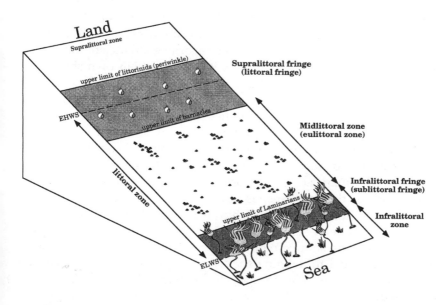

Fig. 2.1 Stephenson and Stephenson's universal zonation scheme, with Lewis' (1964) alternative terminology given in parentheses. Modified with permission from Stephenson and Stephenson (1972).

The Stephensons later qualified some aspects of their scheme, calling it widespread rather than universal and emphasizing that it applied to wave-beaten but not extremely exposed shores. The general features of the scheme, however, do seem to apply to most exposed rocky shores. But in many parts of the world (e.g. Britain and Northern Europe, Canada, New England), sheltered shores cannot be fitted easily into the scheme. In this book, we will refer to their three major zones as the **littoral fringe** (high shore), the **eulittoral zone** (midshore) and the **sublittoral fringe** (low shore) (Fig. 2.1), adopting terms used by Lewis (1964) in his book based on an extensive study of zonation patterns around British rocky coasts.

It should be noted that the Stephensons' scheme (and that of Lewis) is based simply on the relative positions of the major community types that they observed along the vertical gradient. The zones were not defined by reference to particular tidal levels or any other physical factor. Although this work was intended primarily to show that common ecological factors probably operate on shores worldwide, the scheme has also made easier the comparison of work carried out on rocky shores in quite different parts of the world. Instead of ecologists having to define the precise tidal level at which their study was carried out, they can merely state that an experiment was carried out, or an observation made, for instance, in the lower eulittoral zone. Reference to a zone rather than a particular tidal level is often much more useful to other researchers, who would then have a fairly good idea of the general characteristics of the habitat and how it relates to their own shores. We believe that this is an important practical consideration, because detailed definition of tidal levels on a shore is a laborious and unrewarding task – particularly as wave action modifies the environment and hence zonation patterns so much. Nevertheless, several authors (e.g. Ricketts *et al.*, 1968 and Swinbanks, 1982), have proposed general zonation schemes based on a limited number of tidal levels and these are used in some parts of North America (e.g. M.J. Foster *et al.*, 1988, 1991).

The work of the Stephensons also served to direct attention away from tides and tidal levels, with which many ecologists of the day were preoccupied (Colman, 1933; Doty, 1946; Evans, 1957). Even in more recent textbooks on shore ecology a disproportionate amount of space is sometimes devoted to defining the terminology of the many levels on the shore that the tide reaches throughout the year. Tides are undoubtedly an important feature of the shore and the upper limits of distribution of many species can be associated with major high and low tidal levels, but, as shown in Chapter 1, tides merely stretch an existing environmental gradient along which species are zoned.

The lack of a clear association between tidal levels and zonation can be seen in Figs 1.16 and 2.2. Here, the major zones shift upshore relative to predicted tidal levels as exposure increases. The littoral fringe gets uplifted the most, the upper levels of the eulittoral zone (the barnacle zone) considerably, and the low-shore zone much less, if at all. The extra splash and spray on wave-beaten shores extend the marine environment to higher shore levels than would be found on a sheltered shore. In the previous chapter, we argued that the distribution patterns of shore organisms along one environmental gradient cannot be properly understood without reference to the other major shore gradients. This is particularly true when the interaction between the vertical intertidal gradient and the horizontal exposure gradient is recognized.

It should be noted that not all shore ecologists share our enthusiasm for the Stephenson–Lewis approach as a useful field shorthand and framework for more localized study. A system championed by Scandinavian workers recognizes only one main zonal boundary, the **litus line**, located at the lower limit of the supralittoral. The litus line has been shown by multivariate analyses to clearly separate assemblages (Russell, 1972; Bolton, 1981) and, like the Stephensons' scheme, it provides a useful benchmark for locating the zones of shore organisms. Similar schemes to the Stephenson–Lewis approach (e.g. Peres, 1967) have been developed for the Mediterranean (Fig. 2.3). Many Mediterranean shore ecologists, however, have adopted a very detailed approach similar to that of the Zurich–Montpellier school of phytosociology, where generic and specific names of different taxa which characterize a 'community' are combined. Clearly, such a scheme will be more geographically restricted compared with the Stephensons' generalized scheme. Finally, shores on the west coast of North America have been traditionally described using four major zones (Table 2.1). The merits and problems of all these schemes are discussed further by Russell (1991).

2.2 ROCKY SHORE ZONATION IN DIFFERENT PARTS OF THE WORLD

The work of the Stephensons spawned a whole series of descriptions of shores in various parts of the world employing a similar approach. Many of these can be found in the excellent books and papers dealing with particular geographical regions (Table 2.2). Some of the patterns are shown diagrammatically in Fig. 2.2. In North America, particularly on

Table 2.1 Zonation schemes for the Pacific coast of North America*

	Ricketts et al. (1985)† Pacific Coast, North America	Pearse (1980) Ano Nuevo Island, California	Ferguson (1984) Big Sur Coast, California	Stephenson and Stephenson (1972) Pacific Grove, California	Raffaelli and Hawkins interpretation
ZONE 1	LITTORINA KEENAE Porphyra Cladophora Enteromorpha	SPLASH ZONE Bluegreen algae Lichens Porcellio scaber Ligia pallasii Littorina keenae	SPLASH ZONE Littorina keenae Ligia Littorina scutulata	SUPRALITTORAL FRINGE Littorina planaxis Ligia Pachygrapsus	LITTORAL FRINGE
ZONE 2	BALANUS GLANDULA Pelvetia Littorina scutulata/plena Tegula funebralis	HIGH ZONE Porphyra spp. Mastocarpus papillatus Endocladia muricata Pelvetiopsis limitata Balanus glandula Chthamalus dalli Mytilus californianus Phragmatopoma californica Pollicipes polymerus Anthopleura elegantissima Collisella scabra Tegula funebralis Littorina scutulata.plena Nutallina californica	HIGH ZONE Porphyra Pachygrapsus Collisella digitalis/ austrodigitalis Collisella scabra Lottia	UPPER MID-INTERTIDAL Balanus glandula Tetraclita	EULITTORAL
ZONE 3	MYTILUS Pollicipes Nucella emarginata Katharina Nutallina	MID ZONE Iridaea flaccida Anthopleura elegantissima Anthopleura xanthogrammica Dodecaceria fewkesi	MID ZONE Pagurus Tegula funebralis Anthopleura elegantissima Haliotis	LOWER MID-INTERTIDAL Chthamalus dalli Tegula funebralis Thais Nutallina	EULITTORAL

ZONE 4	PHYLLOSPADIX Laminarians	*Collisella digitalis* *Collisella pelta* *Collisella scabra* *Notoacmea scutum* *Tegula funebralis* *Tegula brunnea* *Pisaster ochraceus*	*Lottia* *Katharina* *Mytilus/Pollicipes* (only on offshore rocks)	
		LOW ZONE *Phyllospadix* *Laminaria* spp. *Egregia menziesii* Sponges Bryozoans Tunicates	LOW ZONE Sponges Bryozoans Tunicates *Mopalia* *Tonicella* *Leptasterias*	INFRALITTORAL FRINGE *Alaria* *Lessoniopsis*
				SUBLITTORAL FRINGE

*Modified from Foster et al. (1988).

†Ricketts et al. (1985) use biological indicators to define zones and list subsidiary species.

Table 2.2 Selected publications describing rocky shore zonation patterns in different parts of the world

Ocean/Sea	Location	References
Eastern Atlantic	Iceland	Munda (1991)
	Russia	Kussakin (1971)
	Norway	Levring (1987), Svendson (1959), Jorde (1966)
	Faeroes	Burrows et al. (1954)
	United Kingdom	Southward (1958), Lewis (1964)
	Ireland	Ebling et al. (1960)
	German Bight	Janke (1990)
	Baltic	Wallentinus (1991)
	Netherlands	Den Hartog (1959)
	France (Biarritz)	Renoux-Meunier (1965)
	Spain	Donze (1968)
	Portugal	Ardré (1970), Saldanha (1974)
	Mediterranean	Augier (1980), Lipkin and Safriel (1971), Peres (1967)
	Canary Islands	Lawson and Norton (1971)
	Azores	Hawkins et al. (1990)
	West Africa	John et al. (1991), John and Lawson (1991)
	Southern Africa	Branch and Branch (1981), Field and Griffiths (1991)
Western Atlantic	North America	Berrill and Berrill (1981), Mathieson et al. (1991), Orth et al. (1991), Vadas and Elner (1992)
	Caribbean	Dawes et al. (1991)
	Bermuda	Thomas (1985)
	Colombia	Brattström (1980)
	Brazil	Alves-Coelho and Ramos-Porto (1980)
Eastern Pacific	North America	Ricketts et al. (1968), Carefoot (1977), Brusca (1980), M.J. Foster et al. (1988, 1991), Littler et al. (1991)
	Chile	Santilices (1991)
Western Pacific	Japan	Yajima (1978), Yamada (1980), Mori and Tanaka (1989), Takada and Kikuchi (1990)
	Hong Kong	Morton and Morton (1983), Morton et al. (1993)
	Australia	Dakin (1953), Bennett and Pope (1960), King et al. (1991)
	New Zealand	Morton and Miller (1973), Creese and Ballantine (1986)
Indian Ocean	India	Chapgar (1991)
	East Africa	Hartnoll (1976)
	Aldabra Atoll	Taylor (1971)
Red Sea	Israel	Safriel and Lipkin (1964), Lipkin (1991)
Worldwide/ General		Stephenson and Stephenson (1972)

Pacific coasts, zonation patterns on rocky shores have been traditionally described within a four-zone framework (Table 2.1, Fig. 2.4), largely reflecting the influence of Ricketts and Calvin (Brusca, 1980; Foster et al., 1988). Because the four subdivisions of the shore are defined on the basis of tidal levels the zones are not coincident with those of Stephenson. Carefoot (1977) has shown that Pacific shores can be described using the Stephenson scheme and he has pointed out that it was in part their work in British Columbia that laid the foundations for the 'Universal Scheme' in the first place. Perhaps one reason why the scheme has been

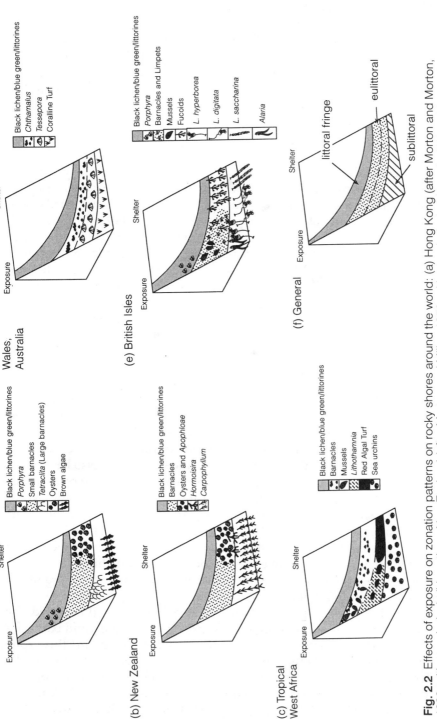

Fig. 2.2 Effects of exposure on zonation patterns on rocky shores around the world: (a) Hong Kong (after Morton and Morton, 1983, with permission), (b) northern New Zealand (after Morton and Miller, 1973, with permission), (c) tropical West Africa (after John and Lawson, 1991, with permission), (d) New South Wales (after King et al., 1991, with permission), (e) British Isles (after Lewis, 1964, with permission), (f) a generalized shore. Does your shore fit this scheme? Horizontal and vertical scales are not necessarily equivalent.

less used in the north-eastern Pacific is the lack of published information on how distributions are affected by the horizontal exposure gradient, especially zonal uplift (Foster *et al.*, 1988), so that the mismatch of zones and tidal levels is not immediately apparent, except for a few studies such as that of the Stephensons.

2.2.1 ZONATION ON ROCKY CLIFFS AND BEDROCK SHORES

Throughout the world and right along the exposure gradient, the littoral fringe is characterized by small littorinid snails, cyanobacteria formerly called 'bluegreen algae' and encrusting lichens. These small snails have attracted a great deal of attention in recent years (Raffaelli, 1982; Reid, 1990), but the bluegreens and encrusting lichens have been largely ignored by marine biologists, although useful information can be found in Whitton (1975) and Fletcher (1980). Nevertheless, it is clear that the lichen genera *Verrucaria*, *Caloplaca* and *Xanthoria* have a wide distribution as do the bluegreens *Calothrix*, *Lyngbya*, *Oscillatoria* and *Rivularia* (Russell, 1991). A black zone of *Verrucaria* and/or bluegreens is an almost worldwide feature. Many different kinds of small invertebrates also occur in this zone, including isopods, mites, nematodes and chironomids (Kronberg, 1988). In warm temperate and tropical areas, highly mobile crabs (e.g. grapsids) exploit the high-shore algae.

The littoral fringe also supports dense swards of ephemeral algae during the winter in temperate areas. These are bleached by the sun in most summers. At higher latitudes, however, they may only occur during the summer, because of ice scour during the winter. A large section of the eulittoral zone of high-latitude shores may also be scoured off each year by ice. The midshore is then characterized by opportunist algae and mobile animals in the spring, as found on many North American shores (Vadas, 1990), and in parts of Scandinavia, the Arctic and Antarctic there may be a bare zone for most of the year (Johannesson, 1989). In the Antarctic, limpets (*Nacella*), which overwinter in the sublittoral fringe, migrate upshore to exploit this rich growth. Bluegreens seem to be absent altogether from the Antarctic supralittoral (Heywood and Whittaker, 1984).

On the most exposed shores the eulittoral is dominated by encrusting, sessile suspension feeders, principally mussels and barnacles (Figs 2.2–2.4). Mussels of the genus *Mytilus* occur in this zone throughout the Atlantic and Pacific, but they are not always the dominant group and they also occur in the sublittoral. *Perna* spp. are characteristic of parts of Australasia, East and West Africa, and Venezuela, whilst *Aulacomya* occurs in Chile, South Africa and the

Kerguelan Archipelago (Suchanek, 1985). On all these coasts, acorn barnacles cover much of the eulittoral and many species are represented. The genera *Balanus, Semibalanus, Elminius, Chthamalus* and *Tetraclita* are especially common. Stalked or goose barnacles, in particular *Pollicipes* spp., characterize the eulittoral in warm-temperate regions, such as New Zealand, the north-west Pacific coast of America, Spain, Portugal and south-west France. The eulittoral also supports mobile herbivores such as true limpets (*Patella, Acmaea, Cellana*), keyhole limpets *Fissurella*, pulmonate limpets *Siphonaria*, and several species of periwinkle (littorinids) and topshells (trochids). Predatory whelks are also common worldwide.

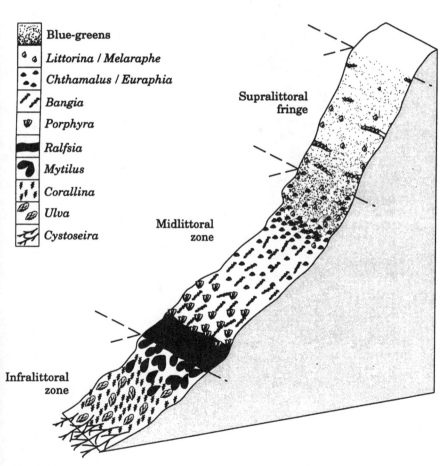

Fig. 2.3 A typical zonation scheme for a Mediterranean shore (modified with permission from Southward, 1965).

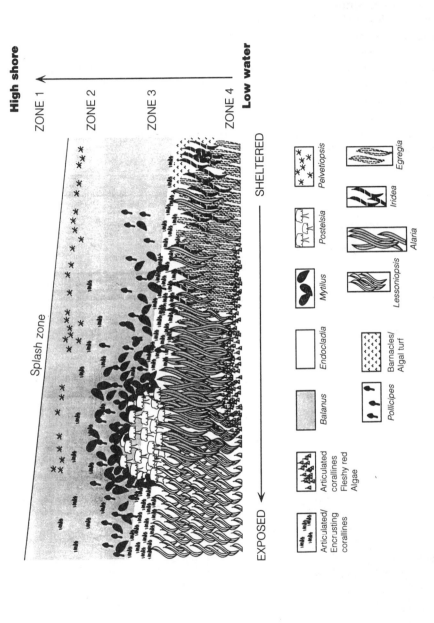

Fig. 2.4 An example of zonation at Monterey, California (adapted with permission from M.J. Foster *et al.* (1991) *Ecosystems of the World 24. Intertidal and Littoral Ecosystems*, Elsevier Science, p. 245).

On the most exposed shores, seaweeds in the eulittoral tend to be ephemeral or short turf forms. In cooler temperate regions, larger seaweeds become more prevalent with increasing shelter. In the North Atlantic, the fucoid seaweeds are particularly evident, occurring in patches on moderately exposed shores, but in shelter they may cover the entire eulittoral. Zonation is usually clear (Fig. 2.5) but the exact sequence depends on location. In the British Isles, *Pelvetia canaliculata* occurs at the top of the eulittoral, followed in sequence by *Fucus spiralis*, then an *Ascophyllum nodosum* zone, often with patches of *F. vesiculosus*, giving way to *F. serratus* and then kelps, usually *Laminaria digitata* but *L. saccharina* in very sheltered conditions. Sometimes a narrow *F. vesiculosus* zone occurs above the *Ascophyllum*, sometimes it occurs just beneath it but above *F. serratus* (see Lewis, for example, in Stephenson and Stephenson, 1972). Further north, in Norway and Iceland, various species in the *F. distichus* complex occur. On the eastern coast of Canada and in New England, the zonation pattern is different again. Here *F. spiralis* gives way lower down to *F. distichus* or *F. vesiculosus*, followed by *Ascophyllum* and then *F. evanescens*, and usually a *Chondrus/Mastocarpus* zone (Vadas and Elner, 1992). The Pacific coast of North America may also have the kelp *Hedophyllum sessile* and the sea palm *Postelsia palmaeloris* at exposed locations. Not all regions, however, have these dense midshore growths of seaweeds. For instance, on the cool sheltered shores of New Zealand the only equivalent to the midshore fucoids is *Hormosira banksii*, but this species never attains the abundance of fucoids in other regions (Figs 2.2–2.4). Oysters may characterize the eulittoral of moderately exposed and sheltered shores in many parts of the Pacific and Indo–Pacific. Pacific oysters recruited from culture also form a zone in some parts of south-west France.

In the low eulittoral and the sublittoral fringe, algal turfs are a common feature. The large kelps are present throughout the entire exposure gradient on colder and nutrient-rich shores. These kelps vary widely in form, but generally have a sizeable holdfast, a long pliable stem or stipe and a broad, sometimes divided, lamina (Fig. 2.6). Many species belong to the laminarian group and these are the true kelps. Non-laminarians, such as the massive fucoid *Durvillea*, may dominate this zone in the South Atlantic and Pacific. In warmer regions, the sublittoral fringe may support a dense covering of ascidians, such as the sea tulip or red-bait *Pyura*, in addition to a red algal turf. In the warmest seas, this zone may also embrace the upper edge of the corals. All these are essentially sublittoral species which just extend into the intertidal and are usually continually swashed by waves on all but the calmest days.

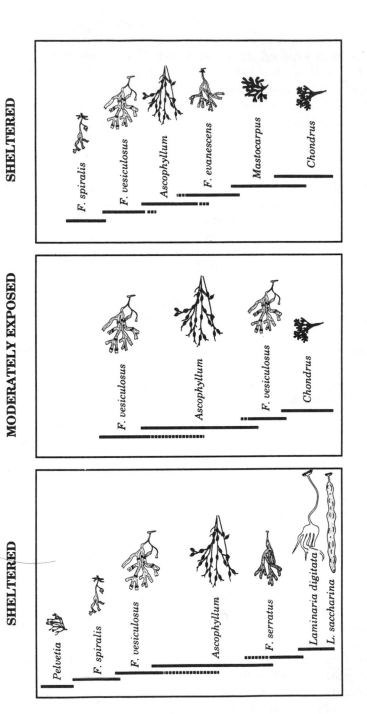

a) British Isles, Southern Norway

b) New England / Canada

c) New England / Canada

SHELTERED

MODERATELY EXPOSED

SHELTERED

Pelvetia

F. spiralis

F. vesiculosus

Ascophyllum

F. serratus

Laminaria digitata

L. saccharina

F. vesiculosus

Ascophyllum

F. vesiculosus

Chondrus

F. spiralis

F. vesiculosus

Ascophyllum

F. evanescens

Mastocarpus

Chondrus

Fig. 2.5 Zonation of fucoids on a range of shores (based on Lewis, 1964, and Vadas and Elner, 1992).

2.2.2 WHY ARE THESE FEATURES SO UNIVERSAL?

It is clear from the above that the biological characteristics (or rather the lack of them) of the infralittoral fringe reflect the harsh physical environment of the high shore; few taxa can cope with such an environment. In some parts of the world, like the north-east Atlantic, the consistency in structure of midshore communities seems more likely to be due to the effects of herbivores, especially grazing limpets, enabling space-occupying animals to dominate the shore (Hawkins and Hartnoll, 1983a; Hawkins *et al.*, 1992a). Domination of the sublittoral fringe and the sublittoral itself by large canopy-forming kelps or tunicates like *Pyura* is possible because growth is not limited by desiccation, and nutrients and food are in ample supply. These species may also have a refuge from sea-urchin grazing from below because the urchins avoid the high-wave-energy environment around low water mark on exposed coasts. However, it should be noted that this does not seem to hold for some regions such as the north-east Pacific (M.J. Foster *et al.*, 1991; M.S. Foster, 1992). Limpets may not be able to move down from the eulittoral into the dense algal growth because they are swept away if they move over plant surfaces. In the temporary absence of these plants, limpets will become established at low shore levels (Underwood and Jernakoff, 1981; Hawkins *et al.*, 1992a), but they cannot keep pace with the subsequent rapid algal colonization and growth and eventually starve to death in the midst of fully grown plants.

Patterns in the sublittoral

Although it is convenient for many shore ecologists to restrict their studies to the region above low water, the system does of course extend into the subtidal. The depth of this sublittoral zone is usually taken as the point at which macroalgal growth ceases. This is usually within a few tens of metres below low water mark, but the record stands at 250 m in clear oceanic water (Russell, 1991) – an unnerving prospect for the shore ecologist!

Structurally, sublittoral assemblages can be divided into single-layered or multilayered systems. The former consists chiefly of coralline encrusting red algae (e.g. *Lithophyllum*) and short turf-forming species such as *Ceramium*, *Ectocarpus*, *Laurencia* and *Dictyota*. These are common in warm temperate regions. Multilayered systems are characterized by a canopy of kelp species such as *Macrocystis pyrifera* or *Laminaria hyperborea*. In some cases the limit of kelps may be set by fields of the anemone *Metridium* (Vadas and Steneck, 1988). Antarctica is unique amongst cold-water sublittoral assemblages in lacking kelps, although the massive fucoid seaweed *Durvillea antarctica* occurs in the upper sublittoral of southern ocean shores. In warm temperate seas, low canopies can be formed by other species such as *Cystoseira* in the Mediterranean. In the atidal Baltic, a sublittoral forest is created by large canopy-forming *Fucus vesiculosus*, and there are many functional parallels between this system and the more familiar North American and South African kelp forests (Wallentinus, 1991). Below this *Fucus* forest the deeper sublittoral (> 20 m) is dominated by mats of the mussel *Mytilus edulis*, an unusual pattern due perhaps to the lack of molluscan, crustacean and echinoderm predators in the low-salinity eastern Baltic. Subtidal *Mytilus* communities also occur in extreme exposure (Hiscock, 1983) and are the common fouling assemblages on offshore structures, especially early in the successional sequence. They are also common in the reduced-salinity environment of disused dock basins in the UK (Russell *et al.*, 1983; Hawkins *et al.*, 1992b).

Kelp forests are very diverse assemblages and those in the north-east and south-west Pacific may have the highest species richness and productivity of all temperate systems (M.J. Foster *et al.*, 1991). More than 30 species of macroinvertebrates and 125 fish species have been recorded from rocky reefs and kelp forests in southern California, together with seals and sea otters. In these vast forests of the north-east Pacific, different species of canopy-forming algae characterize different locations. Thus, in south-western Alaska *Alaria fistulosa* dominates, from eastern Alaska to northern California *Macrocystis integrifolia* and *Nereocystis luetkeana* and from Santa Cruz to Baja California, Mexico, *Macrocystis pyrifera*. Large kelp forests also occur in the Southern Hemisphere, including Chile, the Falklands and Australia. Probably the most detailed kelp studies in the Southern Hemisphere have been made in the southern African *Nereocystis* forests in the Benguela upwelling system (see Field and Griffiths, 1991 for an overview, and Chapter 6). Large beds of seaweeds do not occur in the nutrient-poor waters of the tropics, although interesting tropical kelp beds occur in the seasonal upwelling areas off the coast of Oman in the Arabian Sea (Fig. 2.6).

2.3 DISTRIBUTION PATTERNS ON OTHER SHORE TYPES

2.3.1 BOULDER SHORES

Further along the particle size gradient described in Chapter 1 are boulder shores composed of very large particles weighing many tens of tonnes which will be disturbed only by the heaviest wave action. The distribution patterns described above for bedrock or cliff habitats apply to such shores, with the added biological richness provided by the

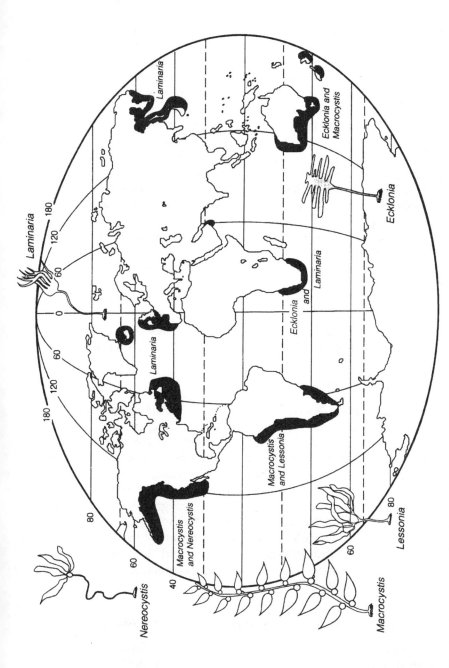

Fig. 2.6 Distributions of major kelp species (after Mann, 1973, and various sources). NB: In southern Europe and North Africa the main laminarian kelp species becomes *Sacchoriza polyschides*.

increase in microhabitats such as overhangs and crevices. Of course, where tidal range is significant, in excess of a few metres, the individual particles are not sufficiently large to each span the entire intertidal gradient. The faunal assemblages underneath the more stable boulders are protected from desiccation and often include many organisms more commonly found in the sublittoral, such as tube worms and bryozoans. The large crevices formed by the spaces between adjacent boulders provide an ideal habitat for gastropods and crabs. These interstices also trap sediment, adding yet another habitat dimension, for burrowing forms. The high species diversity of this habitat may also be enhanced by the periodic removal of species crushed and ground off by boulder movement. This resets the successional clock so that opportunists such as ephemeral green seaweeds appear alongside later successional species (Chapter 4).

Small boulders are often found as islands within sandy flats so that the beach is a mixture of two very different particle sizes. These shores are rich in a variety of species, especially in warmer regions, such as the Sea of Cortez (Steinbeck and Ricketts, 1941; Brusca, 1980), where it is even possible to find several species of small octopus under boulders. On Atlantic coasts, small cobbles are commonly bound together by mussel byssus threads or by the polychaete *Sabellaria*, increasing both substratum stability and habitat complexity (Wilson, 1971; Gruet, 1986).

2.3.2 FROM GRAVELS TO COARSE SAND

The relationships between exposure and zonation patterns on rocky shores, as described by the Stephensons, can be appreciated because the fauna and flora occupy surfaces and their patterns of distribution can easily be seen. This is not the case as we move along the particle size gradient towards depositing shores such as sandy beaches and mudflats. Here most of the organisms are hidden below the surface. Intermediate between boulders and sand are shingle and gravel beaches which are essentially abiotic (Chapter 1). Moving further along the particle size gradient, to sands, particles are now sufficiently small to retain some water at low tide through capillary action and slow drainage. Also the risk of physical damage is lower because the organisms are large relative to the size of particle.

2.3.3 GENERALIZED ZONATION SCHEMES FOR FINE-PARTICLE SHORES

Distribution patterns within a sandy beach or mudflat can only be derived through destructive sampling of the habitat. Samples are usually taken along a transect from high to low shore levels, followed by sorting and identification of the organisms present, often back in the laboratory. Only then are the zonation patterns revealed. Ecologists interested in these small-particle shores have therefore been unable to adopt the Stephensons' broad qualitative approach in their search for general features of zonation.

Nevertheless, some general features have emerged. The Scandinavian marine biologist Dahl (1952) recognized three major zones on beaches. In the north-east Atlantic these are characterized by different kinds of small crustaceans: a supralittoral fringe populated by talitrid amphipods (e.g. *Talitrus saltator*); a midshore region characterized by cirolanid isopods (e.g. *Eurydice pulchra*); and a low-shore region typified by amphipods belonging to the families Haustoriidae (within which Dahl included *Bathyporeia*, now placed in the family Pontoporeiidae) and Oedicerotidae (Fig. 2.7). In other biogeographical areas these zones were occupied by different species of Crustacea or perhaps different kinds of organisms altogether, but there appeared to be an overall pattern. For instance, in West Africa, John and Lawson (1991) recognized an upper zone of ocypodid crabs (in the tropics) or the isopod *Tylos granulatus* (in temperate parts); a midshore zone of the isopod *Excirolana* and the polychaetes *Nerine* and *Scolelepis*; and a lower zone characterized by the bivalves *Donax* or *Gastrosaccus*. In Chile, *Tylos* and the amphipod *Orchestoidea* characterize the upper zone, the middle zone has *Excirolana*, the polychaetes *Euzonus* and *Scolelepis*, and the lowest zone is characterized by the crab *Emerita*, the bivalve *Mesodesma* and various polychaetes (Santelices, 1991).

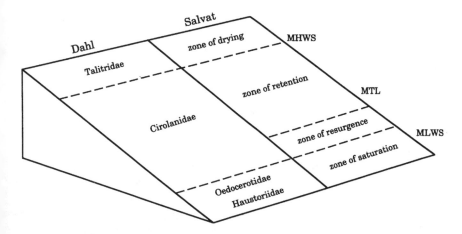

Fig. 2.7 Zonation schemes of Dahl (1952) and Salvat (1964) for sandy shores. MHWS, mean high water of spring tides; MLWS, mean low water of spring tides; MTL, mean tide level. (Reproduced with permission from Raffaelli, D. *et al.* (1991) Zonation schemes on sandy shores. *J. exp. Mar. Biol.*, **148**, Elsevier Science.)

In Britain, Eltringham (1971) observed a similar three-zone pattern on muddy shores: an upper zone characterized by either a complete absence of fauna or by the polychaete *Hediste* (= *Nereis*) *diversicolor*; a mid-tide zone of the bivalve molluscs *Scrobicularia plana*, *Cerastoderma* (= *Cardium*) *edule* and *Macoma balthica*, and, where some sand is present, the lugworm *Arenicola marina*; a low-shore zone with the bivalves *Mya arenaria* and possibly *Angulus* (= *Tellina*) *fabula*, depending on the sediment particle size, along with a number of polychaete species.

An alternative scheme has been proposed by the French ecologist Salvat (1964, 1967). In his scheme there are four major zones on beaches. These are defined not by the fauna itself (cf. Dahl's scheme), but by how the hydrodynamics of interstitial water change with shore level (Fig. 2.7). Zone 1 at the top of the beach is characterized by **dry sand**.

Downshore from this is a zone of **water retention** (zone 2), which is damp (not wet) at low tide. Next comes an **area of resurgence** (zone 3) where interstitial water flows in and out of the sediment with the tide. The lowest zone (4) on the shore is **permanently saturated** and there is little exchange of water over the tidal cycle.

When Dahl's and Salvat's zonation schemes have been applied to beaches it has often not been possible to distinguish between them on the basis of the biological assemblages occurring within each of the zones (e.g. McLachlan, 1990; Raffaelli *et al.*, 1991). This may in part be because the two schemes are not in fact radically different – Salvat's zones 2 and 3 are included within Dahl's midshore – and partly because the sampling and descriptive techniques usually employed produce faunal distribution patterns that are sufficiently imprecise that they may fit either scheme or neither. Also zonal boundaries themselves are probably less well defined in sedimentary environments compared with those seen on rocky cliffs. McLachlan (1990) suggests that Salvat's zone 1 is characterized by air-breathers; zone 2 by intertidal species which can tolerate damp sand during the low tide period; zone 3 species require wet sand whilst the tide is out; and zone 4 species are basically sublittoral forms which extend only a little way up the intertidal gradient.

An important feature of Salvat's scheme is that it should be possible to recognize the major zones solely from the physical appearance and wetness of the sediment surface at low tide. This is possible on some beaches, but on most shores the zone of resurgence is difficult to locate and usually only three boundaries can be seen: the driftline, the water table outcrop, and the low tide swash zone, corresponding roughly to Dahl's three zones (McLachlan, pers. comm.). These features could provide the beach ecologist with the same kind of universal benchmarks as the Stephensons' major zones, facilitating between-shore comparisons. However, much more work needs to be done in other parts of the world to properly establish the generality of these features, especially in areas with restricted tidal ranges, such as the Baltic and the Mediterranean.

Other schemes which define broad zones on the shore include Brown and McLachlan's (1990) more rigorous division of the beach into an upper zone of air-breathing animals and a lower zone of truly aquatic forms. In Swinbanks' (1982) scheme the shore is divided into three major zones: an upper **atmoszone**, a midshore **amphizone** and a low shore **aquazone**. These zones are not defined by any obvious biological or physical features, but by particular tidal levels and as such are not likely to be useful as a reference framework for between-shore comparisons (see also Chapter 3).

Finally, it should be remembered that the intertidal region is only a part of the sandy beach system. The beach is created by dynamic processes which originate and operate beyond the low tide mark. Indeed, some authors have suggested that zonation on beaches should be more properly considered along the gradient running from the backing dunes down into the surf zone and beyond, because of the significant physical and biological interactions between these habitats (Brown and McLachlan, 1990). Not surprisingly, major zones can be recognized below the low water mark on these sandy beaches. As one moves beyond the low tide mark, the first zone encountered is the **inner turbulent zone**, probably an extension of Salvat's saturation zone, and seen as the surf zone on exposed beaches (Fig. 2.8). Further offshore the **wave break point zone** is one of extreme turbulence and few animals are found in this area. Lastly, there is an **outer turbulent zone** where the physical stability of the habitat increases rapidly and there is a high faunal biomass and diversity (Fig. 2.8).

Wave disturbance certainly has a strong influence on the offshore zonation patterns of crustaceans and sedentary polychaetes in the north-east Pacific: tube-dwellers and their commensals are unable to maintain themselves in disturbed sediments (Fig. 2.8). Whilst wave disturbance seems to be an important factor controlling sublittoral and low intertidal beach assemblages, sediment physico-chemical conditions of the kind described above by McLachlan (1990) will be more important along the upper beach.

With decreasing exposure and increasing sediment stability, species richness, abundance and total biomass increase to reach a maximum on sheltered shores (Fig. 2.9). Interestingly, individual body size may increase with exposure (Fig. 2.10), so that the total faunal biomass on

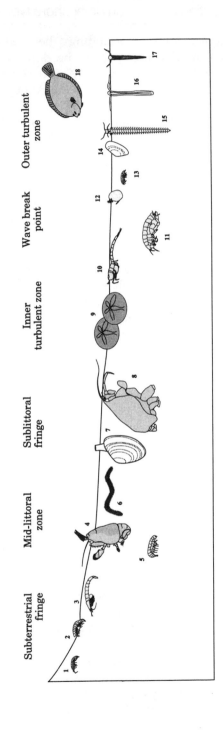

Fig. 2.8 Major intertidal and sublittoral zones on a sandy beach at Monterey, California (reproduced with permission from M.J. Foster et al. (1991) *Ecosystems of the World 24. Intertidal and Littoral Ecosystems*, Elsevier Science, p. 245). Species names: 1, *Megalorchestia* spp.; 2, *Excirolana linguifrons*; 3, *Archaeomysis grebnitzki*; 4, *Emerita analoga*; 5, *Grandifoxus grandis*; 6, *Nephtys californiensis*; 7, *Tivela stultorum*; 8, *Blepharipoda occidentalis*; 9, *Dendraster excentricus*; 10, *Hemilamprops californica*; *Euphilomedes longiseta*, *Olivella pycna*; 11, *Eohaustorius secillus*; 12, *Euphilomedes carcharodonta*; 13, *Rhepoxynius abronius*; 14, *Tellina modesta*, *Nephtys caecoides*; 15, *Northria elegans*; 16, *Magelona sacculata*; 17, *Prionospio pygmaeus*, *Mediomastus californiensis*; 18, *Citharichthys stigmaeus*.

exposed and semi-exposed beaches may not differ as much as e? (Bally, 1981; Fig. 2.10). Biomass increases with shelter (McIntyr and particle size (Fig. 2.10) in some regions, but in other parts of the world the relationship is less clear. It seems that it is the interaction between exposure, particle size and slope (and probably other physical features) that is important. Composite measures of these physical variables, such as Dean's parameter or a modified version, the beach stability index (BSI) which takes into account tidal range (McLachlan, 1992), are better predictors of biological features than is any single variable (Eleftheriou and Nicholson, 1975; Allen and Moore, 1987; Brown and McLachlan, 1990) (Fig. 2.10).

The most obvious change occurring along the gradient from exposed to sheltered conditions is the progressive addition of species and the loss of those unable to tolerate the less turbulent and more reducing sediment environment of fine-particle mudflats (Fig. 2.9). Mobile and robust forms, such as the cirolanid isopods, characterize the most exposed beaches, other groups being unable to maintain themselves in these highly unstable habitats. For instance, at Village Bay in St Kilda, an island group far out into the Atlantic west of Britain, the entire sandy beach may be removed offshore during winter storms to reveal an underlying rocky shore (Scott, 1960). The beach is gradually replaced over the summer when wave action is less severe. Only the cirolanid isopod *Eurydice pulchra* is frequent, polychaetes and amphipods being few in species and numbers (Fig. 2.9). Where exposure is even more severe, as on the south coast of Iceland, the beaches are completely devoid of macrofauna (Ingolfsson, 1975). Interestingly, *Eurydice* can attain high densities on less exposed beaches when sediment properties are modified through disturbance by burrowers, so that the beach develops 'exposed' characteristics (Tamaki and Suzukawa, 1991).

Polychaete worms such as *Nerine* and *Ophelia* become more frequent in areas of moderate disturbance, along with haustorid amphipods. On sheltered beaches and mudflats burrowing bivalve molluscs, such as tellinids and large clams, and epibenthic mussels are abundant and are responsible for much of the high biomass of sheltered shores (Fig. 2.9). Some bivalves, such as *Donax*, cope with the turbulent conditions found on more exposed shores by being able to burrow very rapidly when dislodged by the surf, a characteristic of many amphipods and polychaetes living on such shores. Indeed, *Donax* and the gastropod *Bullia* exploit exposed conditions, surfing up and down the shore with the tides to feed (Fig. 2.11). Other polychaetes may avoid dislodgement by simply burrowing deeply or by coiling their body into an anchor shape (Tamaki, 1987).

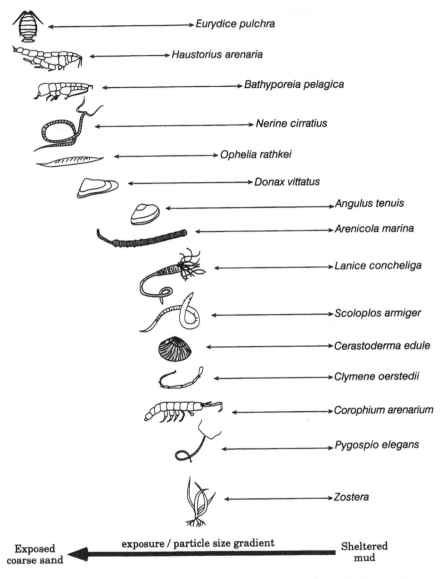

Eurydice pulchra

Haustorius arenaria

Bathyporeia pelagica

Nerine cirratius

Ophelia rathkei

Donax vittatus

Angulus tenuis

Arenicola marina

Lanice concheliga

Scoloplos armiger

Cerastoderma edule

Clymene oerstedii

Corophium arenarium

Pygospio elegans

Zostera

Exposed coarse sand ← exposure / particle size gradient → Sheltered mud

Fig. 2.9 Example of changes in macrofauna along exposure/particle size gradients on sandy shores in Britain. Note how progressively more species are added towards more sheltered, fine particle shores. Community-level trends are shown in Fig. 2.10. Based on personal observations and Allen and Moore (1987).

The fauna of these fine-particle shores characteristically consists of large numbers of deposit- and filter-feeding invertebrates (Fig. 2.9) fuelled by the large amounts of material settling onto the flats, much of which is retained in the sediment and used by the large microbial

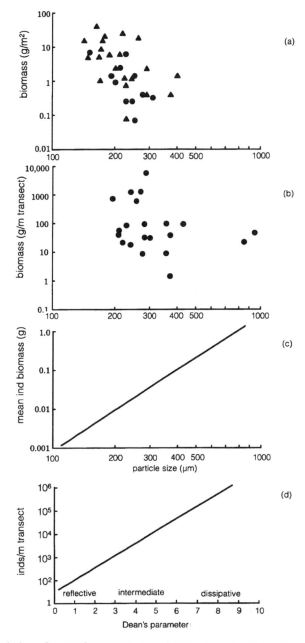

Fig. 2.10 Variation of macrofauna with particle size (a–c) and beach morphodynamic state defined by Dean's parameter (d). (a) Scottish east (▲) and west (●) coast beaches, modified with permission from Eleftheriou and McIntyre (1976) and Eleftheriou and Robertson (1988). (b) Australia, South Africa and Pacific North America, modified with permission from McLachlan (1990). (c and d) South African beaches, modified with permission from Brown and McLachlan (1990).

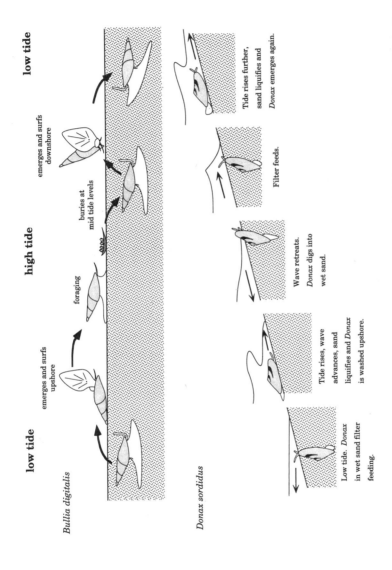

Fig. 2.11 Behaviour of two surfing species on exposed sandy beaches in South Africa. Migrating with the tide probably allows these species to feed in optimum conditions and avoid predators, few of which would be able to forage in the heavy surf. The gastropod *Bullia digitalis* uses its flattened foot as an underwater sail, whilst the bivalve *Donax sordidus* surfs with both the foot and the siphons extended. (Adapted with permission from Branch and Branch, 1981, *The Living Shores of Southern Africa*.)

low tide

high tide

low tide

Bullia digitalis

emerges and surfs upshore

foraging

buries at mid tide levels

emerges and surfs downshore

Donax sordidus

Low tide. *Donax* in wet sand filter feeding.

Tide rises, wave advances, sand liquifies and *Donax* is washed upshore.

Wave retreats. *Donax* digs into wet sand.

Filter feeds.

Tide rises further, sand liquifies and *Donax* emerges again.

populations. Invertebrate biomass is high (McIntyre, 1970), especially in estuarine areas, although diversity generally declines on shores affected by low salinities.

2.4 THE MEIOFAUNA

So far we have concentrated on the larger organisms occurring on rocky, sandy and muddy beaches. Inevitably, shore ecologists focus on those species which are conspicuous and relatively easy to identify from available guides and keys with minimal time and skill required in specimen sorting and preparation. Smaller and less conspicuous species are often ignored altogether. This is regrettable because small organisms may be ecologically all-important. One of these smaller-sized groups is the meiofauna, often extremely abundant in fine-particle beaches and sometimes exceeding the macrofauna in biomass in coarser sands.

The term 'meiofauna' translates roughly as 'the less-than fauna', reflecting their artificial classification on the basis of their small size. For instance, the sediment meiofauna are the organisms in samples that are lost through a 1 mm or 0.5 mm mesh during the process of separating the larger fauna from their sediment environment. Fortunately, this division on the basis of size effectively separates quite different taxa. Although it is appropriate to introduce the meiofauna in the context of fine particle beaches, the group is not in any way restricted to sediments. They also inhabit complex surfaces and the spaces between them such as are found on seaweeds and other structures, and a good review of the ecology of these species can be found in Hicks (1985).

An astonishing diversity of taxa can be found within the meiofauna (Fig. 2.12). All of them are small and many have a worm-like shape. Intertidal sediments tend to be dominated by nematodes which can reach very high densities on finer-particle beaches. Nematodes are usually long and thin, a shape well suited to moving through interstices, although many species are either burrowers or 'sliders'.

The other most widespread taxon is the harpacticoid copepods. These are common on all shores: slender species inhabit the large interstitial spaces found on sandy beaches and bulkier epibenthic and shallow burrowing forms are more common in finer-sediment habitats. Small oligochaetes, archiannelids, acarine mites, acoels and rhabdocoels (turbellarians), together with strange, wholly or predominantly meiofaunal phyla such as gastrotrichs, kinorhynchs, gnathostomulids and tardigrades, will be less familiar to the general intertidal ecologist. The systematics of most of these groups are not easy, especially the soft-bodied taxa, many of which can only be reliably identified from living specimens. However, the identification guides for the most abundant and frequently encountered groups are of exceptional quality and one can now embark on meiofaunal studies with some confidence (Higgins and Thiele, 1988, and references therein).

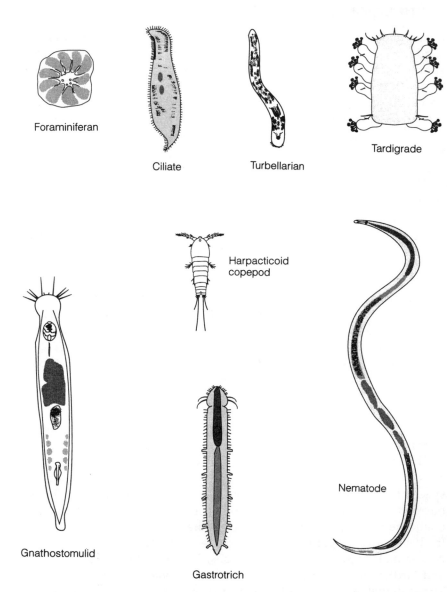

Fig. 2.12 Major meiofaunal taxa found in intertidal sediments (adapted with permission from Higgins and Thiel, 1988).

Sandy beach meiofauna also show zonation patterns in response to tidal height and wave action. Zonation with depth in the sediment is also especially obvious because of their small size. McLachlan (summarized in Field and Griffiths, 1991) recognized four major meiofaunal zones associated with beach physico-chemical conditions (Fig. 2.13).

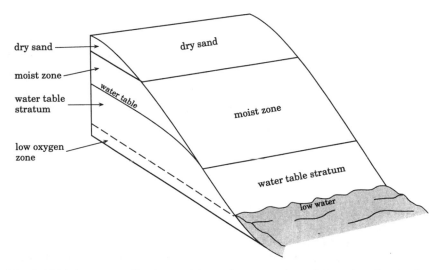

Fig. 2.13 Major zones affecting meiofauna in exposed sandy beaches (see text).

In the upper shore there is a **dry sand zone** which extends down into the sediment to about 15 cm. Here temperatures are often very variable and water saturation can be as low as 50% at low tide. Small nematodes and oligochaetes characterize this zone. A **moist sand zone** extends below the dry sand zone to the permanent water table which reaches the surface at low water mark. Temperatures are fairly constant here and the oxygen saturation is always more than 50% . Harpacticoid copepods, mystacocarids, nematodes, oligochaetes and turbellarians are the common groups here. Underneath this zone is a **water table stratum**, permanently saturated and with oxygen saturations of 40–70%. Nematodes and crustaceans are co-dominant. Deeper still there is a **low oxygen zone** with a sparse population of nematodes extending to depths of a metre or more.

2.5 MACROPHYTES ON SHELTERED SHORES

Macrophytes, absent from a large section of the particle size gradient, begin to reappear on the more sheltered flats (Fig. 1.12), along with many epibenthic animals that are able to move over the stable sediment surface. These are mainly plants with terrestrial evolutionary affinities. Some seagrasses, such as *Zostera* (Fig. 2.14), occur intertidally, but are often more abundant in the sublittoral. In the high intertidal of sheltered flats, salt-tolerant land plants can establish themselves. At high latitudes with a cool climate, these plants form the **salt-marsh**, with the most tolerant species found furthest down the shore. The salt-marsh flora comprises mainly grasses and small shrubby herbs, with the genera *Salicornia*, *Spartina*, *Anthrocnemum* and *Juncus* being widespread in both hemispheres (Chapman, 1977). Salt-marsh is a prominent feature of the high intertidal zone along the east coast of the United States, with over half a million hectares between Rhode Island and Georgia. These assemblages are dominated by *Spartina alterniflora* and have been the focus of much research effort (Chapter 6). In sheltered locations the shore profile is often flat and the land–sea gradient is correspondingly shallow. In consequence, the few salt-tolerant terrestrial species capable of colonizing the high shore may occupy extensive single-species zones up to several hundred metres in width. The Baltic supports a comparable system characterized by beds of the reeds *Phragmites australis* and *Scirpus* spp. (Wallentinus, 1991), although in the low-salinity Baltic, salt tolerance plays less of a role.

At low latitudes, where the environment is warmer and more humid, large trees and shrubs penetrate the higher intertidal, to form the **mangrove** or **mangal** (Chapman, 1977) habitat, adding much structural complexity and biological diversity to the shore (Fig. 2.15). Mangroves do best where the temperature never drops below 20 °C, although some species of tree can tolerate temperatures as low as 10 °C. The tree genera *Avicennia*, *Rhizophora* and *Bruguiera* are widespread and are characterized by extensive modifications of the root systems for support and gas exchange.

Two main groups of mangal vegetation can be recognized: Old World mangal, containing about 60 species of plants, and the less rich New World mangal with only about 10 species (Mann, 1982). These large trees provide a surface for attachment of sessile marine invertebrates whilst the organically rich sediments around the tree prop-roots and pneumatophores support large numbers of crustaceans, molluscs and polychaetes. Fish and crustaceans move in and out with the tide and these areas are important nurseries for the young stages of commercial fish and shellfish (Chapter 6). The tree canopies support a diverse insect fauna as well as roosts and nest sites for insectivorous and piscivorous birds. Larger mammals and crocodiles are a feature of some mangrove forests, particularly in the Indo-Pacific region.

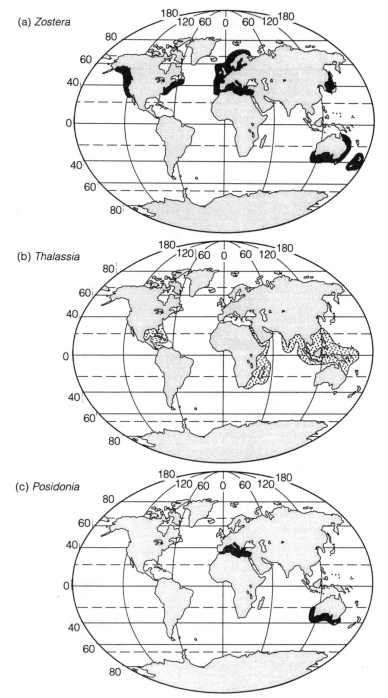

Fig. 2.14 Examples of worldwide distribution of seagrasses: (a) *Zostera*, (b) *Thalassia*, (c) *Posidonia* (data from Den Hartog, 1970).

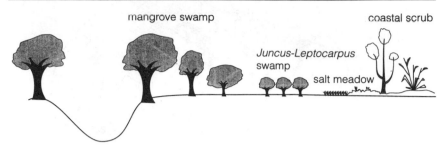

mangrove swamp

coastal scrub

Juncus-Leptocarpus
swamp

salt meadow

Fig. 2.15 Zonation in a New Zealand mangrove habitat (after Morton and Miller, 1973, with permission).

2.5.1 ZONATION IN SALT-MARSHES AND MANGROVES

Many textbooks give generalized accounts of zonation patterns in salt-marshes and mangroves. They emphasize monospecific zones associated with different degrees of tidal inundation and hence salinity stress. (Remember that the salinity stress gradient for terrestrial plants runs in the opposite direction to that of marine organisms colonizing the intertidal from the sea.) These zones are often thought of as temporal stages in a successional sequence generated by pioneer species advancing seawards and stabilizing the sediment. Sediment builds up, allowing less salt-tolerant but more competitive species to colonize. At some point a balance will be reached between the ability of the marsh to stabilize the sediment and the ability of the sea to erode the leading edge of the marsh. This can be dynamic and the mosaic observed may be caused by rare chance events such as very large storms which only occur once every 50 years. Thus, the emphasis is often on a temporal sequence rather than spatial zones, in contrast to views of intertidal seaweed zonation. In many places this simplification does not hold (Fig. 2.16). For instance, in European salt-marshes, a mosaic of assemblage types can often be found and in the mangroves of southern Florida large areas are often dominated by a mix of mangrove species (Snedaker, 1989). In such areas it is probably more appropriate to recognize major vegetation types rather than particular species zones (Fig. 2.17).

2.6 ZONATION ALONG THE ESTUARINE GRADIENT

Zonation along the length of an estuary reflects biological responses to the salinity gradient and its associated gradient of particle size (Chapter 1). Carriker (in Lauff, 1967) provides a broad classification of ecological groups with respect to salinity ranges of the 'Venice system'. In his scheme, the **oligohaline** group is represented by a few species (e.g. some oligochaetes) that are essentially freshwater but which can tolerate

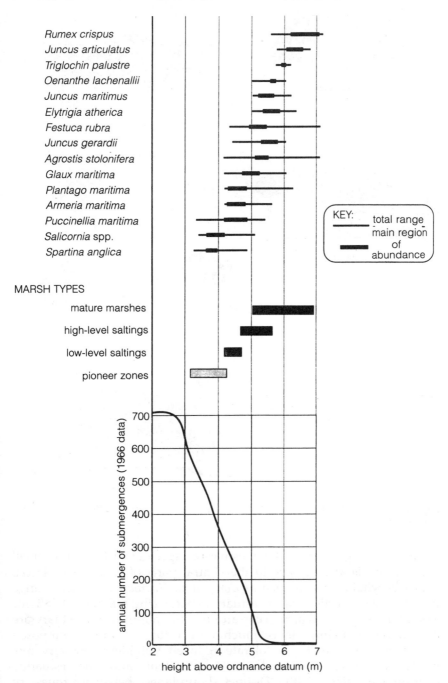

Fig. 2.16 Zonation in a British salt-marsh. (Reproduced with permission from Gray and Scott, 1987, *Morecambe Bay: an assessment of present ecological knowledge.*)

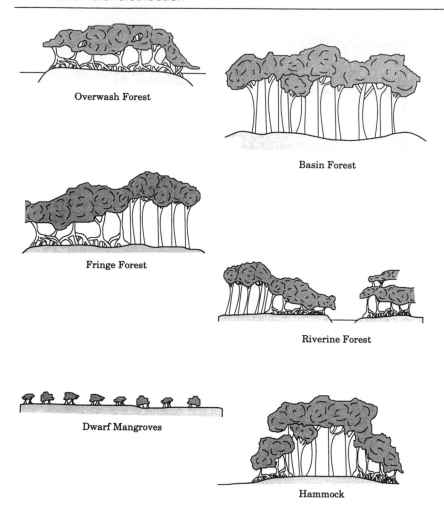

Fig. 2.17 Major vegetation types in North American mangrove systems (adapted from Snedaker, 1989, with permission).

salinities of up to 5‰. True estuarine species, such as the small polychaete *Manayunkia*, live in the central parts of estuaries (5–18‰), whilst **euryhaline** forms such as *Corophium, Hydrobia* and *Nereis* range from the sea to sections of the estuary experiencing salinities of 18‰ or less. Marine species which occur around the mouth of the estuary are classed as **stenohaline**. Hence, much of the estuarine fauna is composed of marine forms that can tolerate reduced salinities, and very few freshwater species penetrate to any extent into the estuarine environment (Fig. 2.18). Distinct boundaries between zones of individual species are usually not possible to locate along the length of an estuary, because of the secondary effects of sediment grade on the distribution and abundance of the fauna and the dynamic nature of the salinity gradient itself.

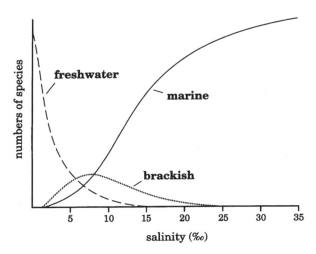

Fig. 2.18 The relative numbers of freshwater and marine species along an estuarine salinity gradient.

On a much larger scale, the Baltic Sea resembles a gigantic estuary or fjord (Wallentinus, 1991). However, the Baltic differs from true estuaries in the almost complete absence of tides so that the salinity variation at any one location is within the range 0.5–1‰ over much of the area, except in places influenced by significant river run-off or meltwater. Zonation patterns are much clearer (Fig. 2.19) because the salinity gradient is stationary. Perhaps because of this stability, most Baltic species seem to live close to their lower salinity tolerance limit (Wallentinus, 1991).

OVERVIEW

Zonation patterns are most striking on rocky shores and the broad patterns of zonation are apparent on many shores worldwide. These change with exposure, with zonation being uplifted with increasing wave action. The Stephensons' three-zone scheme holds reasonably well for more exposed shores worldwide and provides a useful descriptive framework if not followed slavishly. The neat horizontal bands of virtually monospecific stands of several fucoids on sheltered North Atlantic rocky shores are one of the best examples of zonation. Elsewhere on rocky shores, zones may consist of a mixture of species (for example, the highly diverse low-shore red algal turf communities found in many parts of the world) and there can be considerable overlap of species

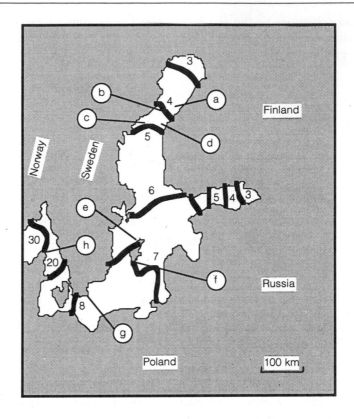

Fig. 2.19 Large-scale zonation in the Baltic, showing examples of inner distributional limits of marine species: a, the isopod *Idotea emarginata*; b, the barnacle *Balanus improvisus*; c, the mussel *Mytilus edulis*; d, the seaweed *Fucus vesiculosus*; e, the flatfish *Pleuronectes platessa*; f, the seaweed *Fucus serratus*; g, the kelp *Laminaria saccharina*; h, the limpet *Patella vulgata*. Thick black lines are surface isohalines (‰) (redrawn with permission from Wallentinus (1991) *Ecosystems of the World 24. Intertidal and Littoral Systems*, Elsevier Science, p. 84).

distributions giving diffuse zonation on many shores. Zonation patterns are less apparent on other types of shore but still occur in response to the various environmental gradients. In many cases (e.g. salt-marshes, mangroves) the zones consist of multispecific groups or assemblages rather than dominant single species. In general, zonation is most apparent on steep environmental gradients where microhabitat variations are unimportant. Microhabitat variability, temporal fluctuations, biological interactions and chance events can modify all these patterns which in some places can vary considerably with time. The reasons for these distribution patterns are explored in the next chapter.

Causes of zonation 3

Zonation is one of the most obvious features of a rocky shore and, with a little effort, it can also be recognized on sandy beaches and mudflats. Each species has an upper and lower distributional limit along the vertical intertidal gradient, and the factors which set those limits have long attracted the attention of shore ecologists, particularly for plants and sessile animals. In this chapter we review the various ideas put forward to explain zonation. We start with the early ideas which emphasized the direct effects of physical factors. We then go on to describe how more recent manipulative experiments have revealed the significance of biological factors for setting distributional limits. Because the majority of studies have been on rocky shores, we will deal with these first. We then consider the causes of distribution on depositing shores and other environmental gradients.

3.1 VERTICAL ZONATION ON ROCKY SHORES

Almost all the plants and animals inhabiting rocky shores are marine in ancestry and character. The number of terrestrial species that permanently live in the intertidal zone is very small. The intertidal zone is therefore characterized by aquatic organisms that require some degree of wetting. Much of the early work on the causes of zonation demonstrated that species at different tidal levels had differing abilities to survive out of water: species at high shore levels have appropriate morphology, physiology and behaviour to allow them to survive for long periods without being submerged in sea water; whereas low-shore species are essentially fully marine and can only cope with brief periods in air (emersion). Most of this work was done by subjecting different species of plants or animals to similar stress regimes of high temperature and low relative humidity in the laboratory (e.g. Baker, 1909; Gowanloch and Hayes, 1926; Broekhuysen, 1941; Biebl, 1952; Southward, 1958; review: Newell, 1979). Not surprisingly, there is a general correlation between degree of tolerance to physical factors and shore position for the majority of intertidal species (Fig. 3.1).

This kind of work was done in parallel with a whole series of studies attempting to link zonation directly with physiological tolerance of the emersion/immersion regime at different tidal levels. These studies are worth briefly recounting as they show how our ideas have changed, although some of the ideas still persist in textbooks and papers.

3.1.1 OLDER IDEAS ABOUT THE CAUSES OF ZONATION

Colman (1933) suggested that certain **critical levels** occurred on the shore where rapid changes in the duration of emersion or immersion were experienced. These levels were apparent on emersion curves constructed by calculating from tide tables the percentage of time that a particular part of the shore spends in air over a year (Fig. 3.2). He showed that several species reached their upper and lower limits in the region of these critical levels, as follows:

I   – the lower limits of many intertidal species occurred between extreme low water of spring tides (ELWS) and mean low water of spring tides (MLWS);
II   – sublittoral fringe species had their upper limits between MLWS and mean low water of neap tides (MLWN);
III   – the upper limits of several species occurred at the extreme, lowest high water level of neap tides (E(L)HWN).

These levels represent the **upper limits** of most midshore and sublittoral marine species respectively. The second and third critical levels also set the **lower limits** for several higher-shore species. Colman suggested that some parts of the intertidal gradient were critical in that conditions change so rapidly within these sections that many intertidal species cannot cope. His approach was a valuable attempt to match distributions with a measure of the stress along the environmental gradient and has been adopted by several other authors, such as Doty (1946) for North American shores and Evans (1947) in Europe (review: Swinbanks, 1982).

Forty-five years after Colman's ideas were published, Underwood (1978) quantitatively re-examined the question. The emersion curve he produced from tide table data was much smoother than that derived by Colman, with no suggestion of critical levels (Fig. 3.2). Furthermore, the claim that the lower and upper limits of several species coincided at

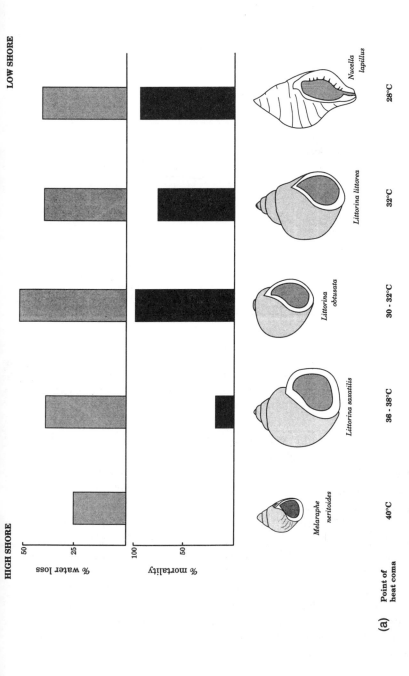

Fig. 3.1 Examples of water loss/mortality rates in high- and low-shore species. (a) Snails occupying different parts of the intertidal gradient on British shores (based on data in Lewis, 1964).

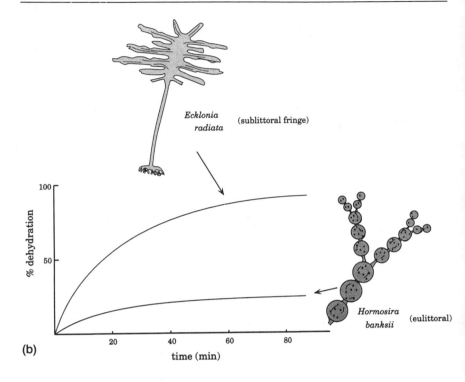

Fig. 3.1 (b) New Zealand eulittoral and sublittoral fringe seaweeds (after Morton and Miller, 1973, with permission). In both cases, species from higher shore levels are more tolerant than those from lower on the shore.

distinct points along the intertidal gradient was not supported by a more rigorous statistical analysis of several UK rocky shores. Underwood showed that the lower and upper limits appear to occur haphazardly along the shore. Thus, there are in fact no physical or biological grounds to support the idea of critical levels (Fig. 3.2). Hartnoll and Hawkins (1982) presented tidal emersion curves from recorded, as opposed to predicted, levels for several locations in the UK. On one or two shores there was some evidence for regions with steep sections to the curve, but it was not possible to make any generalizations concerning the occurrence of critical levels. After much discussion, Lewis (1964) concludes 'even when we have discounted the great distortions of zonation caused by wave action, the grounds for postulating the existence of recurring critical levels are very dubious'. We concur with this view.

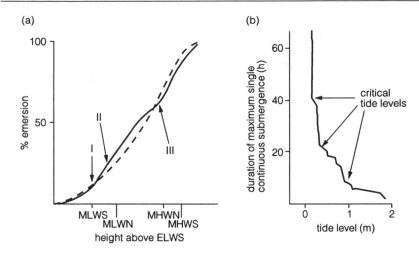

(a)   (b)

Fig. 3.2 (a) Emersion curves for Devonport (Plymouth) calculated by Colman (solid) and Underwood (broken). Critical levels I–III as in text. Modified with permission from Underwood (1978). (b) Critical tidal levels defined by Doty (1946) for San Francisco Bay, USA (after Carefoot, 1977, with permission).

Another early idea linking distribution patterns with the physical environment was that proposed by marine botanists in the 19th century (Engelmann, 1884; Oltmann, 1892). As light penetrates water, its quality changes and its quantity diminishes. Red and blue wavelengths are filtered out most rapidly, and green wavelengths penetrate the furthest. If different wavelengths are available at different depths then this might be reflected by the occurrence of green seaweeds towards the top of the shore, reds at the bottom and browns in the middle, because their photosynthetic pigments are complementary to the colour of the transmitted light (Lüning and Asmus, 1991). At first sight this seems to be the case: on European shores greens tend to dominate the high shore, browns the midshore and reds the low shore and subtidal. However, although the physical appearance of the shore implies that green seaweeds do best at higher levels and reds do best at lower shore levels, this may simply reflect the high abundance of one or two species within these zones. The association between seaweed colour and shore level is less convincing when the relative numbers of species of the three seaweed groups are examined in relation to shore position (Table 3.1). Also, if seaweeds are thick, not allowing light penetration, their colour will be irrelevant because they will absorb all the incident light energy anyway, explaining why thick green seaweeds such as *Codium* can live at depth (Ramus *et al.*, 1976). Only for species like *Ulva* and *Porphyra*, which are thin and membranous so that light passes through the lamina, will the nature and amount of pigment closely reflect ambient light conditions – yet *Porphyra* is a red alga found high on the shore and *Ulva* can be found deep in the subtidal. This generalization also does not appear to hold up to close scrutiny.

Table 3.1 Percentage of species within each of the main seaweed groups that occur within different depth ranges (from Chapman, 1979)

Depth range	Greens	Browns	Reds
Intertidal only	57	30	15.2
Intertidal and sublittoral	28.5	30	18.5
Sublittoral only	14.5	40	66.3

3.1.2 A SYNTHESIS OF MORE RECENT IDEAS ON ZONATION

These older studies linking the physical environment with zonation patterns emphasized the direct effects of physical factors – that is species had distributions directly set by their physical tolerances. In the rest of this section the partial fallacy of these ideas is exposed. It is best to consider upper and lower limits separately and to draw the distinction between plants and sessile animals on the one hand and those animals which can move on the other. Once a plant or sessile animal becomes attached, it cannot change its position and hence regulate the conditions it experiences. A mobile animal can move elsewhere to avoid adverse conditions – either to a different zone or to a more favourable microhabitat such as a crevice or pool.

Causes of upper limits of plants and sessile animals on rocky shores

Over the years, many laboratory experiments have shown that the tolerance to aerial exposure of intertidal plants and animals is greater in higher-shore species than in those from lower on the shore (reviews: Newell, 1979; Norton, 1985). So many of these kinds of experiments have been carried out that it is generally assumed that physical factors such as desiccation or thermal stress directly set the upper distributional limits of any rocky shore species. However, if the periods survived are looked at carefully, these are often much longer than the duration of emersion likely to be encountered in the field. An exception is very high on the shore (e.g. the zone above neap tides in the North Atlantic), where several days can pass before inundation occurs again and organisms can indeed reach their physiological tolerance limits. Here we critically assess the evidence that physical factors set the upper distributional limits of rocky shore species.

Field observations have shown that physical factors can directly kill the adults or young stages of plants and sessile animals. For instance, on sheltered shores in the British Isles, high-shore *Pelvetia* and *Fucus spiralis*, and more rarely the midshore *Ascophyllum nodosum*, all suffer bleaching and die during periods of hot weather (Schonbeck and Norton, 1978; Hawkins and Hartnoll, 1985; Norton, 1985). This does not occur in every year, only in occasional hot summers, and effects are exaggerated during periods of rapid change, such as a warm spring. Fucoids are particularly vulnerable to sudden changes in temperature and relative humidity, although they can increase tolerance ('drought-hardening') in response to gradual change (Schonbeck and Norton, 1979a). On moderately exposed shores, *Laminaria digitata*, along with many low-shore red seaweeds, is also susceptible to spells of hot weather and extreme cold (Todd and Lewis, 1984; Hawkins and Hartnoll, 1985). Juveniles of the barnacle *Semibalanus balanoides* sometimes settle above the adult zone and die during a hot spring (Connell, 1961a; S.J.H. pers. obs.). In a cool spring, a slight range extension may be possible if the juvenile phase survives, because the adults are much less vulnerable. This happens when settlement is heavy and the more favoured sites are already occupied, so that cyprids have to make do with less favourable habitats (Crisp, 1974). Every few years a hot summer will kill these 'out-of-zone' adults, and in these years they will not be replaced by a new settlement. These two examples clearly show that upper distribution limits may be trimmed by occasional extreme abiotic conditions.

Various experimental studies confirm these field observations. These have included upshore extensions of species following the creation of artificial run-offs to make the shore wetter (Dayton, 1975) and transplantation experiments where small pieces of rock are chipped off and cemented to the higher shore using cement or resin glues. The latter approach was first used by Hatton (1938) in France who transplanted barnacles upshore and showed that younger individuals died quicker than older individuals. Similar experiments have been repeated with barnacles by Foster (1971b) and with fucoids by Schonbeck and Norton (1978, 1979a,b,c; Norton, 1985, for a review). In each case species transplanted higher up the shore died. These field results have been backed up by laboratory culture experiments with fucoids, showing that growth is slow or ceases altogether for low-shore species kept in conditions simulating those found higher on the shore.

A more dramatic illustration comes from observations made on the shore following earthquakes in Chile (Castilla and Olivera, 1990) and Mexico (Bodin and Klinger, 1986). Species 'transplanted' upshore by uplifting died and zonation patterns took some time to re-assert themselves. More sinisterly, similar effects have been noted after nuclear bomb testing (Lebednik, 1973) – one experiment where both authors would not like to see any replication!

The upshore transplantation protocol has been rightly questioned by Underwood and Denley (1984). As they point out, 'would the death, through excessive heat, of polar bears transplanted to the Sahara really reveal the causes of the southern geographical limit of the bears?' Whilst it is likely that physical factors of some kind may set upper limits, the precise mechanism involved might be quite complex, and one should not assume that desiccation is entirely responsible. Quite subtle sublethal factors could operate cumulatively towards the upper limits of a species, leading to reduced growth and eventually to death. In intertidal macroalgae, what seems to matter most is the critical water content to which they can be dried and still retain their photosynthetic ability upon re-immersion (Chapter 5).

We have laboured these points because many reviews and other textbooks firmly state that upper limits of intertidal species are generally set directly by physical factors. Whilst this may be the case for some species, particularly those from the upper shore and midshore, it is clearly not true for all shore organisms. Several mid- and low-shore species (e.g. *Fucus vesiculosus*, *F. serratus*) have not been shown to die at their upper distributional limits, even during extremely hot weather when some higher-shore species were badly affected (Schonbeck and Norton, 1978; Hawkins and Hartnoll, 1985). For such species a biological explanation may be more appropriate and this possibility has been explored on seaweed-dominated shores in the British Isles, where canopies are dense and competition important. Thus, *Fucus vesiculosus* has been induced into the *F. spiralis* zone by removal of the species zoned immediately above; *F. serratus* can be similarly induced upward into the *F. vesiculosus* and *Ascophyllum* zones when the species zoned immediately above has been removed (Hawkins and Hartnoll, 1985). Juvenile *Laminaria* can grow further upshore when *Fucus serratus* is experimentally removed (Hawkins and Hartnoll, 1985), but these plants will not survive many years because of occasional hot, dry summers (Hawkins and Hartnoll, 1985; Hill, 1993). Hill (1993) has recently shown that *Laminaria hyperborea*, normally a subtidal species, can extend upwards into the intertidal if *Laminaria digitata* is removed. Although the design of many of these experiments has been criticized (Underwood, 1991), it seems that for most of the time, competition sets the immediate upper limits of these species, although any plants that manage to establish in chance storm-created clearings or experimental patches are killed sooner or later by physical factors.

Grazing has also been shown to be important in setting the upper distributional limits of seaweeds. After the *Torrey Canyon* oil spill, much of the limpet population was killed by the detergents applied to the beaches (Chapter 7). The reduced grazing pressure led to an upward extension of low-shore reds and kelps (Southward and Southward, 1978). The first small-scale experimental demonstration of grazers setting the upper limits of seaweed was done by Underwood (1979, 1980) and Underwood and Jernakoff (1981) in Australia. In their studies, removal of grazers allowed foliose algae to extend further upshore. Similar limpet-removal experiments in the South West of the UK (see photographs in Hawkins and Jones, 1992) allowed a low-shore red algal turf to extend upshore, and, on the Isle of Man, *F. serratus* has been induced to grow far above its normal zone in limpet-removal areas (Hawkins and Hartnoll, 1985; unpublished data). In Chile, the upper limit of *Codium dimorphum* seems to be partly set by a combination of the direct effects of hot, dry summer weather leading to bleaching and grazing (Ojeda and Santelices, 1984). In tropical areas, upper limits are more likely to be set by physical factors, although fewer studies have been made than in temperate regions (Williams, 1993).

Causes of lower limits of rocky shore plants and sessile animals

It might be thought that many intertidal species will have become so highly adapted through evolution to living in a drying environment that they can no longer cope any more with wetter, more marine conditions. Whilst there is much evidence that upper limits are set largely by the direct action of physical factors, it is now clear that lower limits are unlikely to be set directly in this way. As far back as 1909, Baker was led to this conclusion from the results of her experiments where she cultured intertidal brown seaweeds in jars simulating different emersion/immersion regimes. Connell (1973) quotes a prophetic statement by her: 'On the whole it seems as though the greatest competition has been called into play in the lowest zones, the dry and uncongenial regions of the upper shore being left to the most tolerant forms, which, if left to themselves, are able to grow anywhere on the shore'. Supporting evidence for this idea came from the observation that intertidal species could thrive in the subtidal zone: ephemeral algae, such as *Enteromorpha*, and barnacles, such as *Semibalanus*, foul the bottoms of ships, and mussels grow on permanently submerged culture ropes. In special circumstances such as in the Baltic where the normal dominant species do not occur in the subtidal zone due to the low salinity, normally midshore species, such as *Fucus serratus*, grow luxuriously in the shallow subtidal (Chapter 2). All these examples are of aquatic organisms which respire with gill-like structures or, in the case of algae, take in nutrients via a thallus and it is perhaps not surprising that they do so well when permanently submerged. Finally, it has long been appreciated that low-shore species are usually much larger and grow more rapidly than similar species higher on the shore (Stephenson and Stephenson, 1949; Southward, 1958; Lewis, 1964). These observations led many of these early workers to suggest that biological factors such as competition, grazing and predation might be responsible for the lower limits of intertidal organisms.

Clear evidence of the importance of biological factors came in a series of classical experiments in which shore organisms were manipulated in the field. Experiments by Connell (1961a,b) and Paine (1966, 1969) had considerable international impact, and have since found their way into most general ecology textbooks (e.g. Odum, 1971; Barnes and Hughes, 1982; Begon et al., 1986). The experimental analysis of biological interactions was made easy because the most dominant species are sessile and permanently attach to the rock surface. Thus, the resource in greatest demand on rocky shores is the surface of the rock itself – primary space. On a great many shores primary space appears limiting: every square centimetre of the rock surface is occupied, especially towards the lower shore levels (Chapter 1). This finite resource can only be renewed by the removal or loss of organisms already occupying it, through grazing, predation or physical disturbance. This renewal process can be easily quantified and the effects of interactions clearly seen.

Connell (1961a,b) studied the zonation of barnacles at Millport on the west coast of Scotland (Fig. 3.3). High on the shore there is a zone of *Chthamalus* (called *Chthamalus stellatus* by Connell, this species has since been split into *Chthamalus stellatus* and *C. montagui* (Southward, 1976) and at Millport only *C. montagui* is present). Below this species a larger, faster-growing barnacle occurs (*Semibalanus balanoides*, then called *Balanus balanoides*). Connell found that the larvae of *Chthamalus* could normally settle below their main zone, but when *Chthamalus* were transplanted downshore on rocks they would only survive and grow if *Semibalanus* were removed. If this was not done, *Semibalanus* would undercut and crush the *Chthamalus*, excluding them from lower parts of the shore by interference competition. Turning his attention to the effects of predation, Connell excluded the dogwhelk *Nucella lapillus* from the lower part of the *Semibalanus* zone using small, wire mesh cages. This experiment demonstrated that whelk predation normally prevents *Semibalanus* from extending its distribution further downshore, although competition from turf-forming seaweeds and the large low-shore species *Fucus serratus* was also suggested as important and this has subsequently been confirmed (Hawkins, 1983). Similar studies on the high intertidal *Semibalanus glandula* in North America

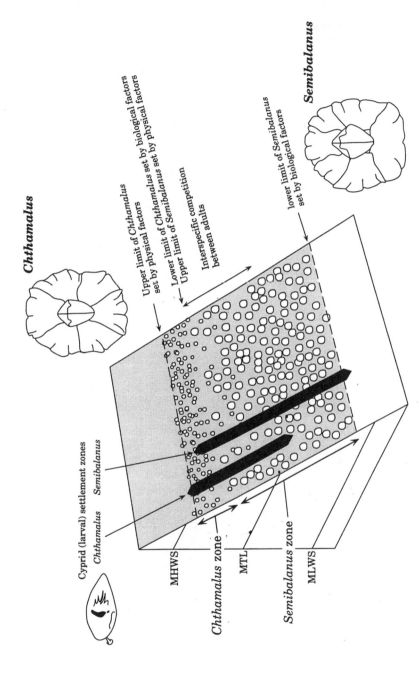

Chthamalus

Semibalanus

Upper limit of *Chthamalus* set by biological factors
Upper limit of *Chthamalus* set by physical factors
Lower limit of *Chthamalus* set by physical factors
Upper limit of *Semibalanus*
Interspecific competition between adults

lower limit of *Semibalanus* set by biological factors

Cyprid (larval) settlement zones
Chthamalus   *Semibalanus*

MHWS

Chthamalus zone

MTL

Semibalanus zone

MLWS

Fig. 3.3 Zonation of *Chthamalus montagui* and *Semibalanus balanoides*, Isle of Cumbrae, Scotland (based on Connell, 1961a,b).

showed that predation by the whelk *Thais* (now called *Nucella*) *lamellosa* was largely responsible for preventing this barnacle from occupying lower shore levels (Connell, 1970).

An equally impressive demonstration of the importance of predation in setting the lower limits of sessile marine invertebrates was carried out by Paine in Washington State on the north-west coast of the USA. He cleared the large starfish *Pisaster ochraceus* from below the zone of the mussel *Mytilus californianus* over a five year period and achieved a downward extension of *Mytilus* of nearly 1 m at one site, and 2 m at another (Paine, 1971, 1974). Paine repeated this work in New Zealand (Fig. 3.4) where he removed both a predator (the starfish *Stichaster australis*) and a plant competitor (the large bull-kelp *Durvillea antarctica*) beneath the green mussel zone (*Perna canaliculus*). This allowed the mussel to extend its range downshore considerably in just nine months (Paine, 1971).

In the 1970s, attention turned to the major zone-forming algae. A series of experiments on intertidal fucoids on the Scottish west coast

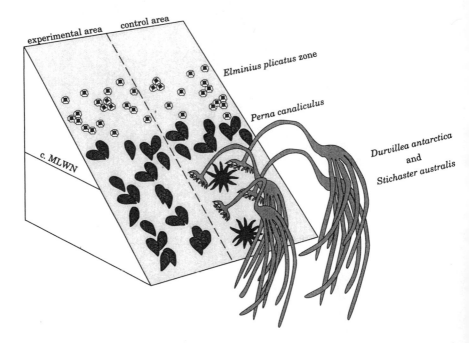

Fig. 3.4 Zonation of the green mussel *Perna canaliculus* on the lower shore at Piha, northern New Zealand, nine months after experimental removal of the kelp *Durvillea antarctica* and the starfish *Stichaster australis* (from data in Paine, 1971).

showed that the lower limits of *Pelvetia canaliculata* were set by competition with *Fucus spiralis* which occupies the zone immediately below. When *Fucus spiralis* was removed *Pelvetia* extended its range downshore. Transplants of these species to lower levels showed that physical factors were unlikely to be responsible as *Pelvetia* and *Fucus spiralis* grew better than in their usual zones (Schonbeck and Norton, 1980). Culture experiments revealed that lower-zoned species of *Fucus* performed better than those higher on the shore when cultured under conditions simulating the low-shore: *Fucus serratus* grew better than *F. vesiculosus* which in turn grew better than *F. spiralis*. Further work on the Isle of Man induced *F. spiralis* into the *F. vesiculosus* zone, and *F. vesiculosus* into the *Ascophyllum* zone (Hawkins and Hartnoll, 1985). Similar studies in Nova Scotia indicate that *F. spiralis* is restricted to the high shore by competition with *F. vesiculosus* (Chapman, 1990). Kelps have been shown on the Isle of Man to exclude *Fucus serratus* from the sublittoral fringe and shallow subtidal (Kain, 1975; Hawkins, 1979; Hill, 1993). *Ascophyllum nodosum* is an enigma: it grows slowly but it still dominates the mid-tidal regions. This has been attributed to its unpalatability to grazers and its ability to proliferate vegetatively (Norton, 1985; Vadas and Elner, 1992; Lazo *et al.*, in press).

In New England, Lubchenco (1980) showed that red algal turf-forming seaweeds (*Chondrus* and probably *Mastocarpus* (Vadas and Elner, 1992)) restricted the downshore extension of *Fucus distichus*. Chapman (1990), working in Canada, has more recently shown that *F. distichus* outcompetes the higher-zoned *F. spiralis*. Thus on both sides of the Atlantic, competition has been shown to be important in setting the lower limits of intertidal seaweeds. Similarly, on the Pacific coast of the USA the lower limits of *Iridaea cordata* are set by competition with the lower-zoned *Laminaria saccharina* (Hruby, 1976), and further south, in Chile, the lower eulittoral species *Codium* and *Gelidium* will extend downshore if the kelp *Lessonia nigrissens* is removed (Ojeda and Santelices, 1984). The fast growth and large size of *Lessonia* and *Durvillea* in Chile allows kelps to dominate the low shore, pushing a zone of calcareous crusts (*Mesophyllum* spp.) into the subtidal and displacing the cushion and fleshy crust seaweeds upshore (Santelices, 1990).

Whilst competition is important in setting seaweed lower distribution limits, there are few examples of lower limits being set by grazing (review: Hawkins and Hartnoll, 1983a). One of the most dramatic examples was a large-scale experiment carried out on the Isle of Man (Lodge, 1948; Burrows *et al.*, 1954), which involved clearing all limpets (*Patella vulgata*) and larger seaweeds from a 10 m wide strip running the length of the intertidal gradient. Three years later, the patterns of distribution of the major seaweeds (mostly fucoids) were very different from those seen on undisturbed shores. Species normally only found at high shore levels could now be found within the strip at lower shore levels and vice versa for low-shore species. Where fucoid cover is patchy and discontinuous, the ability to grow quickly and reach a size (> 5 cm) immune from limpet grazing is probably crucial for survival. Species adapted to a particular zone would be expected to grow faster and, on average, escape grazing more frequently (review: Hawkins and Hartnoll, 1983a). Those settling out of zone are likely to be more vulnerable to grazers and are usually removed quickly.

Grazing also seems to be important in setting the lower limits of the larger subtidal kelps. For instance, when the urchin *Echinus esculentus* was removed from a barren area immediately below the lower limit of the subtidal kelp *Laminaria hyperborea*, the substratum became dominated by kelps (Kain and Jones, 1967). In the subtidal of the Gulf of Maine and in Canada, the lower limits of the kelps *Laminaria digitata* and *L. saccharina* seem to be determined by grazing by the sea urchin *Strongylocentrotus* (Witman, 1987). Removal of urchins allowed the kelps to extend several metres deeper into subtidal mussel (*Modiolus*) beds, where they were eventually responsible for the dislodgement of the mussels during storms. Overgrowth and dislodgement by kelps probably prevents the mussels establishing in shallower water. Witman (1987) suggests that the urchin–mussel association is a facultative mutualism: urchins prevent overgrowth by kelps and mussels provide a refuge for urchins. Of course, at greater depths light will become a limiting factor for kelps.

In all the above examples, the lower limits of major zone-forming organisms were set exclusively by biological interactions of various kinds. As is the case for physical factors and upper limits, the role of biological factors in setting lower limits has become something of a dogma over the last 10–20 years.

However, there still remains the possibility that some intertidal plants and animals are stressed by too much submersion. For instance, the high-shore alga *Pelvetia*, which often spends days out of water, does well at first when transplanted downshore, growing more rapidly than in its own zone (Schonbeck and Norton, 1980), but eventually it rots and dies. This appears at first sight to be a direct effect, but it could in fact be due to a pathogen absent from the high shore (Norton, 1985). To the best of our knowledge there are no other good examples, and early work in which *Fucus* died under constant immersion (e.g. Fischer, 1929) may have been a result of poor culture facilities. A tantalizing hint was given by Burrows (1988), who used an artificial tidal aquarium in which newly settled *Chthamalus montagui* (a high-shore barnacle) showed highest mortality under simulated low-shore conditions. We think this is an interesting area for future work, but be prepared for negative results!

Effects of the behaviour of the larvae of sessile animals and of movement by mobile animals on zonation patterns

In many cases the behaviour of settling larvae is the direct determinant of zonation (Chapter 5). A variety of cues ensures that settlement outside the normal zones is minimized (Crisp, 1974; Denley and Underwood, 1979; Kirchman *et al.*, 1981; Rittschof *et al.*, 1984; Pawlik and Hadfield, 1990; Durante, 1991; Pawlik, 1992). As space is filled up, or in years with heavy settlement, larvae arriving late have to settle in less favourable positions, because as energy reserves run out the larvae become less choosy. Many of these larvae die but some colonize suitable new space. Quite often it is difficult to induce larvae to settle outside their normal zone even if free space is opened up. In some barnacle species (e.g. *Semibalanus balanoides*), rocks bearing adult barnacles have to be transplanted downshore to induce out-of-zone colonization.

Some invertebrates only occur on or among a host organism (often a plant, but sometimes animals such as mussels) which provides a habitat and sometimes food (review: Williams and Seed, 1992). Thus the vertical zonation of species is determined by the secondary space created by the host. In invertebrates this close relationship is usually maintained at settlement by larval behaviour. For example, the classic work of Knight-Jones (1951) showed that certain species of spirorbid tube worm were attracted and settled if presented with their host plant (review: Newell, 1979).

Early work on many species of mobile animals showed that high-shore forms were less tolerant than lower-shore forms. Again it was assumed that their upper limits were set directly by physical factors. As far back as 1929, field observations by Orton of limpets during a spell of hot weather showed them to be dying at their upper distributional limits. Wolcott (1973), in a very thorough study of several species of intertidal acmaeid limpets, measured a variety of conditions in the field and carried out realistic tolerance experiments. He showed that only the highest-zoned limpet ever reached its tolerance limits in the field – and then only from desiccation and ensuing osmotic stress. The behaviour of lower-zoned limpets kept them well within their tolerance limits. Wolcott hypothesized that only when a species borders an unexploited resource – in this case a luxuriant lawn of algae which develops in winter above the grazed zone – do animals really push their tolerance limits as the benefits of energy gain offset the risk of death. During spring, when conditions changed rapidly, some limpets were caught out and died. This also happens on British shores (Lewis, 1954; S.J. Hawkins, pers. obs.).

There are many other examples of upper distributional limits being set by physical extremes in mobile animals, particularly sedentary or homing species such as limpets in which there is reluctance to move (Underwood, 1979; Branch, 1981). There is also evidence that the lower limits of grazers can be set by fast-growing algae low on the shore. For instance, in Australia, limpets are swamped by algae and cannot maintain clear areas for their microphageous mode of feeding. Removal of these algae can lead to a downshore extension of limpets (Underwood and Jernakoff, 1981). In the British Isles, removal of red algal turf can also result in modest downshore extensions of limpets (S.J. Hawkins, unpublished).

For much of the time, however, behaviour patterns seem to be directly responsible for setting the distribution patterns of mobile animals. As with settling larvae, their behaviour ensures that they live within a region in which they can survive the physical environment and where the risk of biological interactions is reduced. Competition and predation have been hypothesized as being important in setting lower limits of mobile animals, but firm evidence is hard to find (Underwood, 1979; Underwood and Jernakoff, 1981; Yamada and Mansour, 1987; Stevenson, 1992; Xue, 1992).

3.2 DETERMINANTS OF ZONATION ON OTHER KINDS OF SHORES

The vast majority of studies of zonation patterns come from bedrock areas in temperate systems. This is because the zonation patterns are often strikingly clear and easily quantified. Also, experimental manipulations are easier to conduct and interpret in a two-dimensional habitat. On subtropical and tropical rocky shores, the universal features of zonation are recognizable, but zonation patterns of individual species within these major zones are often far from clear because of the high diversity and patchy distribution of species (Brusca, 1980). On boulder and stone shores, the instability and fragmentary nature of the habitat can obscure zonation patterns, even in temperate regions. Only on larger boulders in more sheltered conditions are biological processes likely to be similar to those seen on rocky cliffs. Where small boulders and stones are continually moved around by wave action, the fauna and flora of the shore will remain in the early stages of succession and will be characterized by opportunists, such as ephemeral green algae (Chapter 4). Biological interactions are unlikely to be important in setting distributional limits on such shores, although physical factors may still determine the upper shore limits of opportunists.

3.2.1 ZONATION PATTERNS ON SANDY BEACHES AND MUDFLATS

Gravelly and coarse sandy beaches that experience a high degree of exposure support few invertebrates (Chapter 2). Only in less exposed conditions, where beaches have finer sediments and zonation patterns are well defined, have the factors setting upper and lower distributional limits been investigated. As on rocky shores, the higher the shore level, the drier the beach becomes, but desiccation is probably not such a limiting factor for organisms protected to some degree by the sediment in which they live (Chapter 5). At low tide, there remain significant amounts of sea water below the sediment surface held by capillary forces. Even where the low tide water table is at some depth, the sand particles nearer the surface are often surrounded by a thin film of water.

In contrast to species of rocky shores, many sediment species are mobile. Damper conditions can be found not only by moving lower down the intertidal gradient, but also by burrowing deeper, although this will usually be accompanied by decreased oxygen concentrations. The fauna can also move along the intertidal gradient with the ebb and flood of the tide. This is especially true for animals like *Bullia* and *Donax* on more exposed sandy beaches (Chapter 2; Fig. 2.11). Their positional behaviour will keep species in shallow areas covered by the tide, but out of reach of larger fish and avian predators. Thus, on exposed sandy shores and surf beaches there is unlikely to be a clear relationship between the zonation patterns of the fauna and particular tidal levels when the tide is out. The physical characteristics of the beach and its profile change seasonally and many of the invertebrates migrate tidally as well as seasonally in response to changes in wave exposure (Chapter 2). Indeed, Brown and McLachlan (1990) state: 'on open ocean beaches, zonation as recorded by the researcher reflects faunal distribution only at the time of the investigation.'

Whilst this may be true for exposed sandy beaches, it is not the case for finer-particle, less exposed areas. Here, the fauna moves around much less and many species construct semi-permanent burrows and tubes in the sediment. On the most sheltered shores, macrophytes become established. Zonation patterns are more consistent on these shores, although the limits of most species are much more diffuse compared with those seen on rocky shores, and distributions of some species span the entire shore gradient (Fig. 3.5). How much of this diffuseness is due to the three-dimensional nature of the habitat, the potential mobility of some species or the destructive nature of our sampling methodology, will vary from beach to beach and with investigator. Whatever its cause, it implies that physical factors do not trim distributional limits of sand and mudflat species in quite the same way as they do on rocky shores. However, Swinbanks (1982) argues that the upper distributional limits of several species coincide with so-called critical tidal levels. These include several large burrowing species such as

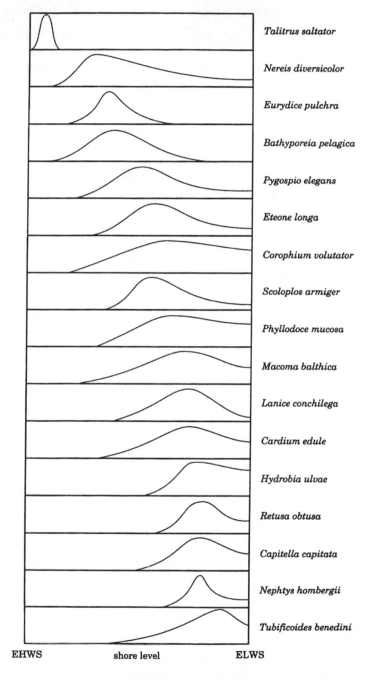

Fig. 3.5 Zonation patterns on a sheltered sandy beach at Newburgh, Aberdeenshire (reproduced with permission from Raffaelli *et al.* (1991) Zonation schemes on sandy shores. *J. exp. Mar. Biol. Ecol.*, **148**, 242).

the lugworm, *Abarenicola pacifica*, thalassinidean shrimps and several species of salt-marsh plants and seagrasses.

Beaches and mudflats are not as amenable as rocky shores to experimental manipulation. Zones can be broad and manipulations might have to be on a large scale. Also, removal of selective infaunal species usually requires significant disturbance, if not destruction, of the sediment habitat and damage to soft-bodied animals. Because the animals are mobile, elaborate precautions have to be taken to ensure that the removed species does not simply re-invade the cleared area. There are many good examples of field experiments which have identified interactions that affect the abundance and the local distribution of sandy beach and mudflat species and these are described in Chapter 4, but few of these convincingly illustrate that such interactions actually limit zonation patterns.

One of the more convincing experiments designed to explore the importance of physical factors in setting upper limits was carried out in Shark Bay, Western Australia by Peterson and Black (1987, 1988). They transplanted low-shore clams (*Circe lenticularis* and *Placamen gravescens*) to higher shore levels and monitored their performance. Both species grew more slowly than in their usual zone and *Circe* suffered higher mortality. It was unclear whether this poorer performance was due to the direct effects of desiccation or whether there was less time available for feeding or aerobic respiration at higher shore levels.

Perhaps the clearest evidence of biological interactions influencing zonation patterns on beaches and mudflats is seen in situations involving a bioturbator or disturber of sediment structure. Ghost shrimps of the genus *Callianassa* live in dense aggregations and rework large amounts of sediment through their burrowing activities. On the sand flats of Elkhorn Slough, northern California, *C. californiensis* excludes the tube-building *Phoronopsis* from higher shore levels by breaking and burying the tubes (Posey, 1986). Fish predation may also set the lower limits of bioturbators, such as ghost shrimps and the large enteropneust *Balanoglossus*. These are preyed upon by the sculpin *Leptocottus armatus*, which comes onto the flats to feed with the tide and which may limit their downshore spread (Posey, 1986). Biological interactions may also be responsible for sublittoral zonation patterns. For instance, in the north-east Pacific the sand dollar *Dendraster* occurs as a broad band in the shallow sublittoral (10 m) and appears to effectively exclude other epibenthic species for space (Morin *et al.*, 1985).

Because of the difficulties of manipulating sediment-dwelling species, intertidal ecologists have had to adopt other approaches for assessing how important interspecific interactions are in limiting distributions. A good example of how the problem might be approached is provided by Croker and Hatfield (1980), who carried out an extremely detailed study of the spatial relationships and tolerances of species of burrowing amphipods in a sandy beach in Maine, USA (Fig. 3.6). A significant correlation was found between the upper limits of the small low-shore *Acanthohaustorius millsi* and the presence of the larger, low-shore *Haustorius canadensis*, but *Acanthohaustorius* seemed capable of tolerating the harsher physical environment at higher shore levels. Croker and Hatfield argued that biological interactions between the two species set the *upper* limits of the lower-shore species, contrary to many of the rocky shore examples so far discussed.

Of course, this hypothesis can only really be tested by manipulating the densities of the two amphipods and monitoring their subsequent performances. Although this approach is not feasible in the field, it can be done in the laboratory or in mesocosms, employing artificially assembled systems in which single- and mixed-species populations can be maintained and monitored. Underwood (1986) provides a good account of designs and analyses, including their pitfalls, for such experiments. Using this approach, Croker and Hatfield (1980) found that when *Acanthohaustorius* was kept with *Haustorius*, the former showed increased mortality and negligible reproductive output compared with its performance when alone, implying some kind of interaction (competition?) between the two species.

A further complication in the analysis of competition between sediment-shore organisms is the three-dimensional nature of the habitat. Rocky shores are essentially two dimensional and competition for space between sessile species is far from subtle, involving overgrowth, prising off and crushing of individuals. In sediments there is the extra dimension of depth, so that there may be considerable overlap of species in their zonation along the intertidal gradient, but they may remain quite separate in the sand column. Peterson's (1977) studies of southern California lagoons illustrate well the processes that might be responsible for such distributions. In shallow (< 1 m) sandy and muddy sediments there was marked stratification of most species with sediment depth (Fig. 3.7), with little vertical overlap. Those species which overlapped significantly in their vertical distribution (*Callianassa* and the bivalve *Sanguinolaria*) showed horizontal spatial segregation. Removal of the large burrowing *Callianassa* by Peterson or by shrimp fishermen resulted in shifts in the vertical distributions and abundances of some of the remaining species (Fig. 3.7), suggesting that interspecific interactions are important in maintaining the vertical stratification.

Grant (1981) has examined similar interactions between *Acanthohaustorius millsi* and *Pseudohaustorius caroliniensis*, which occupy oxidized and reduced (anoxic) layers respectively on sandy beaches in South Carolina. In laboratory-maintained sediment cores containing both species at low densities, the two species coexisted within the upper oxic layer. At high densities *Pseudohaustorius* shifted its distribution to

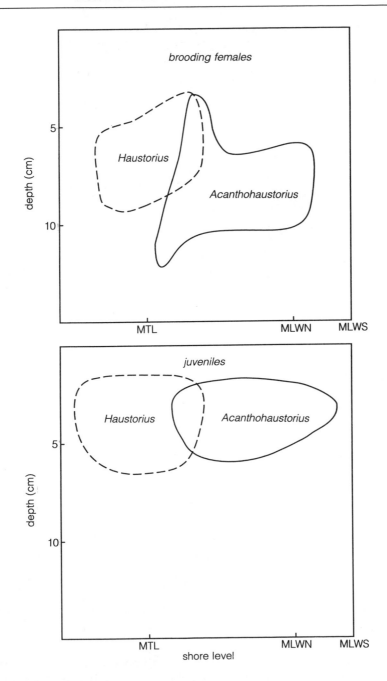

Fig. 3.6 Three-dimensional distributions of *Haustorius canadensis* and
Acanthohaustorius millsi on a sandy beach in summer, Maine, USA (adapted with
permission from Croker and Hatfield, 1980).

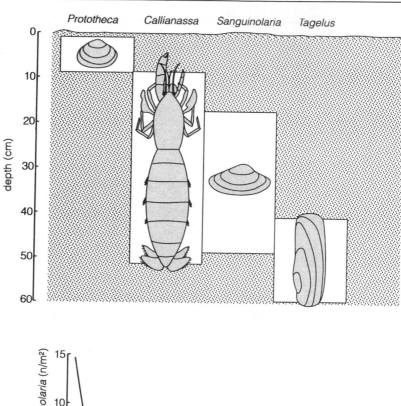

Fig. 3.7 Vertical stratification of infaunal species in Mugu Lagoon, California, USA. Little overlap occurs in vertical distribution of species, except for the bivalve *Cryptomya* (not shown) which appears to be commensal with *Callianassa*, and between *Callianassa* and the bivalve *Sanguinolaria*. However, as indicated in the lower graph, these two species are segregated horizontally. Derived from data in Peterson (1977).

the deeper, more reduced layers. In the absence of *Acanthohaustorius*, *Pseudohaustorius* again occupied the oxidized layers. It would seem that the fundamental niche of *Pseudohaustorius* includes that of

Acanthohaustorius, but it is not realized in the presence of the latter. Clearly, there is potential for significant interaction between these species that could account for their observed relative distributions on the shore. The mechanism of this presumed competitive interaction for living space was not identified.

Recent experimental work on New England salt-marshes has shown that species zonation patterns are caused by processes analogous to those operating on rocky shores (Bertness, 1992). Salt-marshes are not particularly diverse and zones often comprise just one or two species. The environment becomes progressively harsher for terrestrial vascular plants at lower shore levels and extension downshore by any one species appears to be limited by tolerance to physical factors (salt immersion, waterlogging of soil and possibly erosion of the marsh edge by water movement). Landward extension of these plants is limited by competitive interactions with faster growing terrestrial species. Thus there are clear parallels between salt-marshes and rocky shores in the determinants of zonation patterns along the vertical shore gradient – although in saltmarshes the direction of the stress gradient is reversed, with greater stress occurring at lower tidal levels.

3.3 CAUSES OF ZONATION ALONG OTHER SHORE GRADIENTS

In addition to the classic picture of zonation down the intertidal gradient, zonation can be a feature of the exposure, salinity and particle size gradients. The major zonation patterns along these gradients have been described in Chapter 2. In Chapter 1, we suggested that the kinds of processes determining zonal limits along the vertical intertidal gradient might also apply to these other gradients. Thus the zonation of a marine species in an estuary might be maintained by physical factors at its upstream limit and biological factors at its downstream limit. Indeed, McLusky (1989) suggests that marine estuarine species may be constrained to this habitat because of competition with truly marine species, but surprisingly little work has been done on this.

Manipulative experiments are difficult to perform along the exposure and particle size gradients. Some work has been done on the exposure gradient but it is difficult to find similar geologies and topographies at a scale of 100–1000 m necessary to perform experiments. A further problem lies with the nature of the horizontal gradient itself. In contrast to the vertical gradient – essentially a unidirectional stress gradient for organisms of marine evolutionary affinities – stress on the horizontal exposure gradient is more difficult to define (Chapter 1, Fig. 1.7). For some species, exposed conditions are beneficial. For other species, wave action is unfavourable (Denny, 1988). Therefore, it is quite likely that for some species, optimum conditions will be intermediate between exposure and shelter. The assumption that conditions always get more stressful with exposure does not always apply (e.g. Menge and Sutherland, 1976).

Factors operating along the horizontal exposure gradient are well-illustrated by midshore communities in the north-east Atlantic (Hawkins *et al.*, 1992a). At the latitude of the British Isles, sheltered bedrock shores are dominated by dense canopies of fucoids whilst exposed shores are dominated by barnacles, mussels and limpets. This pattern changes with latitude. In the north (Norway, Iceland) seaweeds predominate and extend further into exposed conditions. In the south (Spain, Portugal) limpets and barnacles predominate and fucoids become increasingly restricted to sheltered shores before disappearing except for a narrow band of *Fucus spiralis* high on the shore.

The conventional explanation of this pattern has been that fucoids could not attach on more exposed conditions and, therefore, were excluded from these areas by the direct effects of wave action. This was shown to be untrue when a dense sward of fucoids grew in the absence of grazing after limpets were removed from a 10 m wide strip on the Isle of Man (Jones, 1948; Lodge, 1948) and in subsequent follow-up experiments (Southward, 1956, 1964; Hawkins, 1981a,b). Further confirmation of the exclusion of fucoids from exposed shores by grazing came from the killing of limpets by dispersants during the clean-up operations in the aftermath of the *Torrey Canyon* oil spill. Fucoids completely covered the eulittoral on these shores (Southward and Southward, 1978). However, it is unlikely that fucoids will remain for long after initial establishment in very exposed conditions due to dislodgement by waves, particularly if growing on barnacles which provide an insecure anchorage. Work in New England has suggested, however, that the distribution of *Ascophyllum* may be directly set by wave action preventing settlement of propagules (Vadas *et al.*, 1990; review: Vadas and Elner, 1992). *Patella* are absent in New England and it is possible that this allows *Ascophyllum* to extend into more exposed conditions.

An alternative explanation for fucoid distributions which involves a mixture of biological and physical factors has been proposed (Southward and Southward, 1978; Hawkins and Hartnoll, 1983a). A dynamic balance probably exists between fucoids and limpets plus barnacles, which is analogous to a dynamic chemical equilibrium, and mediated by wave action (Fig. 3.8). In shelter, the balance is tilted in favour of the fucoids and in exposure the balance shifts towards limpets,

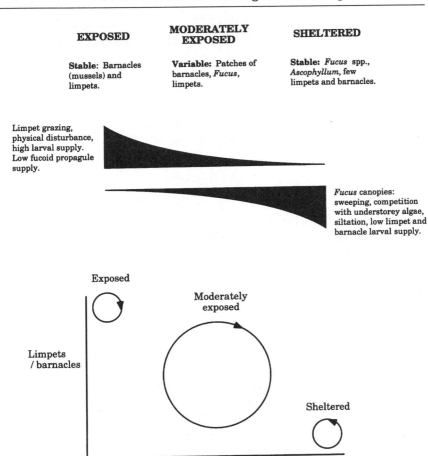

EXPOSED

MODERATELY EXPOSED

SHELTERED

Stable: Barnacles (mussels) and limpets.

Variable: Patches of barnacles, *Fucus*, limpets.

Stable: *Fucus* spp., *Ascophyllum*, few limpets and barnacles.

Limpet grazing, physical disturbance, high larval supply. Low fucoid propagule supply.

Fucus canopies: sweeping, competition with understorey algae, siltation, low limpet and barnacle larval supply.

Exposed

Moderately exposed

Limpets / barnacles

Sheltered

fucoids

Fig. 3.8 The dynamic balance between limpets/barnacles and fucoids on the horizontal exposure gradient in the north-east Atlantic. The lower figure is a reciprocal plot with time showing strong homeostasis in exposed and sheltered conditions. Intermediate conditions show strong fluctuations. (Modified from Hawkins and Hartnoll (1983) *Ocean. mar. biol. Ann. Rev.*, **21**).

barnacles and mussels. Clearly limpets normally prevent fucoids from extending onto exposed headlands, but how fucoids maintain their dominance on sheltered shores and what prevents barnacles and limpets extending into sheltered conditions has yet to be established. Nevertheless, any escape of fucoids on an exposed barnacle-dominated shore will tend to be short lived; whereas any gaps amongst the fucoid canopy on sheltered shores will also be quickly filled up. Shores of intermediate exposure tend to be patchy and highly variable (Hartnoll and Hawkins, 1985). On such shores the various interactions are delicately poised and fluctuations in physical conditions or recruitment variation can push the community one way or another (Chapter 6). The balance of interactions between fucoids and limpets plus barnacles also changes with geographic location. Warmer conditions (e.g. Spain, Portugal) favour greater penetration of limpets and barnacles into shelter. Colder conditions (e.g. Norway) favour expansion of fucoids into exposed conditions at the expense of limpets and barnacles (Ballantine, 1961).

OVERVIEW

For sessile organisms living at high shore levels on rocky shores, and for many midshore species, it seems likely that the upper distributional limits are set directly by physical factors. In contrast, the upper distributional limits of many mid- and low-shore species are set by competition, grazing and predation, although in the absence of these biological constraints, upshore extensions are ultimately restricted by adverse physical conditions. With very few exceptions, lower limits are set by biotic factors.

Unlike sessile organisms, mobile animals have dynamic upper and lower limits which reflect the behaviour of individuals. No doubt these behaviours have evolved and continue to operate with respect to both the physical conditions (upper limits) and biological interactions (lower limits). Most sediment-dwelling organisms are also mobile to some degree and they can move away from desiccating conditions at the sediment surface. However the three dimensionality of the sediment environment and the consequent difficulty of carrying out experimental manipulations means that the relative importance of physical and biological factors for zonation patterns is not well understood for species inhabiting the sediments.

The relative importance of physical and biological factors along the horizontal wave-exposure gradient is even less clear; both biotic and abiotic factors can exclude species from exposed rocky shores and sheltered areas. Whatever the gradient, for each species the underlying physical environment affects its distribution either directly or, more often, indirectly by mediating competitive ability or the presence or absence of predators. In other words the physical environment affects the intensity and outcome of biological interactions. For a long time in the history of intertidal ecology the importance of physical factors was considered paramount. Emphasis then switched to biological interactions as the results of manipulative field experiments became apparent, and perhaps these often dramatic results in turn became over-emphasized. Clearly both physical (direct and indirect) and biological factors interact to set distribution patterns.

Community dynamics 4

Many ecological communities are complicated, often bewilderingly so, and the processes underlying their dynamics are likely to be just as complex. Yet intertidal ecologists have made significant contributions to a general understanding of these processes and much of current theory in community ecology is based at least in part on intertidal studies. This is largely because of the relative ease with which experimental manipulations can be carried out on rocky shores coupled with their accessibility and relative simplicity. In Chapter 3, we describe how these powerful techniques have allowed identification of the relative importance of the physical environment and biological interactions in restricting the distribution patterns of some species along the vertical and horizontal environmental gradients on the shore. But at any point on the shore gradient there will be many species, especially towards the middle and low shore where diversity tends to be higher. In this chapter we examine what causes the patterns of species diversity on both rocky and sediment shores – in other words what structures communities and generates patchiness and fluctuations?

We begin by looking at rocky shores characterized by space-occupying sessile species and conspicuous consumers like starfish, sea urchins and limpets. Grazing and predation seems to be an important organizing force on at least some shores. We then ask how general such interactions are, especially for other kinds of shore, such as sandy beaches and mudflats. Here, the fauna is more mobile and there is great scope for animals to modify the sediment environment and hence the living conditions for other species. Finally, we look at the processes involved in ecological succession on both hard (rocky) and soft (sedimentary) shores.

4.1 ROCKY SHORES

4.1.1 THE ROLE OF GRAZING AND PREDATION

Systems characterized by large starfish

The mid- to low-shore community of semi-exposed, temperate rocky shores consists of a few tens of large species of easily recognized plants and sessile invertebrates – mussels, barnacles, algae – and their consumers, such as whelks and limpets. In some regions, notably the Pacific, large predatory starfish may also be common. In a series of experiments, Paine has demonstrated that predation by these starfish can play a major role in community organization (Paine, 1966, 1969, 1974). Much of Paine's work has focused on the shores of Washington State in the north-east Pacific, where the starfish *Pisaster ochraceus* is a conspicuous component of the food web and feeds on mussels and other sessile invertebrates. Paine cleared *Pisaster* by hand to maintain starfish-free areas on the shore and noted a dramatic change in the rest of the midshore community (Fig. 4.1). At Mukkaw Bay, species richness declined from 15 to 8 species of conspicuous space-occupiers after only 30 months. After 5 years the mussel *Mytilus californianus* and the goose barnacle *Pollicipes polymerus* completely dominated the midshore (Paine, 1974). Starfish are clearly a vital component of the normal community at Paine's sites; remove the starfish, and the community changes. The basic mechanism underlying these changes is that *M. californianus* is a superior competitor for space, but its numbers are normally kept in check by the starfish, which prefers to eat mussels. This allows the coexistence of mussels with the many other competitively inferior sessile species, all of which have a requirement for space.

To determine whether these kinds of interactions are general, Paine carried out similar starfish removal experiments in other parts of the Pacific, notably in northern New Zealand and Chile. In both these regions, the midshore community is similar in structure to that in Washington, although the majority of the species are different. At Anawhata, near Auckland, Paine and New Zealand co-workers removed the starfish *Stichaster australis* and recorded an increase in the abundance of the green mussel *Perna canalicula* (Paine *et al.*, 1985). The decline in species richness was not as dramatic as in Washington, from 20 to 14 space-occupying species in 9 months (Fig. 4.1), but this could have been due to problems in keeping all starfish off the shore. On Chilean shores, the starfish *Heliaster helianthus* is a generalist predator on barnacles, solitary tunicates and mussels, especially *Perumytilus purpuratus*. Removal of *Heliaster* led to an increase in cover by *Perumytilus* from 1% to 46% (Paine *et al.*, 1985).

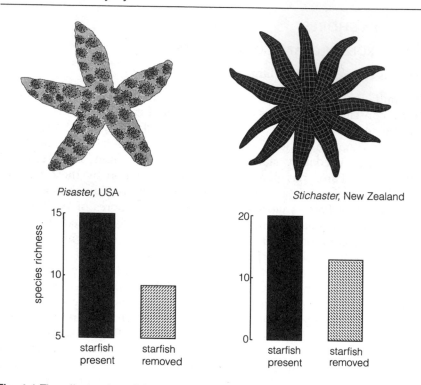

Fig. 4.1 The effects of starfish removal on the species richness of primary space-occupying organisms in the north-east Pacific and northern New Zealand, after Paine and co-workers.

On the North American Atlantic coastline, exclusion of predators (dogwhelks, but also starfish and crabs) resulted in the development of a virtual monoculture of mussels at the expense of barnacles and seaweeds, but the outcome of these experiments was less predictable than those in the Pacific (Menge, 1976). On exposed Atlantic headlands, predators were not so effective in controlling competitive dominants like mussels, compared with more benign habitats, probably because of wave dislodgement when foraging (Menge, 1978a,b). In such places, competition between sessile species, such as the mussel *Mytilus edulis* and the barnacle *Semibalanus balanoides*, may be the dominant biological interaction (Menge, 1976; Menge and Sutherland, 1987; review: Mathieson *et al.*, 1991).

The time for recovery following cessation of the perturbation (starfish removal) is variable. In Washington, recovery is slow, perhaps 10–15 years, because the mussels grow so large in the absence of starfish that they have effectively escaped predation by the time the starfish return.

Having reached a size refuge, the mussels form a dense matrix into which young mussels can recruit and are afforded protection from starfish. It is then difficult to shift the community back from this mussel-dominated, low-diversity system to the normal high-diversity system characterized by starfish. Paine's experiments illustrate well how a community can exist in alternate stable states. Shifting from one state to the other requires a massive perturbation, either the removal of starfish (by Paine) or the removal of the large mussels (by fierce storms). A similar example comes from the shallow sublittoral of several islands off the coast of South Africa. Here, predatory rock lobsters have disappeared, perhaps through overexploitation, and this has allowed their prey, carnivorous whelks, to grow to large sizes. The whelks are now not only too large to be preyed on, but are sufficiently large to attack rock lobsters, thus preventing their re-establishment on these shores (Barkai and Branch, 1988; Barkai and McQuaid, 1988). Not all starfish removal experiments result in a shift to an alternate community state. In Chile, recovery of the shore was fairly rapid (about 4 years), because returning *Heliaster* were able to feed on both large and small mussels. In the North Atlantic, starfish are smaller and are less common in the intertidal except when occasional plagues occur.

Does predation by *Pisaster* increase or reduce diversity?

Whilst the presence of *Pisaster* maintains a higher species richness of sessile primary space occupiers, community-wide diversity may be locally reduced by starfish predation. Beds of *M. californianus* in the north-east Pacific are structurally complex, comprising a matrix of one or several layers of dead and live shells with accumulated sediments and organic material (Suchanek, 1992). Associated with this matrix is a diverse assemblage of plants and animals taking advantage of the secondary space and microhabitat provided. The mussels block out solar radiation, and temperatures within the bed can be reduced by 5–13 °C. Relative humidity is increased by about 15% and wave action is reduced within the matrix. Over 300 species have been recorded from these beds (Suchanek 1992), but this richness varies with shore level and exposure. Within 0.1 m^2 areas, one might find about 25 species at high-intertidal protected sites and about 135 species at low-intertidal exposed sites. Also, the thicker the mussel bed, the more diverse the community. Gaps in the mussel bed caused by predation, wave action and floating logs have a much lower diversity of only about 15 species of primary space occupiers. Thus, depending on viewpoint and the scale of observation, predation by *Pisaster* can be seen as a mechanism for either increasing or decreasing species diversity on the shore. Nevertheless, this starfish clearly plays an important role in organizing the community (Fig. 4.2).

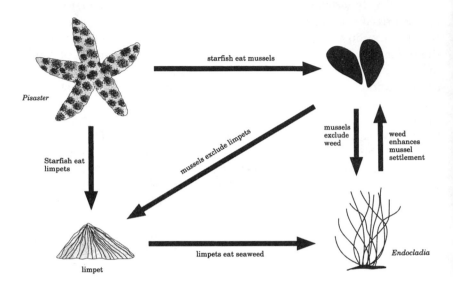

Fig. 4.2 Positive and negative interactions between midshore species in the north-east Pacific.

Grazer–algae interactions

Similar interactions between a consumer and a competitively superior resource occur between sea urchins and seaweeds (Paine and Vadas, 1969). These authors removed the sea urchins *Strongylocentrotus purpuratus* from intertidal pools and *S. fransciscans* from sublittoral areas, in the north-east Pacific. Normally these heavily grazed areas are covered by the pink encrusting coralline algae collectively called 'Lithothamnia', but urchin removal resulted in an immediate increase in foliose seaweeds, several of which were new to the pools. Over the next few months, larger, canopy-forming brown seaweeds dominated these areas, mainly *Hedophyllum* in the intertidal and *Laminaria* in the subtidal. After two to three years, the sublittoral experimental areas had an understorey flora poor in species, suggestive of competition for light. Paine and Vadas predicted that the large browns would eventually competitively exclude the understorey species, the process taking longer at higher shore levels. Dawes *et al.* (1991) describe similar canopy–understorey interactions revealed by manipulation of grazing fish in the Caribbean shallow sublittoral zone.

The urchin–kelp interactions described above are, however, probably more complex than previously thought (Paine, pers. comm.), because the effects of another grazer, the chiton *Katharina*, were not taken into account. This chiton is capable of grazing down the kelp *Alaria*, allowing its weaker competitor, *Hedophyllum*, to become dominant. When either chitons or urchins are excluded, it is *Alaria* that dominates, not *Hedophyllum*. Indeed, when urchins alone are removed, there is an initial flush of *Alaria* before *Hedophyllum* takes over (Paine, 1977, 1980).

In the eulittoral region on shores of the eastern Atlantic, grazing by large patellid limpets is an important structuring agent. This is the case in both the north-east Atlantic (Europe and N. Africa) and in southern Africa. One of the first demonstrations of the importance of a trophic interaction on shore communities was performed by removing limpets from a 10 m wide strip down the shore on the Isle of Man, UK. This early work (Jones, 1948; Lodge, 1948; Burrows and Lodge, 1950) and subsequent experiments (Southward, 1956, 1964) were ignored for a while because they were published in obscure journals, but they clearly showed how important limpet grazing was in preventing the growth of ephemerals and the domination by fucoids of the midshore region of moderately exposed shores. Confirmation came when massive kills of limpets occurred due to widespread dispersant application after the *Torrey Canyon* oil spill (Chapter 7), when seaweeds proliferated on many shores in west Cornwall (Southward and Southward, 1978; Hawkins and Hartnoll, 1983a), and in subsequent experiments (Hawkins, 1981a,b; review: Hawkins *et al.*, 1992a).

On south and east coasts of South Africa, the influence of limpet grazing acts in two ways. In the mid and high eulittoral, generalist grazers are important as in the North Atlantic. Lower down there are zones dominated by 'gardening limpets'. A characteristic patch of algae is associated with each individual limpet and this is vigorously defended. *Patella longicosta* has a garden of the encrusting brown algae *Ralfsia*, and *P. cochlear* a garden of filamentous reds. When these limpets are removed a completely different assemblage of algae grows in their place (Branch, 1981; Branch and Branch, 1981; Branch, 1985 for reviews).

Experiments like these have shown that some shores, at least, have **keystone predators** or **grazers**, like *Pisaster*, *Strongylocentrotus* and *Patella*, which are all-important in maintaining community composition and structure. Like the keystone in a building or an arch, these animals may not be particularly conspicuous, but their removal leads to rapid, cascading changes in the structure they support. An important feature of keystone species is that their significance can only be convincingly gauged through experimental manipulations.

Provision of space: physical versus biological disturbance

A high species diversity occurs on these shores because of the continued reduction of superior competitors and the renewal of the major resource required by the sessile fauna and flora – space on the rock surface. Physical processes are also important in freeing space, especially in removing older and larger prey which have escaped predation by virtue of their size, but which often become more vulnerable to dislodgement by waves as they age. This effect is often amplified when waves move large objects, such as floating logs. Along many parts of the north-east Pacific, trees grow down to the edge of the shore, so that a considerable amount of timber is carried onto the shore in breaking waves. Logs weighing many tonnes make a significant contribution to the provision of space, smashing into mussels and barnacles to leave a bare patch available for colonization by other species (Dayton, 1971). Similar wave action can be exaggerated by pebbles and even large stones being moved around by waves (Shanks and Wright, 1986). Scouring by sand or gravel in suspension can also have a major effect on rocky shores. Reefs jutting between stretches of sand are often inundated by sediment which kills all the plants and animals. Ice scour in colder regions can have a similar devastating effect. Unlike predation, this disturbance is unselective with regard to species and size of prey removed, but the effect is similar – the creation of bare space within the community. However they are formed, such spaces will be further enlarged by wave action, particularly in winter, because the individuals immediately adjacent to the cleared area are more susceptible to dislodgement.

In the previous examples, consumers increase the diversity of the primary space occupiers by feeding preferentially on the superior competitor. But predators can also lower species diversity. We have shown above how this can happen if the superior competitor also provides secondary space for other species (Suchanek, 1992). It can also occur when at high predator densities even the less preferred prey are eaten and only a few unpalatable species are left. A third reason is that the predator may actually prefer to eat the competitive inferiors, not the superiors. In New England, the dominant herbivore over much of the intertidal of sheltered and moderately exposed shores is the snail *Littorina littorea*, which was introduced into the USA in the 19th century. In feeding trials, this snail consistently prefers smaller ephemeral seaweeds, such as *Enteromorpha* and *Ulva*, to the tougher fucoids which can dominate these shores (Lubchenco, 1978; Chapter 5). The preferred smaller seaweeds can only thrive in disturbed areas where fucoids have been dislodged by wave action, or in tide pools which are not colonized by fucoids. Experimental exclusion of *Littorina* from the open shore (non-pool areas) leads to an increase in seaweed diversity, because the smaller species are no longer grazed down (Fig. 4.3). In contrast in tide pools, in the absence of fucoids, the dominant seaweed is the fast-growing *Enteromorpha*, a preferred food of the snail. The effect of *Littorina* on seaweed diversity in these tide pools depends entirely on snail density. At both high and low densities, diversity is low and only at intermediate densities is diversity high (Fig. 4.3). Under low grazing pressure, *Enteromorpha* dominates and smothers other seaweeds; at high *Littorina* densities, overgrazing occurs. With intermediate levels of grazing, a mosaic of various less competitive species can occur in the gaps left by selective removal of *Enteromorpha*. Lubchenco's work underlines the importance of knowing the food preferences of predators and grazers when interpreting the outcome of exclusion experiments.

The intermediate disturbance hypothesis (Caswell, 1978) suggests that at low levels of disturbance, certain competitive species will predominate and hence diversity will be low. At intermediate levels of disturbance, no one species will predominate and diversity will be high. As disturbance increases further, only a few highly tolerant or very opportunistic species will occur. This hypothesis has been tested in boulder fields (Sousa, 1979) where stable boulders have a low-diversity community of dominant plants (red turf-forming algae such as *Laurencia* spp., *Gelidium* spp., *Ceramium* spp., *Mastocarpus stellatus*); highly unstable boulders have a limited suite of opportunistic ephemeral species (*Ulva*, *Enteromorpha*); highest diversity occurs on boulders which get turned by storms occasionally, as succession is halted and restarted but proceeds beyond the early pioneer species phase.

Many of the results of predator or grazer removal experiments can be accommodated within Caswell's generalized intermediate disturbance hypothesis, predation being a biological disturbance. If predators prefer competitive dominants, then at low predation pressure diversity will be low, at intermediate levels it will be high as no species will win, and at very high levels it will be low due to overgrazing or overpredation and only a few highly resistant or strongly defended species will be present.

Hierarchical and non-hierarchical interactions

On some rocky shores, predators have been shown to mediate the interactions between competitors, which would otherwise lead to competitive exclusion of all but a few species. The interactions can be complex; species may have positive as well as negative effects on others (Fig. 4.2; see also below), and there is likely to be a hierarchy of competitive and predatory relationships within the community. This is well illustrated by Dayton's (1971) work on rocky shore communities

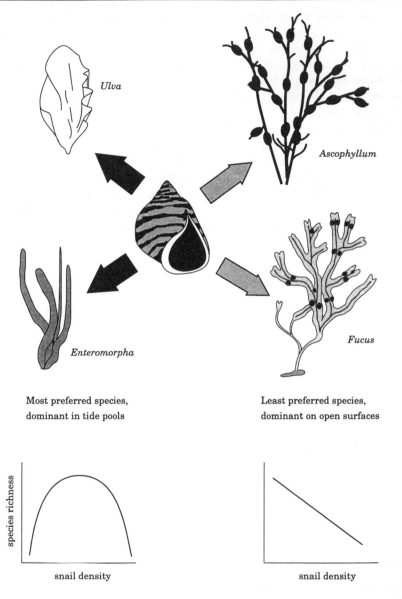

Fig. 4.3 Effects of prey preference on the outcome of *Littorina littorea* removal experiments on New England shores (based on Lubchenco, 1978).

similar to those on which Paine worked. Through a series of carefully controlled small-scale predator exclosure and enclosure experiments, Dayton was able to rank the competitive abilities for space of the main sessile species as follows: *Mytilus californianus* was the best competitor for space; next was *Balanus (= Semibalanus) cariosus*, which could outcompete *Balanus glandula*; *B. glandula*, in turn, outcompeted *Chthamalus dalli*, and seaweeds were the least competitive (Fig. 4.4). Coexistence between these species, and hence community diversity, was maintained by several species of predator, each acting at a different level in the competitive hierarchy. Thus, *Mytilus* can be kept in check by the starfish *Pisaster* (and floating logs), *Balanus* spp. by dogwhelks (*Thais* spp., now called *Nucella*) and acmaeid limpets. The smaller size of adult *Chthamalus* make them less profitable to the whelks which prefer *Balanus*, whilst the limpets bulldoze off the larger post-settlement cyprids of *Balanus* but tend to skate over the smaller cyprids of *Chthamalus* (Fig. 4.4). Acmaeid limpets graze down algae.

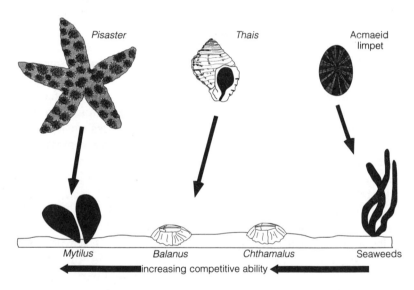

Fig. 4.4 Hierarchy of competitive and predatory interactions between species in the shore community described by Dayton (1971).

Dayton's work revealed the complexity of interactions between species on rocky shores. Sometimes these interactions are sufficiently complex that manipulation of only one of the species involved could lead to erroneous conclusions. This is well illustrated by examples from Australia (Underwood *et al.*, 1983) and North America (Dungan, 1986). In both cases, the significance of the interactions was only revealed by painstaking experimental manipulations of each of the species using factorial experimental designs.

In New South Wales, Australia, the major midshore species found on shores of intermediate exposure are the barnacle *Tesseropora rosea*, the limpets *Cellana tramsoserica* and *Patelloidea latistrigata* and the predatory whelk *Morula marginalba*. High densities of *Cellana* had a negative effect on the settlement and survival of the barnacle, through bulldozing, but a positive effect at lower densities because they graze algae which would otherwise pre-empt barnacles for space. Once established, however, barnacles had negative effects on *Cellana*, probably because of the reduced space available for grazing. In contrast, barnacles have positive effects on the other limpet *Patelloidea* by providing a refuge from a harsh physical environment and its superior competitor *Cellana*. The whelk had negative effects on *Patelloidea*, on which it preferentially feeds. Not surprisingly, the outcome of these interactions, and hence the structure of the midshore community, varied dramatically according to shore level, local weather and the timing and intensity of larval settlement of limpets and barnacles (Underwood *et al.*, 1983).

In Sonora, northern California, Dungan (1986) manipulated densities of the limpet *Collisella strongiana*, the barnacle *Chthamalus anisopoma* and an encrusting brown alga *Ralfsia* sp., and was able to tease out the direct and indirect interactions between these species. The direct effects were as follows: barnacles pre-empt space for both the limpet and *Ralfsia*; *Ralfsia* pre-empts space for barnacles; the limpet grazes on *Ralfsia*. The importance of indirect effects is highlighted by the sole removal of the barnacle. This manipulation did not lead to an increase in *Ralfsia* cover because of the compensatory effect of providing space for limpets. These grazed back the *Ralfsia* and in doing so enhanced the recovery of the barnacle. Without knowledge of this indirect effect, one might have erroneously concluded from the barnacle removal experiment that barnacles and *Ralfsia* do not interact. The lesson to be learned is that ecologists should be prepared to examine a range of alternative hypotheses when interpreting the results of their field experiments plus tease out direct and indirect effects.

Patchiness

Patchiness is a fundamental feature of most communities (Pickett and White, 1985) and is easily studied on rocky shores because of their two-dimensional nature. Many shores can be seen as a patchwork or mosaic of species or assemblages on various scales in different phases of succession, from cleared bare rock to complete cover by a dominant species. The dynamics of these patches vary with wave exposure, time of year and patch size. For instance, in Paine's Washington sites, 1–5% of mussels were removed every month during the winter at the exposed Tatoosh Island, but in summer and on the mainland, the removal rates were an order of magnitude less (Paine and Levin, 1981). Small and intermediate bare areas (< 0.3 m²) can be obliterated rapidly by lateral movement or by peripheral mussels leaning over and creeping at a rate of about 0.5 mm per day. Large areas can only be replaced by recruitment from the plankton, and recovery typically begins about 2 years after initial patch creation. The turnover time for mussel beds in this part of the world ranges from 8 to 35 years, depending on location.

The dynamic nature of patchiness has been investigated in detail for smooth limestone ledges on semi-exposed shores on the Isle of Man in the Irish Sea (Burrows and Lodge, 1950; Hawkins and Hartnoll, 1983a,b; Hartnoll and Hawkins, 1985; review Hawkins et al., 1992a; Fig. 4.5), where grazing by the limpet *Patella vulgata* is an important structuring agent. Local reductions in limpet density allow clumps of the seaweed *Fucus vesiculosus* to establish, especially on barnacle shells. The continual sweeping of the *Fucus* fronds over the rock surface dislodges a large proportion of the settling barnacles *Semibalanus balanoides*, and the damper conditions under the plants encourage aggregations of limpets, especially juveniles, as well as the dogwhelk *Nucella lapillus*, a predator of barnacles. The clumps of fucoids disappear through loss of insecurely attached plants growing on barnacles and, eventually, ageing. The plants are not replaced locally because of the dense aggregations of grazing limpets and the sweeping action of the fucoids themselves. Once their shelter has disappeared, limpets and dogwhelks disperse and in their absence barnacles can now settle successfully. Barnacles settle better in the gaps between fucoid clumps, which also have fewer limpets as these tend to be grouped under the seaweeds. The grazing efficiency of limpets is poor in stands of older barnacles and new escapes of *Fucus* occur in these areas. A patch of *Fucus* lasts about 3–4 years and the community functions as a series of cycling patches, usually out of phase with each other.

There are many positive and negative interactions between the various elements in the mosaic, and some species moderate other interactions. Thus, limpets prevent algal growth but fucoid patches encourage the recruitment of juvenile limpets; newly settled barnacles are reduced in number by limpets, but are probably permitted to settle due to the removal of competitively superior ephemeral algae; barnacles reduce limpet foraging efficiency allowing algal escapes; dogwhelks thin-out barnacles, allowing limpets to more effectively reduce algal cover; the sweeping by fucoid fronds reduces barnacle settlement.

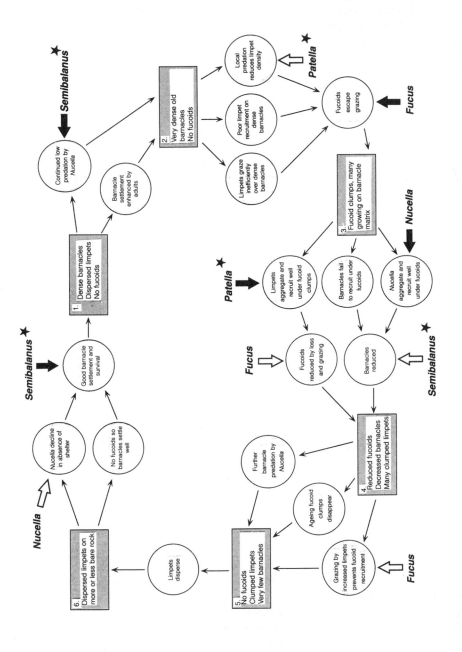

Fig. 4.5 Qualitative model of the dynamics of patches on semi-exposed shores on the Isle of Man, UK. Recruitment inputs are shown by arrows (open arrows inhibit and closed arrows promote the cycle). Asterisks indicate recruitment from the plankton. (Reproduced with permission of Ophelia Publications from Hartnoll and Hawkins (1985) *Ophelia*, **24**, p. 62.)

Recruitment variation also helps initiate events. Thus a combination of poor limpet recruitment and good fucoid recruitment will increase the likelihood of patches of fucoids arising in escapes from grazing. This likelihood will be further enhanced if a good barnacle settlement has occurred. Conversely, good limpet and poor barnacle recruitment will reduce the probability of fucoids growing. This variability is often due to seemingly random events in the plankton affecting the number of larvae and whether or not they come inshore in large numbers. Wind direction at the time of settling can affect the supply of barnacle larvae (Hawkins and Hartnoll, 1982a). The supply of larvae will also depend on the vagaries of British weather as this will affect reproductive output and early growth, particularly in seaweeds which will be stressed by desiccation in rare good summers. These chance factors are called **stochastic events**. The cycle described following an escape from grazing is reasonably predictable and can be called a **deterministic process**. The stochastic events are external to the cycle; the deterministic processes are internal. The effect of the stochastic events will depend on where in the mosaic cycle they occur. Some will speed up cycling, others will slow it down or brake it altogether.

4.1.2 HOW COMMON ARE KEYSTONE SPECIES ON ROCKY SHORES?

Predator removal or exclusion experiments have made a major contribution to our understanding of the nature and consequences of biological interactions on rocky shores. However, the danger with all new and exciting concepts, and this applies particularly to the keystone model, is that they are apt to become generalizations or paradigms which are used uncritically to explain and interpret the patterns and changes in species abundance in all kinds of communities.

An obvious feature of keystone species is that they exert a very strong controlling influence on their prey. However, this cannot be deduced from simple observations, through intuition or even by measuring the amount of energy consumed: only manipulative experiments can reveal unequivocally the presence of a keystone predator (Paine, 1980; Chapter 6). This point is often overlooked in discussions about the roles of keystone species in marine systems, particularly where changes in the system are associated with the disappearance of particular species. The problem is well illustrated by the debate surrounding changes in the abundance of large kelps in the Pacific and the Atlantic sublittoral.

In the Pacific, the geographical range of the sea otter, *Enhydra lutris*, extended in historical times from Southern California up through the Aleutian Islands to the Japanese Archipelago. Sea otters were hunted for their fur as early as 6750 years ago on the Californian Channel Islands (Tegner, 1989), but they were driven almost to extinction in the 18th and 19th centuries. It has been suggested that the decline in sea otters probably allowed an increase in their large herbivore prey, such as the abalone *Haliotis* spp. and sea urchins. Certainly, the recent recovery of sea otters following protective legislation is associated with declines in the abalone fishery in some areas (Tegner, 1989). Comparisons of the ecology of the Aleutian Islands with and without sea otters reveal that islands with otters have lower populations of grazing invertebrates, especially the sea urchin *Strongylocentrotus*. They also have better-developed kelp communities (*Alaria, Laminaria, Agarum, Thalassiophyllum, Desmerestia* and various reds) compared with islands lacking otters (Estes and Palmisano, 1974; Estes and Duggins, 1995). On the basis of these associations, Estes and Palmisano suggested that the sea otter plays a keystone role in this system through its presumed ability to control the grazers of kelps. A similar case has been made for the South American otter *Lutra felina* which eats herbivorous gastropods such as the whelk *Concholepas concholepas* and the clingfish *Sicyases sanguineus*, which are important consumers of primary space users on Chilean shores (Santelices, 1990).

On the Atlantic seaboard of North America, there is a clear association between the abundance of kelp (*Laminaria* spp. and *Agarum*) and densities of sea urchins. Lobsters (*Homarus americanus*) prey on the urchin *Strongylocentrotus droebachiensis*, and overfishing of lobsters (the presumed keystone species) is thought to have contributed to an urchin population explosion as a consequence of which large areas of the kelp beds were overgrazed, producing barren grounds (review: Mathieson *et al.*, 1991).

Urchins are known to graze kelps and to be functionally important predators in other locations (e.g. Kain and Jones, 1967; Kitching and Ebling, 1961; Paine and Vadas, 1969), so that the circumstantial evidence for their predators – sea otters and lobsters – being keystone species is impressive. But, as Elner and Vadas (1990) and Chapman and Johnson (1990) have pointed out, there have never been any carefully controlled manipulations carried out on lobsters to test rigorously the keystone proposition. Furthermore, urchins are not taken readily by lobsters as once thought (Elner and Campbell, 1987). Predation on the early life-history stages of urchins may be a key factor: substantial numbers of urchin larvae are preyed upon by jellyfish and larval fish, whilst post-settlement urchins are taken by a wide variety of fish and invertebrate predators (Hooper, 1981). Disease is also a likely factor controlling sea urchin populations, with epidemics of parasitic amoebae obliterating populations when they reach high densities (Liddell and Ohlhurst, 1986; Scheibling, 1986; Chapman and Johnson, 1990; Vadas and Elner, 1992).

The manipulation of sea otter densities would be difficult and controversial so a test of the keystone hypothesis for this species is unlikely. Also, the supporting evidence for their keystone role is not as clear as first supposed. Factors other than sea otter predation on sea urchins, such as storms and incidence of upwelling, affect kelp abundance in central California, and in Southern California where sea otters do not occur, kelp forests persist in the face of often dense aggregations of sea urchins (Schiel and Foster, 1986; Foster, 1990). In cases like this, where the answer to the keystone question has important management or conservation implications, but experiments are not feasible, ecologists can only advise as best they can, bearing in mind the possibility that the removal of, or failure to protect, that predator could lead to cascading and undesirable changes in the rest of the system.

Finally, it is debatable whether the term 'keystone' should be applied to species like sea otters and lobsters at all. The original concept was of a predator which mediated competition between prey species (Paine, 1969). Whilst sea otters may be capable of regulating urchin densities, there is no suggestion that otters or lobsters mediate competition between urchins and other species. The urchins merely overgraze kelp beds in the absence of sea otters.

The need for intertidal ecologists to be careful in uncritically attributing keystone status to particular species is supported by a growing literature showing that many rocky shores lack keystone predators. For instance, on Australian shores, there are numerous examples where predator removals (including starfish, Keough and Butler, 1979) do not lead to increased competition between prey and the exclusion of species (review: Underwood and Denley, 1984). Neither is there much evidence for keystone predators on some tropical American shores (Menge et al., 1986). We must also be cautious in assuming that similar interactions will occur on shores with similar communities. Whereas Paine (1969) observed a dramatic response to starfish removal within 2 years, Dayton (1971) observed no response after 3 years, the possible reasons for which are discussed below. Paine and Suchanek (1983) drew comparisons between the sea squirt *Pyura praeputialis* in Chile and the mussel *M. californianus* in the north-east Pacific, suggesting that their ecological roles might be similar (although they did not manipulate this system). This sea squirt also occurs on Australian shores, but when its main predator, the whelk *Cabestana spengleri*, was removed there were no effects of the kind anticipated in Chile (Underwood and Fairweather, 1986). Also, sea urchins amongst Chilean *Macrocystis pyrifera* forests do not play the same role as urchins do in the otherwise ecologically similar North American kelp forests (Santelices, 1991). Thus, the same type of community (e.g. a kelp bed) in different geographical areas may be structured and maintained by quite different processes.

Even within a region, the importance of predation can vary significantly between shores. The removal of the whelk *Morula marginalba* from a number of similar shores in southern Australia has shown that the effects on prey abundance can vary tremendously between shores (Fairweather and Underwood, 1991), underlining the danger of generalizing about interactions from the results of a single study. These authors argue that only by carrying out many replicated experiments on different shores in an area can the 'interaction norm' be established.

Finally, it is perhaps unfortunate that in systems where predation or grazing seems an important organizing force, the spotlight has fallen on the predator by giving it a special status, 'keystone'. It must be remembered that such species owe their status to the competitive abilities of prey species, rather than any inherent characteristics they themselves possess. In other words, these consumers are only recognized as being keystone species because a competitively superior prey emerges following their removal. Perhaps instead of asking the question why it is that some shores possess keystone species whilst others do not, we should be asking why some shores have competitive dominants, whilst others do not appear to. One of the important differences between apparently similar shores may be the recruitment patterns of the prey species (Menge, 1992). When prey recruitment is sporadic and unpredictable, as is the case for many sessile species with planktonic larvae, the response of the community to predator removal may be quite different in different years and in different places. If intertidal ecologists really wish to understand fully how shores behave, then they will have to extend their interests offshore to include those factors affecting the supply of algal propagules and the larvae to the shore (Gaines and Roughgarden, 1985).

4.1.3 SUPPLY-SIDE ECOLOGY

In recent years there has been a growing realization amongst shore ecologists that variation in the arrival of new propagules or larvae will be a major influence on the abundance of species. This variation will affect the outcome of biological interactions, thereby affecting community dynamics. In part this has been a re-discovery – or as Underwood and Fairweather (1989) put it, 'new wine in old bottles' – because these ideas have been around for a long time. They first surfaced in the fisheries ecology literature in the early 1900s to help explain the tremendous variation in recruitment to fished stocks from year to year. Later, Southward and Crisp (1954) noted the importance of larval supply in both the geographic distributions and the relative abundance of barnacles. In the south-west of England, warm years favoured *Chthamalus* recruitment whereas colder years favoured *Semibalanus balanoides*. These ideas were developed further by Jack Lewis and co-workers between the 1960s and early 1980s (Lewis, 1964, 1976; Kendall *et al.*, 1985; see papers in Moore and Seed, 1985). They were also central to the patch dynamics model developed for Isle of Man shores in the early 1980s (see above).

The recruitment of species with larvae that spend considerable time offshore in the plankton will depend on the vagaries of water movement and the uncertainties of larval food supply. Recruitment will also be from remote locations, and so adults will not produce the next generation for the same locality. Thus, chance events offshore are likely to influence new inputs into the population. In the mid-1980s the term 'supply-side' ecology was coined (Lewin, 1986). In contrast to the view that hierarchies of interactions occurred on the shore and that the removal of key species would lead to predictable outcomes (deterministic processes), studies on the effects of larval supply emphasized the importance of chance (stochastic) events.

Chance fluctuations sometimes just lead to noise, but sometimes two or three independent chance events combine to drive a system in a particular direction. Taking the patch dynamics model (Fig. 4.5) as an example, a calm spring could lead to an early phytoplankton bloom and a matched early release of barnacle larvae (Connell, 1961b; Hawkins and Hartnoll, 1982a), resulting in a high barnacle settlement and reducing limpet foraging efficiency. If frosts had occurred two winters previously, limpet recruitment to the adult population, which occurs in the second year after settlement, could have been reduced (Bowman and Lewis, 1977). These events in themselves could lead to fucoid escapes in late summer, especially if it was a cool damp summer which may favour early fucoid growth.

Levels of recruitment and their fluctuations will affect community structure and dynamics: interactions will be intense in high recruitment areas and much less severe where recruitment is low (Gaines and Roughgarden, 1985; Menge *et al.*, 1985; Menge and Olsen, 1990; Menge, 1992; Gaines and Bertness, 1992), and variable recruitment may generate instability. If two species compete for a limiting resource, such as space, and one grows faster than the other, then eventually the slower-growing species will be outcompeted. Competitive exclusion will only occur if there is always an adequate supply of propagules of the dominant species. If this supply failed for any reason, then it would give the inferior competitor the opportunity to recruit into any gaps created by death, disturbance or predation.

This point is well illustrated by barnacles on UK shores. *Semibalanus* and *Chthamalus* have coexisted in the midshore on the Devon coast for many years (Southward, 1991). Further north in Scotland, where Connell worked (Chapter 3), *Semibalanus* wins on the mid and low shore and excludes *Chthamalus*. The continued coexistence in Devon can be explained in various ways: sometimes space is undersaturated due to low recruitment (Burrows, 1988; Southward, 1991); in warm years the reproductive output of *Chthamalus* is favoured and is usually reflected a couple of years later in higher counts in the adult population; in contrast, the reproductive output of *Semibalanus balanoides* is favoured in cool years. Both have refuge populations from which they can recruit: *Semibalanus* does well in estuaries, *C. montagui* high on the shore and *C. stellatus* on exposed headlands.

Occasionally, dense settlements of particular species will lead to severe intraspecific competition. In dense settlements of barnacles this competition can lead to whole sheets of individuals being lost (Barnes and Powell, 1950), which will open up space for less competitive colonists. Alternatively, dense settlements will swamp the ability of predators to eat them, leading to escapes (e.g. Sebens and Lewis, 1985).

4.1.4. HOW DOES THE IMPORTANCE OF BIOLOGICAL INTERACTIONS VARY ALONG THE VERTICAL AND HORIZONTAL SHORE GRADIENTS?

The extensive literature on manipulative experiments on rocky shores has prompted a number of generalizations about the relative importance of predation, competition and physical factors along major environmental gradients. Menge and Sutherland (1976) produced a synthesis applicable to communities in general, but which was largely based on studies on rocky shores. These ideas can be summarized as follows.

Predation is the dominant interaction structuring communities in physically benign environments. As environments become harsher, predation becomes less important and competition is a major structuring force. In even harsher environments, competition becomes less important and physical factors assume major importance. On the basis of this we might expect the following: predators will be important structuring agents at low shore levels and on sheltered shores; competition will be more important at higher shore levels and on more exposed coasts; physical factors will be the major organizing force in high shore and very exposed coasts. Intuitively, these generalizations seem to make sense, but more and more exceptions have come to light (Underwood and Denley, 1984).

The early ideas of Menge and Sutherland did not take into account the importance of intensity of recruitment, in which interest was re-awakened in the 1980s (see above). They revisited their model (Menge and Sutherland, 1987; Menge, 1991), incorporating recruitment. This model assumed that mobile organisms (grazers, predators) are affected more by environmental stress than are sessile organisms, and that the complexity of feeding interactions decreases with increasing stress.

If recruitment is high, three predictions can be made on the basis of the model. Firstly, in physically stressed environments, grazers and predators will have little effect as they are absent or not very effective anyway. Physical disturbance will also affect competition for space. Both mobile consumers and sessile organisms are regulated directly by physical stress. Secondly, in intermediate environments, grazers and predators are still not very effective, but sessile organisms are less affected by stress and thus can frequently attain high densities leading to competition for space. Thirdly, in the least stressful environments, grazers and predators prevent competition for space, unless space occupiers can escape to reach a high abundance.

At lower levels of recruitment, the importance of competition is reduced for a given level of environmental stress. Low recruitment of predators or grazers will slow their population increase and hence their effectiveness, even in benign conditions. In stressful conditions the effect on consumers will be the same irrespective of density of recruitment (see above). However, abiotic stress will have a greater importance over a wider range of environmental conditions if recruitment and hence numbers are low. Low recruitment will render competition between space occupiers less intense, even in benign conditions. Under low recruitment, space-occupying organisms should be regulated by physical factors at the severe end of the environmental gradient and by predation at the benign end.

In contrast to these rather complex theoretical considerations of Menge and Sutherland, a summary based more on natural history has been developed in Europe (e.g. Southward and Southward, 1978; Hawkins and Hartnoll, 1983a). These ideas deal primarily with different forces organizing communities on the vertical gradient on more exposed shores (Figs 4.5, 4.6) and along the horizontal gradient in the midshore (eulittoral) (summarized in Hawkins *et al.*, 1992a). We start by taking a slice down a moderately exposed shore.

High up on the shore in the splash zone the community tends to be highly seasonal in occurrence. Many ephemeral algae occur over the autumn, winter and early spring, but they are killed by hot weather in late spring and summer. No doubt competition occurs between ephemeral algae, but it has not been investigated. Grazing is also important but is restricted to the vicinity of refuges from which small littorinids can forage. Only on highly pitted surfaces with many refuges will grazing be important. There is a convincing argument that physical factors are paramount in this zone, which broadly equates with the littoral fringe in the Stephensons' three-zoned system (Chapter 2).

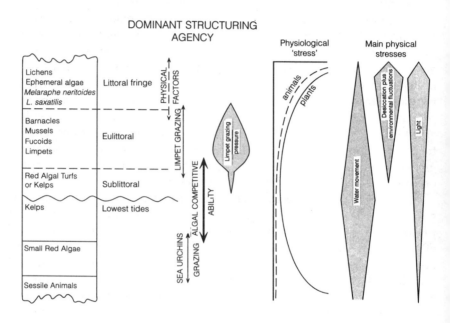

Fig. 4.6 The relative importance of various structuring agencies on the vertical gradient on moderately exposed shores in the north-east Atlantic. This is also an explanation of the Stephenson and Stephenson universal zonation scheme (Chapter 2).

At the top of the eulittoral on moderately exposed shores there is often a thin band of *Pelvetia* and *Fucus spiralis* (Ballantine, 1961). Here limpet grazing seems to be incapable of preventing fucoid growth. Few limpets are present and they seem to be dependent on the *Fucus* canopy, rapidly disappearing when this cover is removed (Hawkins and Hartnoll, 1983b). This band disappears in more exposed conditions as limpets become more abundant. Thus there seems to be a refuge from grazing for fucoids high on the shore. Beneath this zone, grazing by limpets becomes very important, with patches of seaweeds occurring primarily as escapes from grazing. The removal of limpets leads to the whole shore becoming dominated by fucoids. The limpets remove fucoids and probably allow barnacles to establish. Predation by dogwhelks is not thought to be an important structuring agent, with the possible exception of very broken shores having many refuges from which dogwhelks can forage and eat barnacles and mussels (Hughes and Burrows, 1993).

Towards the bottom of the eulittoral, limpet grazing becomes less effective and algal turfs, *Himanthalia* and large kelps dominate the shore. On some shores there is a marked switch between the ability of limpets to graze and the ability of seaweeds to grow (Hawkins and Jones, 1992). The plants can actually swamp grazing activity. In this zone, where physical conditions are probably optimal for plants, the ability of algal turfs to dominate space, or that of kelps to dominate the canopy, is probably the major structuring agent. In the *Laminaria digitata* zone, for example, removal of kelp leads to a proliferation of understorey plants (Hawkins and Harkin, 1985) which swamps any limpets present. 'Lithothamnia'-covered feeding areas for limpets are presumably kept clear by sweeping kelp fronds. The organization of the sublittoral fringe therefore seems to be dominated by plant competition. This continues down into the sublittoral proper, where the canopy of *Laminaria hyperborea* dominates. Removal of this kelp leads to a proliferation of small plants. With depth or increased number of refuges on broken substrata, sea urchin grazing increases in importance and probably becomes the major structuring agency. Physical factors re-assert themselves as light fades and bedrock often gives way to sand or cobbles.

Moving onto sheltered shores, the littoral fringe is very narrow but is probably still largely physically dominated. The whole eulittoral zone is covered by large macroalgae which can occur in contiguous zones forming virtually 100% blanket cover. These algae seem to be largely free of grazers. Competitive interactions probably structure the shore as the removal of the canopy leads to the proliferation of algae, and grazers seem to have little impact. Predators are rare and do not seem to be very important. Perhaps this situation and that in the sublittoral fringe is analogous to the escape from predation bottleneck under benign conditions suggested by Menge and Sutherland (1987).

On very exposed, barnacle-dominated shores, limpet grazing remains important, as shown by the proliferation of seaweeds following the *Torrey Canyon* oil spill. This also happened on mussel-dominated shores. Patellid limpets occur in very high densities on exposed rocky shores in both the North and South Atlantic and grazing is therefore important even in exposed conditions, contrary to the Menge and Sutherland

(1987) model. On very exposed shores, physical disturbance is likely to be important in generating the mosaic of mussels and barnacles observed, but no work has been done on this yet in Europe. In very exposed conditions, mussels can even intrude into the sublittoral fringe and supplant seaweeds (Hiscock, 1985). Again, competitive interactions would seem to be important.

Recruitment levels will affect the outcome of all these interactions. On moderately exposed shores where the balance between limpet grazing and fucoid growth is delicately poised, variations in recruitment are likely to generate fluctuations. On exposed shores, high recruitment of mussels or barnacles will also generate fluctuations because dense populations may self-destruct if intraspecific competition leads to insecure anchorage (Barnes and Powell, 1950). On sheltered shores, there is always sufficient recruitment of fucoids, but in many places the supply of limpet and barnacle larvae seems low, perhaps due to lack of water movement. The exceptions to this are sheltered straits with strong tidal currents which bring in large numbers of larvae; but even here they settle under fucoid canopies and do not pre-empt space.

Southward and Southward (1978) suggested a simple conceptual model of the relationship between fucoids and grazers/barnacles on the gradient between sheltered bays and exposed headlands. In shelter, a clearance in the canopy is rapidly repaired and any localized increase in limpets or barnacles which may occur is short lived. Conversely, in exposed conditions any escape of fucoids is similarly short lived. This idea was extended by Hartnoll and Hawkins (1985), who added a broadly fluctuating region on moderately exposed shores where the balance between limpets and fucoids was less stable (Fig. 3.8).

In conclusion, variability in recruitment can be just as important as the intensity of recruitment and can drive fluctuations in patchy communities. Similarly, habitat complexity and the provision of refuges can be crucial for avoiding physical extremes and predation (Menge *et al.*, 1985). The arrangement and nature of refuges will also affect spatial patterns. Highly topographically complex shores will have a static patchiness caused by microhabitat variation and by predictable constraints on the activity of consumers (see also Menge and Lubchenco, 1981; Menge *et al.*, 1986). Smooth surfaces, where the biota provide refuges such as on the limestone ledges investigated on the Isle of Man, will be much more prone to fluctuations and spatial relocation of mosaics.

4.2 SEDIMENT SHORES

4.2.1 PREDATORS OF SEDIMENT COMMUNITIES

On less exposed sandy beaches and tidal flats the range of predators is expanded, compared with that on rocky shores, to include large numbers of highly mobile forms, notably epibenthic crustaceans, intertidal and juvenile fish and shorebirds. Waders and passerine birds also feed in the intertidal zone of rocky shores (e.g. Feare and Summers, 1985; Marsh, 1986) but their numbers never approach those feeding on mudflats. During the autumn, these shores often experience a massive influx of migratory shorebirds, which utilize the enormous prey densities present in the flats (Chapter 6). For example, during spring migration over one million shorebirds occur within Delaware Bay, USA, on their way from South America to the Arctic (Orth *et al.*, 1991). With the exception of a few species, such as ducks, most shorebirds feed on invertebrates when these are uncovered by the tide. On the flood tide, and at all times in areas where the tidal range is negligible, fish and mobile crustaceans are the main epibenthic predators of sediment invertebrates. Shrimps, prawns, juvenile crabs, gobiid fish and juvenile flatfish, may use these areas as a nursery ground, attaining high densities and consuming large numbers of smaller invertebrates. Large flatfish, including flounders, plaice, dabs and the elasmobranch rays, feed extensively on sandy beaches and mudflats. More detailed accounts of the fish assemblages on sheltered sediment shores, the importance of beaches, mudflats and mangroves to these fish, and the energetic relationships between fish and sediment invertebrate assemblages can be found in Chapter 6.

How important are shorebirds and flatfish in organizing communities?

Because of the difficulties of maintaining exclosure or enclosure cages on exposed sandy beaches, most manipulative studies have been carried out on muddy tidal flats and sheltered sandy shores. There is a risk of losing cages in areas where heavy wave action occurs from time to time, or where tidal currents are strong, ruining weeks of hard work and proving very expensive in materials. Consequently, there is little experimental evidence of the importance of predation in exposed sandy habitats (Brown and McLachlan, 1990).

On beaches and mudflats that experience significant tides, shorebirds are often the most conspicuous predators and their feeding rates can be prodigious (Baird *et al.*, 1985; Chapter 6). Yet where experiments have been carried out to exclude shorebirds from their invertebrate prey, this exclusion has not, in general, led to the kinds of dramatic changes in prey community structure and composition that are apparent when predators are removed from rocky shores (e.g. Kent and Day, 1983; Quammen, 1984; Reise, 1985; Raffaelli and Milne, 1987; Raffaelli *et al.,* 1990b). Some shorebirds are quite capable of seriously depleting densities of particular prey, especially in muddier habitats, for example sandpipers and redshank feeding on *Corophium* (Goss-Custard *et al.,* 1977; Boates and Smith, 1979), and eiders feeding on *Mytilus* (Goss-Custard *et al.,* 1982; Raffaelli *et al.,* 1990b), but these effects do not cascade through the prey community. Also, effects seem to vary from study to study, probably reflecting temporal and regional differences in shorebird feeding pressure (Schneider, 1985 and references therein). An effect is most likely to be seen if experiments are done in late autumn/early winter, when shorebird numbers are highest and prey production is lowest. At this time, shorebirds may hasten the seasonal decline of prey, particularly the dominant species, thereby increasing the evenness component of diversity (Schneider, 1978).

Large numbers of several species of fish regularly move up into the intertidal zone to feed when the tide is in, moving off the flats with the ebbing tide (e.g. Wolff *et al.,* 1977, Raffaelli *et al.,* 1990b). When these fish were excluded from intertidal sediments, only relatively small changes in the densities of one or two species occurred – certainly not the dramatic changes documented for some rocky shores (Van Blaricom, 1982; Kent and Day, 1983; Quammen, 1984; Reise, 1985; Thrush *et al.,* 1991; Raffaelli and Hall, 1992).

Experiments on smaller epibenthic predators

It is relatively easy to assess the impact of shorebirds and large fish on sediment communities by using large-mesh exclosures. If smaller predators, like crabs and shrimps, are to be excluded, then the mesh has to be small, usually only 1 mm or 2 mm. Apart from the risk of changing the entire environment (Chapter 8), a fine mesh will also exclude larger predators, like shorebirds and fish, and it then becomes impossible to separately assess the effects of smaller and larger predators.

In a series of experiments using cages with a range of mesh sizes, Reise (1985) progressively excluded smaller and smaller predators from mudflats in the German Wadden Sea (Fig. 4.7). Fortunately, the larger shorebirds and flatfish had little or no effect on his communities, so that the smaller meshes used to exclude gobiid fish (*Pomatoschistus*), shrimps (*Crangon*) and crabs (*Carcinus*) really did test for the effects of these predators alone. Multi-mesh exclosure designs are potentially useful, because they allow an assessment of the cumulative impact of predators, but they have the disadvantage that small meshes could well create sediment artefacts that confound interpretation of the results of the experiment (Chapter 8).

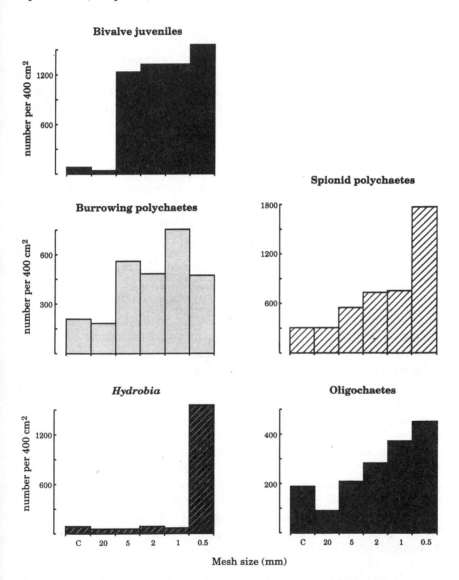

Fig. 4.7 Invertebrate densities inside exclosures with different mesh sizes in the Wadden Sea (after Reise, 1985, with permission). C denotes control (uncaged) area.

An alternative approach for smaller predators is to use enclosures. Here predators are included within cages and the controls are similarly caged, so that the only difference between treatments is the presence or absence of predators. Such experiments have shown that small epibenthic crustaceans, such as the crab *Carcinus maenas*, can have a significant effect on the densities of sediment invertebrates (e.g. Gee *et al.*, 1985; Reise, 1985). Similarly, Woodin (1974) found that when the tube-building polychaete *Platynereis bicaniculata* was excluded from areas of mudflat, the burrowing polychaete *Armanda brevis* increased markedly. Conversely, when the two species were allowed to mix, *Armanda* declined in abundance. However, if small crabs, *Cancer magister*, were added, *Armanda* increased again because the crabs fed preferentially on *Platynereis*, demonstrating that crab predation may contribute to the coexistence of these two polychaetes.

A major problem with the enclosure approach is deciding how many predators to include within the cage. Natural densities of predators can be estimated from trawl data, but the mesh of the net and the speed of towing are likely to affect density estimates markedly. A better method for shallow atidal areas or on intertidal flats at slack water, is to seal off an area of the sediment with a box-like fine-meshed net and then to fish out the enclosed epibenthos (Evans and Tallmark, 1979). Underwater television can also be used to film areas of mudflat throughout the tidal cycle and in this way predator activity as well as density can be assessed, although this can only be done accurately for relatively small areas of 0.5 m^2 or less. With a combination of several techniques it should be possible to obtain reasonable estimates of the natural densities of predators.

However, many enclosure experiments have not used natural densities of predators. Significant effects on prey assemblages have been generated by enclosing predators at several times their background densities, but how meaningful are such results? The rationale for using high predator densities is rarely stated explicitly, but there are often anxieties about having only one or two predators in an enclosure in case one of them dies. A better solution to this problem is to construct larger cages. Ecologists may also be trying to 'force' a response in the prey assemblage because of time constraints. Not surprisingly, where high predator densities have been employed, a significant decline in prey densities often follows. More interestingly, other experiments using high densities of predators have reported few significant effects, as is the case for the sand goby *Pomatoschistus minutus* (Gee *et al.*, 1985; Jaquet and Raffaelli, 1989) and the brown shrimp *Crangon crangon* (Olafsson, 1988; Raffaelli *et al.*, 1989a). Where natural densities of predators have been used, in some instances there have been significant effects (e.g. Reise,

1985), but the overall picture that emerges is similar to that described for shorebirds and fish: small epibenthic predators seem to be significant in some localities, but not in others.

A quite different approach to investigating interactions between small predators and their prey populations is to transfer intact sediment into tank systems to which predators can be added and any interactions carefully monitored. This approach has obvious attractions, but also has its problems. For instance, without a fairly elaborate seawater system, the supply of natural food and recruits to the experimental treatments will be severely disrupted. Also, normal hydrographic events, which appear to be important for some sediment assemblages, are no longer a feature of the system. An up-to-date account of this area of research is provided by Gamble (1991).

Most intertidal ecologists have focused on the direct interactions between predators and their invertebrate prey. In other words, they assume a two-level system. However, a three-level perspective may be more appropriate in some circumstances (Commito and Ambrose, 1984). For example, if shorebirds, fish and crustaceans feed preferentially on larger polychaetes, which may themselves be predators of infauna, then these second-level predators will increase when protected by an exclosure and reduce the other invertebrate populations inside exclosures. Indirect effects like these will confound the interpretation of manipulative experiments, as discussed already for rocky shore systems. Examples of second-level predators include nephtyd, nereid and phyllodocid polychaetes and nemerteans, although their interactions with their prey are not particularly well understood (Commito and Shrader, 1985; Reise, 1985; Davey and George, 1986; Olafsson and Persson, 1986; Ambrose, 1991). If Commito and Ambrose are correct, then numbers of these predators should increase relative to those of other infauna in exclosure experiments. Wilson (1986) could find little evidence for this in the literature, but Kneib (1991) provides a number of examples of how indirect effects might lead to misinterpretations of manipulative experiments.

4.2.2 COMPARISON OF THE SIGNIFICANCE OF PREDATORS ON ROCKY AND ON SEDIMENT SHORES

The contrast between the role predators play on sediment and rocky shores appears striking. Even in those experiments on sediment communities where large increases occur in prey numbers following predator removal, the effects are not translated into the dramatic, cascading changes seen in at least some mussel and kelp communities.

We might conclude that sediment communities lack keystone predators. However, if the attribute 'keystone' is conferred by the response of the prey, as suggested earlier, then a more pertinent question might be, 'does intense competition occur between sediment prey species and, if so, why do superior competitors not emerge as they do on rocky shores?'

Competitive interactions certainly do occur in sediment communities and they can be intense. Interference competition between individuals of the spionid polychaete *Polydora paucibranchiata* results in an even spacing of individuals (Levin, 1981) and aggressive interactions have been reported between nereid polychaetes (Roe, 1975). Adults and juveniles of the same species may interact with dramatic effects on local population demography. For instance, larger individuals of the amphipod *Corophium volutator* (Raffaelli and Milne, 1987) and the polychaete *Ceratonereis eythrocephala* (Kent and Day, 1983) exclude juveniles in the absence of shorebirds and flatfish respectively (Fig. 4.8). These predators normally feed preferentially on the larger prey items, thereby promoting coexistence of small and large individuals. Similar adult–recruit interactions within and between species are probably the norm in sediments, although in most cases the actual mechanisms remain to be established. Interspecific interference competition has been reported for *Nereis diversicolor* and *Corophium volutator* (Olafsson and Persson, 1986), and for burrowing and tube-building polychaetes (Woodin, 1974; see above); the interactions affecting zonation patterns of sediment fauna (Chapter 3) are also likely to be of the interference type.

Exploitation competition, for example by pre-empting space, is less evident in sediments (Dayton, 1984). One of the most-often-quoted examples of exploitation competition comes from studies on the mudsnail *Hydrobia* (Fenchel, 1975 a,b). Fenchel argued that interspecific competition between *Hydrobia ulvae* and *H. ventrosa* for particulate food on Scandinavian shores has led to character displacement, the two species partitioning the sediment resource on the basis of differential body size. Fenchel's studies were based on Scandinavian populations, and there seems to be little evidence in other populations (Barnes, 1980). Also, Levinton (1987) has argued that particle size is unlikely to be the basis for resource partitioning, because *Hydrobia* can very effectively scrape the sides of large particles. Although there is some debate as to whether character displacement actually occurs in *Hydrobia*, it is undeniable that exploitation competition for food does occur. Density-dependent growth effects have been reported for several bivalve species (Wilson, 1991), suggesting intraspecific competition.

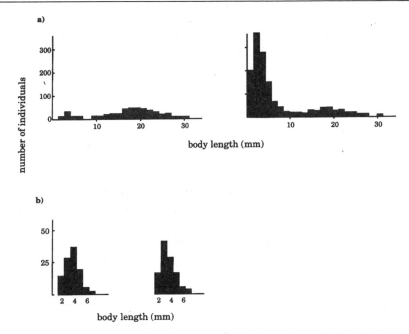

Fig. 4.8 The size distribution of major prey species in exclosure (left) and control (right) areas. (a) Flatfish predation on the polychaete *Ceratonereis* (adapted from Kent and Day, 1983, with permission). (b) Shorebird predation on the amphipod *Corophium* (adapted from Raffaelli and Milne, 1987, with permission).

There is no real equivalent in sediment communities of the kinds of interactions known to occur between space-competing rocky shore species. Sediment invertebrates are rarely permanent fixtures, so that there is no opportunity for the aggressive undercutting, overgrowing and crushing that takes place on rocky shores. In sediments, individuals may simply move away from each other, thereby diffusing any impacts of interference competition. Whilst exploitation competition is likely to be for food rather than space, its effects on growth and fecundity are similarly likely to be diffuse and translated into community-level effects only after a long period. During this period, any such effects may be obscured by major changes in other factors, such as physical disturbance and variations in recruitment patterns.

If this is true, then it raises some interesting problems concerning the design of manipulative field experiments in sediments. Most sediment caging experiments are relatively short term, of the order of weeks rather than months. In contrast, most rocky shore experiments run for months or years. The shorter time course of many sediment experiments is usually adopted to minimize problems with caging artefacts (Chapter 8) which may confound interpretation of any effects recorded. Such time scales may be appropriate for discerning effects through intense interference competition or aggressive exploitation competition for living space, but considerably longer periods may be required to detect any effects of competition for food.

4.2.3 ANIMAL–SEDIMENT INTERACTIONS

Bioturbation: disturbance and stabilization

Whilst exploitation competition seems unlikely to be an important organizing force in sediment communities, disturbance-induced sediment changes may be more important. These can take a wide variety of forms.

The smallest, microbial components can have major effects on sediment chemistry, as described in Chapter 2, but they also stabilize the sediment by secreting quantities of mucopolysaccharides that glue the sediment particles together. Diatoms and bacteria are important in this respect in muddy sediments (Paterson *et al.*, 1986), and work in experimental flumes has shown that these organisms can greatly increase the water velocity needed to erode the sediment surface. Mucus production is a feature of most infaunal species and their effects on sediment binding may be substantial.

The physical presence of larger organisms can also stabilize the sediment, inhibiting erosion by water movement. Many tube-building species fall into this category, especially the larger polychaetes such as the sand mason *Lanice conchilega*. If densely enough packed, the tubes of this and similar species, which may penetrate many centimetres below the surface, can greatly increase critical erosion velocities. However, if tube density is low, the sediment can be destabilized, because the projecting tubes produce swirling vortices in their wake (Fig. 4.9). Other sediment stabilizers include algal mats, mangrove trees, mussel beds and seagrasses (Fig. 4.10). The latter can have pronounced effects on sediment characteristics by creating local hydrodynamic changes (Jones and Jago, 1993), which may also promote larval settlement (Eckman, 1987). They also provide a refuge from predation for small epibenthic species.

Sediment destabilization is mainly caused by deposit feeders which browse, manipulate, sort and process sediment particles. Large deposit feeders, such as the polychaete *Arenicola marina*, act like conveyor belts (Rhoads, 1974), ingesting particles from many centimetres below the surface, passing them through their guts and depositing them as faeces on the sediment surface (Fig. 4.11; Chapter 5). Polychaetes like *Lagis*

(*Pectinaria*) similarly browse upside down in sediments, selectively ingesting small and medium-sized particles. These are repackaged as faecal pellets and deposited at the surface (Fig. 4.12). The net effect of these conveyor-belt species is to change the vertical size distribution of particles. The surface becomes a loose fabric of large particles (faecal pellets) with a high water content, increasing the susceptibility of the sediment to erosion. In deeper-water environments, the incompatible stabilizing and destabilizing effects of suspension feeders and deposit feeders promote spatial separation of these two major feeding types, a phenomenon termed **trophic group amensalism**. However, this phenomenon has not been widely reported for intertidal and shallow sublittoral systems (Hall *et al.*, 1993).

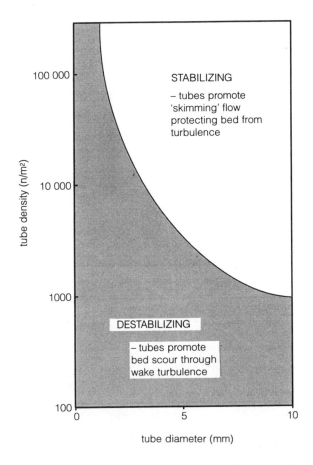

Fig. 4.9 The relationship between worm tube size, density and sediment stability (adapted from Eckman, 1985, with permission).

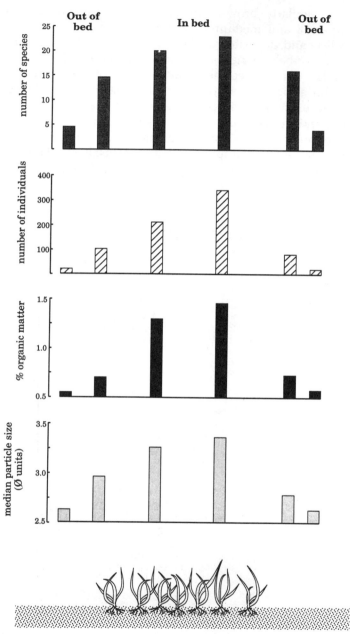

Fig. 4.10 Effects of seagrass beds on sediment characteristics (modified from Orth *et al.*, 1991, with permission).

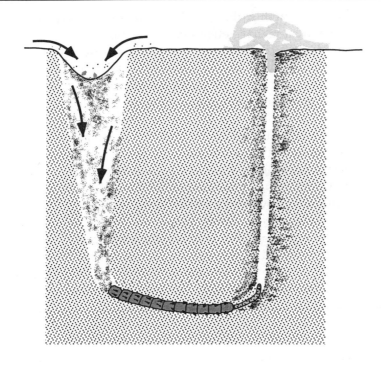

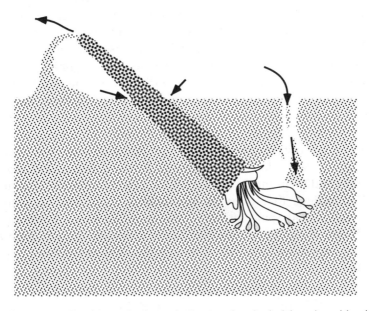

Fig. 4.11 Sediment disturbance by the polychaetes *Arenicola* (above) and *Lagis* (below). Arrows depict direction of water flow.

As well as altering the physical structure of sediments, deposit-feeding bioturbators have effects on sediment chemistry. These include: increasing the exchange rates of compounds across the sediment–water interface by several orders of magnitude; lowering the depth of the redox potential discontinuity (RPD) layer; increasing the penetration of oxygenated water; and bringing to the surface reduced compounds such as methane, ammonia and hydrogen sulphide so that they become oxidized to carbonate and nitrate and hence available for water-column processes (Rhoads, 1974).

Particle processing by intertidal organisms can be substantial (Table 4.1), but its impact is probably exceeded many times by water pumping. Many species occupy permanent or semipermanent burrow structures, extending several centimetres below the sediment surface, through which water is pumped. For instance, the amphipod *Corophium volutator* can pump 25–100 ml h^{-1} through its burrow (Reise, 1985). Such species have the capacity to depress the RPD layer and significantly change sediment chemistry at depth, so that oxic halos are often seen surrounding the burrow in an otherwise black, reduced layer (Reise and Ax, 1979).

In many parts of the world, mudflats, salt-marshes and mangroves support high densities of small crabs (*Uca, Sesarma, Helice, Macrophthalmus*) which excavate burrows extending several centimetres below ground over extensive areas (Table 4.2). Burrow densities can reach hundreds per square metre so that the sediment becomes honeycombed with galleries, greatly increasing the surface area available for exchange of materials such as ammonia, oxygen and sulphide. Burrow maintenance results in significant sediment reworking, surface material being buried and material from deeper layers being brought to the surface. In addition to these bioturbatory effects, several species may have direct trophic effects on meiofaunal and plant populations, although the majority of species seem to be primarily deposit feeders (Table 4.2). Anyone who has seen these populations cannot fail to be impressed by the densities and activities of the crabs and their impact on the physical structure of the sediment. It is hard to believe that they do not play a major role in community organization and some authors have accorded them 'keystone' status (e.g. Smith *et al.*, 1991). However, such generalizations require a more rigorous experimental manipulative approach than has been carried out to date.

Table 4.1 Sediment reworked by sublittoral deposit feeders*

Species	Amount reworked (ml mud individual^{-1} year^{-1})
Clymella torquata (polychaete)	96–274
Pectinaria gouldii (polychaete)	400
Yoldia limatula (bivalve)	257
Nucula annulata (bivalve)	365
Callianassa (crustacean)	75 cm layer transported to surface by population annually

*Source: Rhoads (1974).

Table 4.2 Selected studies on the effects of burrowing crabs on mudflats, salt-marshes and mangroves

Species	Location	No. m^{-2}	Effects	Reference
Helice tridens	Japan	70	Crabs buried surface litter (*Phragmites*). Burrows increase surface area by 59%. In 14 days crabs excavated 3% of the upper 40 cm of soil.	Takeda and Kurihara (1987)
Sesarma spp.	Townsville, Australia	–	Ammonium and sulphide higher in plots without crabs. Mangrove (*Rhizophora*) propagules more abundant in crab-free areas.	Smith *et al.* (1991)
Uca pugnax	Eastern USA	70	Excavated 18% of upper 15 cm of sediment per annum.	Katz (1980)
Uca pugnax	New England, USA	70	Burrows increase drainage, soil oxidation and decomposition of below-ground material.	Bertness (1985)
Uca sp.	New South Wales, Australia	Not stated	Increase microtopography of mangrove forest floor, reduce surface algal cover and change sediment grain size.	Warren and Underwood (1986)
Uca spp.	South Carolina, USA	–	Buried (reduced) material brought to surface, oxidized material fills old burrows. Turned over 18% of sediment in upper 30 cm each year.	Gardner *et al.* (1988)
Uca vocans and *U. polita*	Queensland, Australia	20	Meiofaunal abundance increases 3 to 5-fold when crabs removed, possibly due to absence of intense disturbance.	Dye and Lasiak (1986)

Sediment disturbance by epibenthic predators

All these processes generate small-scale patchiness in sediment habitats, and this must go some way to maintaining community structure and

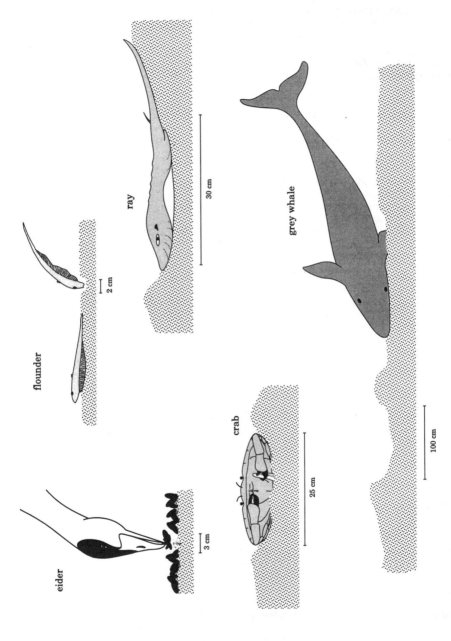

Fig. 4.12 Pits created by different epibenthic predators (adapted from Hall *et al.*, 1993, with permission).

composition in the absence of the strong interactions seen on hard substrata (Dayton, 1984). To these small-scale disturbances must be added those caused by mobile predators. Disruption of the sediment fabric and gross disturbance of the surface layers can be brought about by bulldozers, such as the gastropods *Nassarius* and *Ilyanassa*, with dramatic effects on populations of small metazoans. The effects of such disturbances on meiofaunal community structure and composition are inconsistent. Thus, the amphipods *Monoporeia affinis* (Olafsson and Elmgren, 1991) and *Corophium volutator* (Limia, 1989) both seem to have little impact on overall nematode abundance and, in the case of the former, on nematode species composition. However, they do have an effect on the distribution of nematodes in the sediment, nematodes moving deeper in the presence of the amphipods. Warwick *et al.* (1990a) describe dramatic effects of disturbance by hordes of soldier crabs on meiofaunal populations of intertidal flats. The sediment-reworking activities of lugworms (*Arenicola*) in the Wadden Sea have negative effects on the juveniles of other macrofaunal species, including the polychaetes *Nereis* (= *Hediste*) *diversicolor*, *Nephtys hombergii*, *Pygospio elegans*, *Scoloplos armiger* and *Capitella capitata* and the bivalves *Mya arenaria*, *Cerastoderma edule*, *Macoma balthica* and *Angulus* (= *Tellina*) *tenuis* (Flach, 1992). Effects of *Arenicola* on the amphipod *Corophium volutator* were pronounced and were probably due to this species actively emigrating away from the disturbance.

Other epibenthic species may physically disturb the sediment locally by taking bites of the surface and winnowing the contents, as done by flounders, rays and shelduck, or by trampling, puddling or excavating individual prey items, as done by many shorebirds (Savidge and Taghon, 1988; Cadee, 1990; Raffaelli *et al.*, 1990a). This kind of disturbance creates a discrete depression or pit (Fig. 4.12) that infills, perhaps acquiring sediment and biological characteristics different from those of the surrounding area. In the shallow sublittoral (< 20 m depth), larger predators, such as walrus (Oliver *et al.*, 1985), sea otter (Hines and Loughlin, 1980), larger crabs (Thrush, 1986; Hall *et al.*, 1991) and stingrays (Reidenaur and Thistle, 1981; Van Blaricom, 1982), excavate large pits in pursuit of their prey and these pits may develop populations quite different from those of the background sediment which remain recognizable for many weeks (Fig. 4.13). Perhaps the most dramatic pit-forming process is that involving grey whales *Eschrichtius robustus* winnowing beds of onuphid polychaetes and amphipods seaward of the breaker zone in the north-east Pacific (reviewed in: M.J. Foster *et al.*, 1991).

The pits created by disturbances may become a trap for organic material, attracting amphipods capable of exploiting these high-quality patches (Van Blaricom, 1978), whilst in other instances, they may

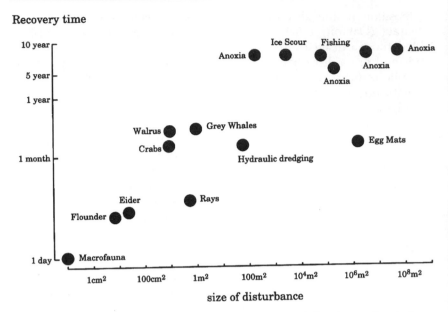

Fig. 4.13 Relationship between recovery time and size of disturbance for shallow water sediments (adapted from Hall *et al.*, 1993, with permission).

provide a refuge for opportunistic poor competitors, such as capitellid worms. Sediment-disturbing predators could therefore generate a mosaic of patches, each in a different stage of recovery. In other words, they could create a patchiness in the habitat, and hence maintain a high prey diversity, by continually resetting the successional clock (see below). Whether succession actually proceeds in a patch depends, to a much greater extent than on rocky shores, on the size of the disturbed area. On a rocky shore, the main space occupiers are not mobile and the surface itself is stable. In sediments, the fauna is more mobile and the substratum is potentially unstable. Unless the disturbed area is large or the sediment is stabilized by biogenic structures, such as amphipod or polychaete tubes, orderly successional processes are unlikely to occur. The sides of the pit will simply collapse and recovery will be more or less instantaneous. The pits made intertidally by birds and fish on sandy beaches and mudflats are usually less than a few cm² in size and infill rapidly. Larger sublittoral pits created by larger predators will retain biological characteristics different from those of the surrounding area for longer (Fig. 4.13).

Perhaps the best way to assess whether disturbances by larger epibenthic predators are of sufficient magnitude and frequency to affect the diversity of sediment communities, is to exclude the disturbing predators from an area. Where this has been done for flounders (Raffaelli and Milne, 1987), eiders (Raffaelli *et al.*, 1990a), and crabs (Hall *et al.*, 1991), any biological changes occurring within individual pits do not seem to be translated into community-wide or landscape-scale effects. However, much more work needs to be done in this exciting area.

4.3 COMMUNITY SUCCESSION

4.3.1 THEORETICAL BACKGROUND

Early theories of succession in ecological communities were derived by the piecing together of different terrestrial plant communities (e.g. lichen-covered rock, scrub and mature forest) into a presumed orderly sequence. These different communities represent temporal stages in an ecological succession from bare rock through to massive forest systems. Because of the enormous time scales over which succession occurs on land it is not practical to observe the whole sequence at any one location. Also, the different successional stages are usually scattered over a large geographical area. Here, the shore ecologist is at a distinct advantage. All the stages of succession might be represented by the patchy mosaic seen on the shore, and documenting the successional process from bare substratum to something approaching a 'mature' community is feasible within a relatively short-term research programme.

Terrestrial ecologists now recognize that the idea that succession proceeds through a series of orderly species replacements and culminates in a single climax community is simplistic. Initial colonization of bare areas and later invasions by species are not as predictable and orderly as once believed, and there may be several possible end points to succession. Also, succession never really stops. The clock is continually reset by disturbance, which may be physical and biological. Indeed the timing of these disturbances in the successional process can have a major influence on the course of subsequent succession. The shorter-term and smaller-scale ecological processes operating on shores allow the intertidal ecologist to experimentally interfere with the course of succession and identify the mechanisms acting at each stage in the process. However, as shown below, one needs to be cautious in extrapolating observations made at a scale of less than a few square metres to whole coastlines.

4.3.2 SUCCESSION ON ROCKY SUBSTRATA

In trying to understand succession on rocky substrata it may be helpful to distinguish between three kinds of successional process (Lubchenco and Menge, 1978; Hawkins and Hartnoll, 1983b), each of which can be mimicked experimentally.

Primary succession, the colonization of virgin substrata created by physical factors such as flaking of the rock surface, by immersion of breakwaters and other human-made structures in the sea and by violent geological activity such as volcanic eruptions, as happened when the new island of Surtsee was created off Iceland. Natural virgin sediment is difficult to find because fresh deposits of this material on the shore are likely to originate from deeper water, and sand grains rapidly – within minutes – acquire an organic film and probably micro-organisms.

Secondary succession, the colonization of encrusting algae or rock possessing an intact microbial film. This type of succession occurs after fairly catastrophic events such as storms ripping off sheets of mussels, barnacle hummocks or algal canopies.

Tertiary succession, whereby new resources are released through the local removal of dominant individuals of, for instance, fucoid seaweeds or mussels, grazers or predators.

Different studies have emphasized these three kinds of succession to different degrees. For instance, the patch dynamics of eulittoral communities described by Hawkins and Hartnoll (Fig. 4.5) refer to tertiary succession, whilst many experimental clearings either deal with primary succession, if an effort was made to sterilize the rock surface, or with secondary succession, if the larger organisms were simply removed. These distinctions need to be borne in mind when comparing different investigations of succession.

Interactions occurring during succession

Three models of succession have been proposed: facilitation, tolerance and inhibition (Connell and Slatyer, 1977). The species involved in these interactions can likewise be viewed as facilitators, inhibitors or tolerant. Facilitators modify the environment or the microclimate, such that it becomes possible for other species with more exacting requirements to colonize later. Facilitators are often outcompeted by these species, the recruitment and establishment of which they promote. Inhibition occurs when an early colonist effectively prevents the establishment of other species, which can only invade when the inhibitor is removed. Tolerant species are usually slower-growing forms which are neither positively nor negatively affected by the presence of initial colonizers, but which eventually dominate the community because they outlast other species.

Different kinds of interactions may occur between different pairs of species at the same stage in the successional sequence. For instance, species A may facilitate species B, but inhibit species C. Alternatively, a successional sequence may include elements of more than one type of process. Several shallow sublittoral fouling studies have stressed the importance of inhibition, including Sutherland and Karlson's (1977) detailed analysis of the colonization of settlement plates at Beaufort, North Carolina. In this study, any species could colonize the plates, depending entirely on when the plates were submerged and what larvae were available to settle at the time. Subsequent colonization by other species could only proceed if the initial colonizer was removed, suggesting inhibitory effects of each species on every other.

Various intertidal studies (review: Sousa and Connell, 1992) have also shown that inhibition of late-arriving algae occurs by early settlers, usually fast-growing ephemerals (e.g. Sousa, 1979; Lubchenco, 1983). This inhibition can be broken by grazers and thus the rate of species replacement is increased. Grazing accelerates succession, particularly if early successional species are preferred. This is usually the case as grazers prefer fast-growing, undefended species such as *Enteromorpha* and *Ulva* (Watson and Norton, 1985). Conversely, if grazers prefer later successional species, succession will be slowed and held at an early stage. This was shown by Sousa (1979): in the absence of grazers, a successional sequence of a mixed assemblage of ephemeral greens (*Ulva*) and filamentous reds plus crustose corallines gave way to a turf of perennial red algae (*Gigartina canaliculata* and *Laurencia pacifica*) and then to large perennial brown algae (*Egregia laevigata* and *Cystoseira osmondacea*). Grazing by crabs and gastropods maintained the assemblage at the early stage, which eventually gave way to more grazer-resistant crustose corallines. Damselfish selectively weed out late successional species from their territories. These are often not consumed, but whether they are weeded out or eaten, the result is that the fish maintain essentially a garden of fast-growing ephemerals as their preferred food (Lassuy, 1980; Montgomery, 1980; Hixon and Brostoff, 1983; Irvine and Northcote, 1983).

In contrast to examples of inhibition, there are few clear examples of facilitation. It is often implicit in the very early stages of succession (e.g. Wahl, 1989), but there have been no successful manipulations of the algal film. On rocky substrata, evidence for facilitation includes Scheer's (1945) observation that bryozoans settle abundantly on surfaces only after a distinct microalgal community has developed; this took about 4 weeks. Also, mussels seldom appear early in the recolonization of rocky shores, and their larvae attach preferentially to filamentous structures or other complex surfaces provided by other biological material. Finally, some kelps will only colonize after filamentous species have established (Connell and Slatyer, 1977). Turner (1983, 1985) showed that an early successional species of brown algae (*Phaeostrophion irregulata*) facilitated the establishment of the surf grass *Phyllospadix scouleri* (a higher plant which has roots). This facilitation was a provision of a microhabitat in which the seeds of the surf grass can attach and germinate. Two other studies (Harris *et al.*, 1984 and Farrell, 1991 – see below) showed that facilitation occurred by early successional species providing a refuge from grazing.

No examples of tolerance have been unequivocally demonstrated. Turner and Todd (1993) have recently carried out a study similar to Sutherland and Karlson's, but on an exposed rocky intertidal site on the coast of Scotland. There was little evidence for any orderly sequence of colonization on their clear acetate panels. All the taxonomic groups involved (sponges, barnacles, bryozoans and ascidians) recruited more frequently onto panels immersed for longer periods, indicating that the prior development of microbial films might be important for their recruitment – evidence for facilitation. In contrast, early successional metazoans did not appear to be a prerequisite for the establishment of 'later' species, all of which could have recruited during the initial immersion period. This suggests that the species which dominate late in the succession are tolerant.

Connell and Slatyer (1977) pointed out the need for a rigorous experimental approach to test whether particular kinds of interactions were central to the successional process, for instance by removing particular species in the presumed sequence, and thereby altering the order of arrival of species. This manipulative approach is lacking in most investigations on rocky shores, which have generally involved clearing and/or sterilizing an area of rock and meticulously describing the community that develops. A notable exception, however, is Farrell's (1991) study of succession in the upper eulittoral barnacle–weed zone of Oregon rocky shores. Farrell cleared areas by scraping the rock surface, leaving barnacle bases, encrusting algae and microbial films in place, and followed succession over a three-year period. He also evaluated the roles of various consumers at different stages in succession, predicting that consumers (limpets *Lottia (Collisella)* spp.) would slow down, or accelerate, or have no effect on succession depending on the mechanisms of species replacement.

In unmanipulated sites, the first colonizer was the barnacle *Chthamalus dalli*, which was later replaced by another barnacle, *Balanus glandula*. The seaweeds *Pelvetiopsis limitata*, *Fucus distichus* and *Endocladia muricata* only appeared after *Balanus* was established. The rate of succession at these sites differed from less than 12 months to 36 months depending on variations in the timing of *Balanus* recruitment. Experimental manipulation of barnacle cover, including the provision of artificial barnacles, showed that barnacle shells were necessary for successful algal recruitment, probably by providing a refuge from limpet grazing. This is an example of facilitation. However, the prior presence of *Chthamalus* had no effect on subsequent *Balanus* establishment, an example of tolerance. Farrell found no evidence of inhibition. Experimental removal of limpets led to an increase in the rate of succession, principally through increased survivorship of *Balanus*, and a more rapid colonization by seaweeds. Thus consumers normally slow down succession in this system.

On UK shores, there can be considerable variation in the sequence of early arrivals depending on the timing of grazer removal (Hawkins, 1981a; Hill and Hawkins, unpublished). If grazers are removed in winter and spring, diatoms precede ephemeral greens which are followed by fucoids. If the same experiments are done during the summer, fucoids can develop directly in barnacle-covered areas, without the need for an intermediate ephemeral alga (or diatom) stage. In the autumn of one year a diatom phase was immediately followed by fucoid development. Clearly the sequence of succession depends on the timing relative to propagule availability.

Succession on different spatial and temporal scales

Studies like Farrell's and that of Hartnoll and Hawkins (Fig. 4.5) reveal much of the fine detail of the interactions between species during succession. However, the experimental approach adopted constrains observations to relatively small spatial scales and, because of funding, to relatively short time scales. Such studies provide an understanding of successional patterns and mechanisms underlying patch dynamics, but do these processes account for successional changes over longer time periods and at larger spatial scales? Although it does not deal specifically with succession, the general point is well made by Dayton's discussion of how our perception of the importance of various disturbance processes depends on the temporal scale chosen (Dayton *et al.*, 1992). For instance, floating logs create primary space on some rocky shores in the north-east Pacific by smashing holes in the sessile communities. Dayton argues that this source of disturbance was probably much more widespread before wholesale deforestation occurred around the world's coastlines. Similarly, shallow-water assemblages on the east coast of the

USA may well have been structured by predation by demersal fish, such as cod and haddock, the stocks of which are now seriously depleted. Some of the patterns and distributions we see today may well be partly a reflection of such processes, which are no longer evident. In other words, those processes which we think are important may not be entirely responsible for the characteristics and distributions of communities we see today.

The effect of spatial scale on our view of succession is best illustrated by considering the changes that have occurred since the *Torrey Canyon* oil-spill disaster in south-west Britain in 1967. The ecological consequences of this are described in detail in Chapter 8. The oil itself probably did little environmental damage, at least much less than the dispersants which were hurriedly applied to clean up the coast. These chemicals were highly toxic and resulted in the widespread removal of limpets and other grazing invertebrates. As a consequence, this coastline bloomed with opportunistic green algae, and it was only later that fucoid seaweeds, limpets, barnacles and mussels reinvaded. The recovery of these shores was monitored by Alan and Eve Southward, who put forward the following succession scenario. A dense cover of fucoid seaweeds, which were normally scarce on exposed British shores, became established early on in the succession sequence, but was eventually trimmed back by a massive population of limpets, allowing domination by barnacles. In subsequent years the fucoids returned, but never attained the high levels seen during the early successional period. Eventually the community settled down to levels of spatial and temporal variation characteristic of undisturbed shores. A damped oscillation model of succession driven by interactions between grazing limpets and fucoid algae and fluctuations in recruitment has been proposed (Southward and Southward, 1978; Hawkins and Southward, 1992).

The *Torrey Canyon* story has been (and is still being) documented over a relatively long time scale (> 25 years) and on a relatively large landscape scale (whole shores spread over tens of kilometres). This is in marked contrast to the spatio-temporal scales on which succession has been investigated on individual fouling plates (usually less than 20 x 20 cm), or in most patch dynamics studies. The time taken for an area to return to its normal state following grazer removal depends on the size of the area affected. Thus for areas of one square metre or less, a return to normality is seen within 3–5 years of limpet removal. In larger-scale manipulations, such as the 2 m and 10 m wide strip removals carried out by Hawkins (1981b and unpublished) and Jones (1948) respectively, the recovery process took about 8–10 years. In the case of the *Torrey Canyon*, recovery took at least 15 years, probably longer. Interestingly, the recovery sequence (ephemeral algae–fucoids–limpets–barnacles) was essentially similar across all spatial scales; the scale-dependent differences in recovery time were due to differences in the length of time for which each sequence lasted.

4.3.3 SUCCESSION IN SEDIMENTS

Much more effort has been put into studying succession on hard substrata compared with effort spent on sediments. This is due in part to the ease with which hard substrata can be cleaned and manipulated, but also reflects the economic significance of biofouling. As a result, relatively little information is available for beaches and mudflats, although our understanding of deeper sublittoral muds is probably much better in this respect. Sediments below the wave shoaling depth are not prone to regular disturbance by normal physical processes, so the interactions between the organisms and the sediments they inhabit can be more clearly appreciated. Similar processes may well occur in very sheltered, fine-particle beaches where the same kinds of organisms are found. It is therefore worth spending some time considering those processes which have been unravelled in subtidal studies.

Conceptual approaches

Three conceptual approaches are apparent in investigations of the disturbance–recolonization process in sediments, and these have been broadly classified as **environmental, life-history** and **biotic interaction** approaches (Zajac and Whitlatch, 1987). These approaches should not be considered exclusive, but simply different perspectives on the same process.

The environmental approach is typified by Johnson's (1973) model where recolonization depends to a large extent on the response or tolerance of species to the altered sediment characteristics. As infilling proceeds, and conditions revert to those of the wider surrounding sediment, then other species can invade the disturbed area. This model assumes close adaptations between organisms and their environment, and the successional process is fairly predictable.

The second approach focuses on the life-history traits of species (Grassle and Grassle, 1974; McCall, 1977; Pearson and Rosenberg, 1978; Rhoads *et al.*, 1978). Those opportunists with high reproductive rates and wide dispersal of larvae (but poor competitive ability) are usually the initial colonizers of disturbed areas. Slower-growing, but competitively superior, species invade later and eventually outcompete the opportunists. In this model, disturbance maintains poor competitors in the system. The successional process is seen as a progressive bioturbation of the sediment which depresses the RPD layer and extends the depth of more favourable conditions (Rhoads and Boyer, 1982). Just how applicable this model is to succession in intertidal and shallow subtidal sediments is unclear. Apart from areas experiencing high organic loadings, denuded intertidal sediments, even muddy ones, do not usually have an obviously hostile physico-chemical environment. Consequently, the ameliorating roles of bioturbators are probably not so important intertidally. Grassle and Saunders (1973) suggested that *r*-selected traits have allowed opportunists to avoid competition with superior competitors, but these traits may simply enable opportunists to respond to locally enhanced food resources, such as organic material accumulating in depressions.

A third approach to understanding successional dynamics has been outlined by Zajac and Whitlatch (1985). Here the biotic interactions between species that facilitate or inhibit the invasion of species into disturbed areas are the focus. Facilitation may occur during the recolonization of sand flats, for example, through small-scale alterations of the local environment by the tubes of polychaetes and crustaceans (Gallagher *et al.*, 1983). These authors also found inhibitive interactions, probably predatory, during succession. Inhibition also seems to be important in some shallow sublittoral assemblages (Santos and Bloom, 1983; Zajac and Whitlatch, 1985), especially where meiofaunal predators are abundant (review: Hall *et al.*, 1992).

Clearly these three approaches have much in common, but they address questions about succession at different levels. Ideally, all three approaches should be integrated. A good example of how this could be done is provided by Zajac and Whitlatch's (1985) exploration of the interactions between the erosive and depositive effects of water flow, type of larval development and interactions of a facilitative, inhibitive or tolerant nature.

Experimental approaches

Experimental manipulation of the successional process is difficult in sediments, but it is possible. For instance, large bivalves can be forced out of the sediment by cooling the base of cores (Rhoads and Boyer, 1982). However, such behavioural manipulations are not easy to carry out and the organism's response can be frustratingly inconsistent and unpredictable. It is perhaps not surprising that experimental approaches to succession in sediments have focused mainly on the recolonization of sterilized or defaunated sediment. Four features are apparent from such studies. First, colonization times are variable, often ranging from several hours to many days. Second, the recolonization areas concerned tend to be small (10 cm² to < 1 m²). Third, recolonization is mainly by adults, not by larvae. Fourth, no consistent functional groups emerge as pioneers, unlike the case for sublittoral sediments, although there is a suggestion that opportunists such as capitellid polychaetes are more abundant at the start of the recolonization process. Unfortunately, in most of these studies, sediment physico-chemical characteristics were not monitored alongside changes in the fauna, so that the mechanisms underlying succession remain obscure. Also, the use of small recolonization areas might underestimate the significance of larvae in the recolonization of larger areas.

In conclusion, it seems that in some intertidal sediments, opportunists are the first to occupy denuded areas, but only for brief periods. These are then quickly replaced by other species, but there is little evidence that the opportunists ameliorate the environment, as has been postulated for sublittoral sediments. Perhaps these later species are best described as tolerant (Connell and Slatyer, 1977), colonizing intertidal sediments despite the presence of 'pioneers', not because of them. Some of the apparent differences between subtidal and intertidal succession may be due to the small spatial scale of investigations of disturbance on beaches and the physical instability of their sediments, both of which would militate against the kinds of interactions documented for physically stable subtidal systems. Finally, very little is known about the part played (if any) in the successional process by microbial and meiofaunal organisms, especially how these might modify or condition sediments for macrofauna, and these aspects would repay further study.

OVERVIEW

Early work on rocky shores showed the importance of grazers and predators in structuring communities by preventing domination by fast-growing competitive superiors. In some communities 'keystone' species are very important but in others their importance has been questioned. In some cases, biological interactions are hierarchical, with cascades of effects occurring down through the community – the so-called 'top-down' control. There is also evidence that many communities are non-hierarchical, with complex suites of positive and negative interactions between species, particularly where interactions are finely balanced, and stochastic factors (e.g. physical disturbance or fluctuations in recruitment) are important in these situations. Recently there has been a growing awareness of the importance of the supply of propagules (larvae, spores etc.) in community ecology. The importance of different physical and biological processes will vary with position on the major environmental gradients on rocky shores. The recruitment regime will also be important in influencing the intensity and outcome of interactions between species and in generating fluctuations. Ideas from freshwater biology have recently been borrowed to explore the role of underlying differences in nutrients and hence productivity in adjacent nearshore waters on the way communities are structured (so-called 'bottom-up' forcing) (see Chapter 6).

In sediment communities, the direct effects of predation on community structure are not so marked as on rocky shores. One reason why predation may not be important in sediments is that competition is probably not intense because of the three-dimensional nature of the sediment. Indirect effects of sediment disturbance by predation can be important. Also, bioturbation by large burrowers will rework the sediments and have major effects on community structure.

Succession has been explored in both rocky and depositing communities to test models of succession. On rocky shores there is limited evidence for facilitation, and there is little scope for environmental modification by early arrivals. Inhibition seems more common in successional sequences and needs to be broken by grazing or predation to allow the next stage to proceed. There are very few good examples of tolerance. Often a sequence will have elements of all stages. The timing of initiation of succession is also important as propagule availability will affect sequences of succession.

On a larger scale, recovery of shores following grazer kills after oil-spill clean-ups has occurred through major oscillations of seaweeds and grazers which eventually dampen to natural levels of fluctuation. It is likely that on most rocky shores succession never stops with small-scale disturbances opening up resources all the time. Within sediment communities there is greater scope for sediment modification and hence facilitation. This certainly occurs in sheltered subtidal locations with limited water movement, but may not be common in many intertidal situations experiencing significant sediment transport.

Coping with the physical and biological environment

5

When first colonized from the sea, shores must have represented an inhospitable, although unexploited, resource. Nevertheless, many species are now clearly able to cope with the physical problems of living in an environment characterized by sharp environmental gradients and fluctuating conditions. As well as the essentially physical problems, there are a host of biological pressures facing intertidal organisms, particularly towards lower shore levels. In this chapter we consider some of the features of shore plants and animals that allow them to survive these abiotic and biotic pressures at different points along the major vertical and horizontal gradients. These features include many of the classic adaptations of shore organisms to the rigours of the physical environment as well as the coevolutionary interactions between consumers and resources which make up so much of the detailed natural history of the seashore. First we consider how the dispersal stages of shore organisms actually find a suitable place to live. Then we go on to describe how, once settled on the shore, these organisms deal morphologically, behaviourally and, to a limited extent, physiologically with the main abiotic pressures. For more on physiological adaptations see Newell (1979), Rankin and Davenport (1981) and Dring (1982). Finally, we discuss a variety of morphological, behavioural and life-history features thought to be the result of predator–prey and herbivore–plant interactions and how these may have developed into mutualisms.

5.1 FINDING A PLACE TO LIVE

5.1.1 ROCKY SHORES

Plants and sessile animals on rocky shores are fixed in one place as adults and it is imperative that the dispersal stages settle in the right place on the shore. A mistake by the larva or algal propagule at this stage of the life cycle cannot be corrected later on. For those species with restricted dispersal, such as fucoid algae, the sea palm *Postelsia*, spirorbid polychaetes and some bryozoans, the chances of settling in the right place are good. For more widely dispersed species, such as barnacles and mussels, the dispersive stage must somehow regain the coast and then find the appropriate place on the shore. Whilst some seaweeds such as *Enteromorpha* have motile spores and many larvae are capable of directed movements, the only way in which these tiny stages can effectively 'navigate' back to the shore is by appropriate positioning in the water column (Crisp, 1976), for instance by changing their phototactic response. Many intertidal seaweed spores are photopositive, behaviour which will to some extent aid in keeping position. Most larvae start life positively phototactic, but as settlement nears, low-shore species can become negatively phototactic whilst mid- and high-shore species remain positively phototactic (review: Newell, 1979).

Early attempts to explain the distributions of sessile species on the shore were based on the idea of random settlement and subsequent differential mortality or movement by juveniles. This parsimonious explanation is no longer tenable following the discovery that the larval stages of many species can select their settlement site to some degree (Newell, 1979; Chia, 1989; Pawlik, 1992). Settlement is now viewed as a sequence of phases during which the dispersive stage responds positively or negatively to a range of physical and biological cues (Newell, 1979). Because most of what we know about this comes from studies on larvae, much of what follows concerns sessile animals rather than plants.

Having regained the shore, settlement occurs in two stages. First, there is a period of **testing**, when the larva alights and temporarily attaches on the substratum, usually by mucus. If the surface is not suitable, then the larva can swim off in a broad-scale search for other surfaces. The second stage is an **attachment** phase, during which quite detailed exploration of the surface occurs before **metamorphosis**. In some sessile species, subsequent detachment and secondary settlement can occur, as in the mussel, *Mytilus edulis* (Bayne, 1964).

The final site of settlement and metamorphosis is determined by precise cues. These include the texture of the substratum, such as the presence of pits, crevices and concavities, which afford protection to larvae and metamorphosed juveniles. Even mobile animals have nursery areas in pools and crevices from which they migrate, often up the shore, when they become larger and less vulnerable (e.g. Raffaelli and Hughes, 1978). Biological cues may include the presence of adults or juveniles of the same or related species, the presence of a host organism in the case of epiphytic species, or species which occur in the same zone.

Gregarious settlement (next to individuals of the same species) ensures successful later reproduction, but can also lead to crowding and increased competition for space and perhaps food, which in sessile organisms can have catastrophic outcomes (Barnes and Powell, 1950). Various spacing-out behaviours have been observed in the final stage of settlement, which go some way towards preventing overcrowding, although these are unlikely to be effective when settlement is very high.

Some molluscs use cues from compounds similar to the neurotransmitter GABA (gamma amino butyric acid) found in encrusting coralline red algae. These include the gastropods *Haliotis* and *Trochus* (Heslinga, 1981) and the chitons *Katherina tunicata* and *Mopalia mucosa*, which are found in low-shore pools (review: Morse, 1992). Such cues can be exact predictors of the likelihood of successful settlement and survival, especially in species such as bryozoans or spirorbids, which attach to host plants or animals (Knight-Jones, 1953, 1955; Ryland, 1959, 1960, 1962; Williams and Seed, 1992). Strathmann and co-workers (1981) have shown that *Balanus glandula* uses the flora from particular tide levels as a cue to settlement. This seems to be a good predictor of the physical conditions in which the barnacle can survive, but can be a poor indicator of predation, because on some shores individuals settling low on the shore beneath the adult zone are devoured by predatory gastropods. Recent work has shown that settling barnacles avoid mucus laid down by predatory whelks (Johnson and Strathmann, 1989) thus avoiding areas with a high predation risk. Purely physical cues must prompt settlement under some conditions, otherwise virgin rock or underwater structures would not be colonized. The presence of a suitable microfloral film is thought to help (Chapter 4), but may not be essential (Wahl, 1989).

By using these physical and biological cues, larvae can accept or reject sites as they are encountered. Their energy reserves are often finite, however, and not surprisingly, larvae become less discriminating with age as their competence decreases (Lucas *et al.*, 1979). This is particularly marked in lecithotrophic larvae (larvae with a yolk sac which do not feed) or larvae with a specialized non-feeding settlement stage having a finite energy reserve, such as barnacle cyprids. In species with highly seasonal dense larval input (e.g. many temperate or boreal species of barnacle), pre-emptive occupation of favourable sites will result in many offspring having to settle in unsuitable places. This is more likely for larvae arriving late in the settlement season. These tend to 'top-up' space made available due to mortality (Connell, 1961b; Hawkins and Hartnoll, 1982a), but late arrivals make little significant contribution to the barnacle population. The lack of success of later-arriving cyprids may be a selective factor, along with their poorer naupliar survival, which helps to ensure early and tight synchrony with the spring phytoplankton bloom (Crisp, 1976; review: Hawkins and Hartnoll, 1982a).

5.1.2 SANDY SHORES AND MUDFLATS

Compared with rocky shore species, much less is known about the settlement behaviour of sediment invertebrates. Very early on, Wilson (1954) showed that the polychaete worm *Ophelia bicornis* was attracted to sediments that had microbial assemblages (particularly flagellates) growing on them. No doubt many of the problems faced by the larvae of rocky shore species are shared by the larvae of sediment dwellers, but the fauna of sandy beaches and mudflats have the potential to move, both as juveniles and as adults, unlike many of the rocky shore organisms discussed above. For instance, polychaetes like *Arenicola* may move their burrows every few days, and this probably also applies to tube-dwelling species and deposit-feeding bivalves (Woodin, 1991). Thus, it may not be quite so critical for the settling larvae of these species to make the right decision the first time. The relatively high mobility of sand and mudflat species makes any larval settlement choice quite risky, because the local fauna and its activities might change within hours of metamorphosis. In other words, at the micro- and meso-scales, sediments do not provide settling larvae with as many predictive clues as do rocky shores.

Given these differences between rocky and sandy shores, we might expect larval settlement behaviour to differ. In particular, the idea of larvae selecting sites on the basis of positive cues sent out by dense monospecific stands of adults, as is the case in barnacles, is probably untenable for most species (but not all; see below) living in sediments, where faunal assemblages are more mixed. Woodin (1985) has considered five basic larval settlement patterns in sediment: **gregarious** settlement, site selection because of **positive** cues, site rejection because of **negative** cues, settlement involving **neither positive nor negative** cues, and **hydrodynamic entrainment** of larvae to a site. Gregarious settlement is reported in the sipunculid *Golfingia*, in the echiurid *Urechis*, in the sand dollar (heart urchin) *Dendraster excentricus*, and probably occurs in the reef-forming polychaete *Sabellaria*. In all cases, some kind of chemical released by the adults is thought to be involved. A positive cue of sulphide has been reported for the opportunistic polychaete *Capitella*, and negative cues, associated with the presence of *Abarenicola pacifica*, in the polychaete *Pseudopolydora kempi*. No cues were evident in the bivalve *Mulinia lateralis* settling around the tubes of the polychaete

Diopatra, suggesting passive entrainment of larvae. Entrainment has been suggested for several species as a result of changes in local hydrography brought about by sediment microrelief and biogenic structures, such as worm tubes (Chapter 4). In practice, however, entrainment is difficult to separate from the effects of positive cues (Snelgrove, 1994). In those situations where juvenile and adult distributions are quite different, for example *Arenicola marina* (Reise, 1985), this might be due to negative cues or post-settlement biotic processes, or to a combination of both.

Woodin (1991) suggests that the role of positive cues is much less important for the larvae of sediment dwellers than for those of rocky shore species. Instead, negative cues are probably as important for informing a larva where not to settle. Negative cues can take the form of the presence of other species or an unattractive sediment physico-chemistry. For example, larvae may react negatively to tactile contact with the tentacles of deposit feeders, such as *Pseudopolydora* (Wilson, 1983), or to chemical cues released around arenicolid burrows (Woodin, 1985). Adults of several species of hemichordates and polychaetes are known to release brominated aromatic compounds, which are thought to reduce the acceptability of sediments to larvae (Woodin, 1991). If a site is rejected, larvae can travel to other areas reasonably economically. Larvae leaving the sediment surface need only to move a few mm into the overlying water before they encounter velocities much greater than their swimming speeds (Chia *et al.*, 1984; Butman *et al.*, 1988). Thus a vertical swim of a centimetre could move a larva laterally by several metres. In fact, the whole process of initial settlement, testing of the sediment, detachment, transport and resettlement, probably occurs over much smaller distances, in the order of centimetres in nereids and *Capitella,* which probably reflects the size of habitat patch recognized by these species.

5.2 COPING WITH THE VERTICAL AND HORIZONTAL GRADIENTS ON ROCKY SHORES

Once settled on the shore, individuals are faced with having to cope with an inhospitable physical environment, principally desiccating conditions and a significantly higher or lower temperature than is found in the sublittoral. Also conditions will vary far more and be much more unpredictable, especially at higher shore levels. Organisms will also encounter various problems at different points along the gradient between shelter and exposure, particularly in the need to attach and to avoid dislodgement.

5.2.1 VERTICAL GRADIENT: DEALING WITH EMERSION

Whilst many experiments have shown that higher-shore species are more tolerant of extreme temperatures and desiccating conditions than lower-shore species (Chapter 3), most organisms are unlikely to encounter their physiological tolerance limits, unless they are mid-to-high-shore species. In particular, plants and animals living above MHWN may remain uncovered for several days, and under such conditions, damage or even death will occur. For many species, morphological features such as a thick epidermis, cuticle or shell will reduce heat gain and water loss, but the general principle is the same in each case: some form of structure seals the organism from the environment (review: Newell, 1979). Many animals withdraw inside a shell when the tide is out, by sealing themselves in with an operculum in the case of gastropods, or by closing the valves in the case of bivalves, or by closing the valve-like tergal–scutal plates in the case of barnacles. In many instances, mucus and mucus-like substances are used to reduce water loss. These substances form an extra layer on the epidermis of some gastropods and intertidal fish, or make a seal around the shell, as in limpets. High-shore littorinid snails reduce contact with a hot substratum by using mucus, often in the form of threads, to hang them away from the surface. Animals also have the option of using evaporative cooling by controlled opening of the respiratory apertures. In the short term this can reduce temperatures, but of course it leads to further desiccation.

Several species of high-shore limpets tend to be higher domed than lower-shore species. This trend also occurs within a species, for example in *Patella vulgata* there is often a change to a more domed shape at higher levels. This domed shape reduces the circumference to body size ratio and hence water loss from the shell margin. Whether the change in shape is due to a phenotypic response or the differential selection of higher-domed morphs remains unknown, but shell shape does change when an animal moves microhabitat, pointing to a phenotypic response. In snails and limpets living at lower latitudes, higher-shore species often have ridged or pimpled shells which are thought to increase re-radiation of heat (Branch and Branch, 1981).

Rocky shore animals also face problems of respiration at low tide (review: Newell, 1979). Under moist conditions, some gas exchange can take place, but continued aerial respiration will lead to water loss through the gills, desiccation and consequent osmotic problems. There are two main ways in which animals cope with the problem of respiration when the tide is out: by having respiratory mechanisms which allow effective aerial respiration but minimize water loss; or by tolerating anaerobiosis when the tide is out, with repayment of the oxygen debt under more favourable conditions. Various mechanisms keep respiratory surfaces wet, allowing aerial respiration, but reducing evaporative loss when the tide is out. This is well illustrated by the structure of the gills of limpets. Keyhole limpets found in the subtidal and on the lower shore have a pair of primary gills with water coming in through the margins and out of the apical hole. Acmaeids have one gill and respiratory currents which go along the side of the body. Patellids, which can cope very well with emersion in air, have a set of secondary gills around the pallial margin. Water is trapped around the edge of the shell allowing the gills to function whether the tide is out or in. In many cases, anaerobiosis is an emergency back-up used when conditions worsen and water loss continues.

In many intertidal species, the consumption of oxygen differs in air and water, and this difference varies with shore level (Branch and Newell, 1978; Innes, 1984; Dye, 1987). Low-shore species, such as *Tectura* (= *Acmaea*) *testudinalis*, have a higher respiration rate in water than in air. In *Littorina littorea*, which occurs higher on the shore, the difference is less marked, but throughout the range of temperatures usually encountered in the environment, aquatic respiration exceeds that in air. In the Mediterranean, the splash-zone limpet *Patella rustica* (= *lusitanica*) has a higher respiration rate in air than in water, but the converse is true for the lower-shore *Patella caerulea* (review: Branch, 1981).

The relationships between temperature and respiration rate are complex and may be related to food availability, as illustrated by South African limpets (Branch, 1981). On these shores, there are at least twelve species of *Patella*. The low-shore *P. cochlear* has a low respiration rate but shows a much more marked response to temperature change than do mid- and high-shore species. In particular, the high-shore *P. granularis* seems able to suppress the rate of increase of respiration in relation to temperature increase. It also suppresses respiration in air compared with water at high temperatures, suggesting conservation of energy. Both *P. cochlear* and *P. granularis* are probably food-limited in the field. In contrast, midshore species with ample food supplies have higher respiration rates, respiring faster in water. *Patella granatina* and *P. oculus* from the cold west coast of South Africa have a high respiration rate at low temperatures. Respiration is faster in air at the temperatures likely to be encountered when the tide is out, but faster in water at the temperatures encountered when the tide is in. It has been suggested that these high rates of respiration allow maximum exploitation of the abundant algal resources.

Like sessile animals, intertidal seaweeds also have to tolerate desiccation and their rate of photosynthesis is progressively inhibited as water is lost (Chapter 3). This has been well documented in fucoids on British shores. All fucoid species lose water at a similar rate and after 4 hours on sunny days only 30–40% of their fresh weight remains. Rehydration occurs rapidly following submersion but recovery depends on shore level. High-shore species like *Pelvetia canaliculata* and *Fucus spiralis* reach normal photosynthetic rates 2 hours after resubmersion, even if the water content has fallen as low as 10–20% (Oliveira and Fletcher, 1977; Dring and Brown, 1982; Beer and Kautsky, 1992; review: Lüning, 1990). *Fucus vesiculosus* plants, from high and low water parts of the range, can recover following water reductions to only 30% and 50% respectively. The lower-shore *Fucus serratus* also only tolerates a 40% reduction in water content and, as might be expected, the sublittoral-fringe kelp *Laminaria digitata* is much less tolerant of water loss. As seaweeds do not have roots they are also dependent on being submerged in water to take up the nutrients needed for photosynthesis, protein synthesis and hence growth (Dring, 1982; Hurd and Dring, 1991).

The rate of water loss also depends on seaweed morphology, particularly the surface area to volume ratio. Membranous, sheet-like forms, such as *Porphyra*, *Ulva* and *Monostroma*, lose water very rapidly (although a species of *Porphyra* from southern Africa has a mucus layer which protects the cells). Sack-like algae (e.g. *Halosaccion*, *Leathesia*, *Colpomenia*, *Hormosira*) retain moisture in central internal spaces and this enables photosynthesis to go on during periods of emersion. Some species retain water amongst their branches and water loss will be less in turf-like forms, or those that grow in dense swards, than in solitary plants. Genera like *Fucus*, *Codium* and various coarse reds all have a small surface area relative to their mass which also helps to reduce water loss (Norton, 1991).

Extreme cold will occasionally be encountered at temperate latitudes, as happened during the hard European winter of 1962/1963 (Crisp, 1964). It is a permanent feature of life at higher latitudes. Some animals simply tolerate freezing (Table 5.1), whilst others, such as fish, manage to avoid it (review: Rankin and Davenport, 1981). Intracellular freezing of the cell contents is almost always lethal, but freezing of extracellular fluids can be tolerated by frost-resistant animals. The actual cause of death at low temperatures is usually osmotic damage rather than ice eventually reaching the cells, and euryhaline species (those capable of living in fluctuating salinities) fare better. As the extracellular body fluids freeze, the remaining fluid becomes more concentrated. The resulting concentration gradient across the cell membrane reduces cell volume by osmosis until a critical point is reached and the organism succumbs.

Most littoral molluscs can survive at temperatures low enough to freeze up to 60–70% of the body water and some seem exceptionally tolerant. For instance, the snail *Littorina littorea* can survive to about –20 °C with around 75% of its body water frozen (Murphy, 1979; Murphy and Johnson, 1980). Some species are more vulnerable at certain times of the year, suggesting that acclimation can take place. For instance, in summer death occurs in the barnacle *Semibalanus balanoides* when 40–45% of its body water is frozen at about –7 °C, whereas in winter this species can tolerate freezing of up to 80% of its body water at temperatures down to –18 °C.

Table 5.1 The percentage of body water of various molluscs frozen at −15 °C*

Species	% Frozen
Littorina saxatilis	67
Modiolus modiolus	65
Mytilus edulis	62
Littorina littorea	59
Crassostrea virginica	54

*Source: Rankin and Davenport (1981).

Intertidal seaweeds can also tolerate very low temperatures (Lüning, 1990; Lüning and Asmus, 1991 and Table 5.2). The lower lethal temperatures are far below the freezing point of sea water (–1.91 °C at 35‰, –1.63 °C at 30‰). As in animals, freezing of extracellular fluids occurs first. This has a desiccating effect because of diminished water vapour above the ice, although the effect is not usually lethal. Lethal tissue damage is caused by intracellular formation of ice crystals damaging the plasma membranes. In low-shore and sublittoral seaweeds with large cell vacuoles, intracellular ice forms very rapidly. Freeze-tolerant species from mid and upper shore levels avoid intracellular ice formation, probably remaining permanently supercooled, because of 'antifreeze' substances in their cells. These are found in much greater amounts in Antarctic green algae compared with temperate species. Some polar species are remarkably tolerant. Arctic *Fucus distichus* can survive temperatures of –40 °C for several months but, as with desiccation, photosynthetic rates are generally inhibited by freezing and obviously in the winter it will be dark as well. In the low-shore species *Chondrus crispus*, the photosynthetic rate recovers after 3 hours at –20 °C, but not after 6 hours. In contrast the higher-shore *Mastocarpus stellatus* recovers after 24 hours at –20 °C. Similar results have been obtained for intertidal fucoids, where the degree of inhibition of photosynthesis following exposure to low temperatures for 3 hours corresponds to their position on the shore.

Table 5.2 The percentage of tissue water frozen in several intertidal brown and red algae*

Species	% Frozen
Subtidal/rock pool	
Alaria esculenta	82.2
Dumontia incrassata	81.2
Laminaria digitata	78.2
Chordaria flagelliformis	73.3
Fucus distichus	71.6
Lower eulittoral	
Chondrus crispus	59.1
Palmaria palmata	51.4
Upper eulittoral	
Fucus vesiculosus	78.9
Mastocarpus stellatus	56.6

*Source: Davison *et al.* (1989).

5.2.2 HORIZONTAL GRADIENT: DEALING WITH WAVES

Staying put: glues, cements and anchors

Both plants and animals have to stick onto the shore when they first settle. Species of green and brown seaweeds with zoospores first attach with their flagella and this is followed by secretion of glycoproteins and mucopolysaccharides for adhesion (Norton, 1981, 1992; Fletcher and Callow, 1992). Fucoid zygotes adhere initially by a thin layer of mucopolysaccharide before producing rhizoids, whilst red algal spores are covered by a sphere of mucilage which adheres to the rock. In all algae, rhizoids are produced which stick to the rock using both mucopolysaccharides and mechanical attachment exploiting surface irregularities. These rhizoids then proliferate to strengthen the bond, either by coalescing to form a disc-like holdfast or by producing a spreading branch-like structure (haptera). Attachment can sometimes be so good that the rock itself gives way rather than the holdfast.

Sessile animals which permanently attach themselves to one spot for life use various cements. Spirorbid polychaetes attach temporarily by mucus secreted from ventral glands. They then spread the contents of the shell gland over the body by rolling from side to side (Nott, 1973). A further secretion from the ventral gland (now on the upper surface) is used to create a primary mucus tube. After metamorphosis this is replaced by a calcareous tube secreted by the collar of the animal. This tube is firmly cemented to the substratum. Mussels anchor themselves to the rock by a number of byssus threads. These threads are secreted by a gland in the foot which can be extended away from the shell up to about one shell length. One end of the thread is attached to the root of the foot while the other has a disc-like byssal plaque about 1 mm in diameter. The thread is secreted by two glands as a 'glue' and a 'hardener' into a groove on the foot. This mixture solidifies and the attachment plaque is released after 10 minutes. The byssus further hardens and browns over the next few days. Although these threads have a tiny diameter (20–30 μm), they are extremely strong and flexible.

After temporary attachment and searching of the substratum using the disc-like adhesive tips of the antennules, the barnacle cypris larva cements itself to the rock. First the cypris antennules fold to form suction cups, and adhesion is reinforced by a cement after folding of the limbs. As long ago as 1854, Charles Darwin indicated that the cement is secreted by a pair of glands located behind the eyes of the cyprid. As in the mussel byssus, there is a glue and a hardener. Unlike most commercially available epoxy cements, these glues work extremely well in an aqueous environment and not surprisingly they have received considerable interest from industry. During metamorphosis the cement glands migrate to the outer edge of the body inside the carapace (shell) and spread cement in concentric circles around the perimeter of the base.

Mobile animals, such as snails, limpets, chitons, sea anemones, starfish and urchins, have to be able to attach firmly during stationary periods. All these animals use mucus for adhesion. The foot of the limpet, *Patella vulgata*, has nine different pedal glands, six of which secrete mucus for use during locomotion, whereas a more viscous mucus is secreted for adhesion at rest (Grenon and Walker, 1982). The most detailed work on the use of mucus has been carried out on several species of limpets in South Africa. Those in exposed conditions have a thick, stiff foot and secrete a very thin layer of mucus. This gives a much faster bond (like a thin layer of water between two glass plates) but allows only slow movement. Those on more sheltered shores secrete more mucus and tend to have a much more flexible foot. They can move around more, but do not have such great powers of adhesion (Branch and Marsh, 1978). Some more recent work has suggested that limpets also adhere by suction (Smith, 1991), with the mucus helping to give a good seal. Starfish and sea urchins use their tube feet as suckers and mucus-like compounds improve their adhesion considerably.

Changes in size and shape along the exposure gradient

Differences in morphology occur along wave exposure gradients in a variety of seaweeds. Narrower blades with several splits along their length typify kelps from wave-exposed areas (*Laminaria hyperborea*, *L. digitata*, *Sacchoriza polyschides*, *Hedophyllum sessile*), whereas plants with broader blades with fewer or no divisions occur in more sheltered conditions, even within the same species. Strong water movement imposes various stresses on the thallus: tension, shear forces, bending and twisting. Narrow, flat blades, especially if divided like streamers, provide a highly streamlined shape compared with broad and undulate blades. Most of these shape differences appear to be phenotypic growth responses. For instance, *Laminaria saccharina* can be induced to take on a more streamlined shape with a narrow blade and greater elongation rates simply by attaching weights to the extreme end of the blade. Similarly, transplanted *Alaria* have been shown to change from a broad-fronded sheltered form to an elongated type characteristic of exposed shores (reviews: Denny, 1988; Norton, 1991).

Different seaweeds may cope with the same wave conditions in different ways, as shown by *Lessonia nigrissens* and *Durvillea antarctica* which both occur in the exposed sublittoral of central and southern Chile. *Lessonia* has a strong, stiff stipe that bends with the flow. In contrast, the elastic stipe and stretchy blade of *Durvillea* allows it to align with the flow completely. Flexible stipes that can be bent parallel with water movements reduce the stress on the thallus and result in the plant being closer to the rock where movement is less. These structures respond best to the chaotic multidirectional water movements typical of exposed shores. In the North Atlantic, flexible elastic stipes are found in species from wave-exposed sites, such as *Alaria esculenta*, whilst *Laminaria digitata* has a more flexible stipe than the deeper and hence less wave-beaten *L. hyperborea*. Presumably the stiff stipe of *L. hyperborea* is more important in maximizing access to light than in resisting wave action.

Whilst avoiding dislodgement by water movement is a major priority for rocky shore species, all seaweeds require some degree of water movement to break down the boundary layer around the thallus. Materials have to diffuse in and out of the plant through this layer of slowly moving fluid, so that the thicker the boundary layer the slower the uptake of materials. In still water, the boundary layer can be several millimetres thick. Many species have a surface of spiny outgrowths (*Macrocystis pyrifera*), wavy margins (*Laminaria saccharina*), or even holes or undulations in the blade (*Agarum cribosum, Laminaria saccharina*) which are thought to enhance the turbulence in water flowing over the lamina and hence allow greater uptake of raw materials for photosynthesis. There is also evidence that fucoids grow hair-like protrusions during periods of low nutrient concentration and that these are involved in nutrient uptake (Hurd *et al.*, 1993).

In molluscs, the shells of species in areas with considerable water movement tend to be flatter, with large apertures providing a greater attachment area to the substratum (Fig. 5.1). Limpets are perhaps the ultimate body plan for clinging to a rock surface, and species found in areas of greater water movement tend to be flatter. Gradients of size and shape also exist within species, especially if they do not have a planktonic stage so that localized selection can occur. In dogwhelks (*Nucella, Thais*) and several species of littorinids, individuals with thin shells and larger apertures occur in more exposed conditions, allowing greater clinging power. In sheltered conditions crab predation selects for thicker-shelled, narrow-apertured forms (see below).

5.3 COPING WITH LIFE ON SANDY SHORES

5.3.1 BURROWING: EXPLOITING A THREE-DIMENSIONAL ENVIRONMENT

Unfavourable conditions can be avoided on sediment shores by burrowing. Most intertidal macrofauna exploit the physical properties of the sediments, in particular **dilatancy** and **thixotropy**, to burrow. When the water content becomes less than 22% by weight, any force applied upsets the close packing of the grains and the interstitial water is no longer able to fill all the spaces between the particles. The sediment becomes hard and resistant to shear forces. This is what happens when you firmly push against a sandy beach with your foot and the sand in the immediate region goes hard and dry. The effect is called **dilatancy**. In contrast, when the water content is above 25%, there is a reduction in resistance with increasing rates of shear and the sediment becomes a liquefied slurry of grains and water. If repeated small movements are made with your feet, a small puddle of quicksand is formed. This is called **thixotropy**. Most burrowing animals use repeated small penetrations to displace sediments, making use of their thixotropic property. Repeated small agitations of the sediment result in the ratio of resistance to penetration decreasing by a factor of ten. Subsequently, anchorage is achieved by dilatancy and the process is then repeated (reviews: Trueman, 1975, 1983; Newell, 1979).

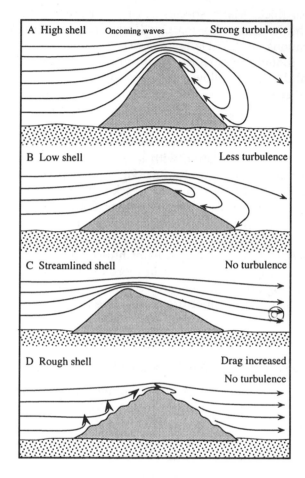

Fig. 5.1 Shell shape variation in limpets in relation to water movement experienced (adapted with permission from Branch and Branch, 1981, *The Living Shores of Southern Africa*).

Burrowing in soft-bodied animals, such as polychaete worms, burrowing anemones, bivalves and gastropods, is essentially similar. There is an initial penetration phase followed by burrowing proper. Penetration is usually made by repeated small movements. Large movements would end up displacing most animals away from the sediment because their weight is low in water. In worms such as *Arenicola*, repeated probing movements of the proboscis are made, each movement applying very little force (Fig. 5.2). This scraping action pushes sediment aside and the head becomes drawn into the initial burrow. The pharynx is then everted, enhancing the probing of the proboscis. Subsequently, waves of contraction of the body wall dilate the mouth and **evert** the proboscis against the sediment. Anchorage is achieved by the worm sticking to the sediment using its chaetae and by flange-like ridges on the front segments which become erect and press against the sediment. After about 20 seconds, sufficient penetration has been achieved for the burrowing cycle to start (see below). In bivalves, initial penetration is achieved by rapid probing of the foot. Once the animal is firmly lodged in the sediment, burrowing proper occurs through a series of powerful digging movements.

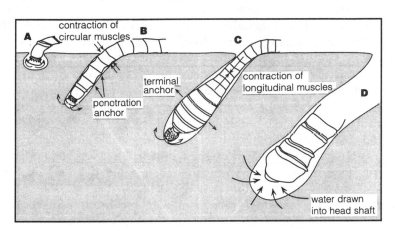

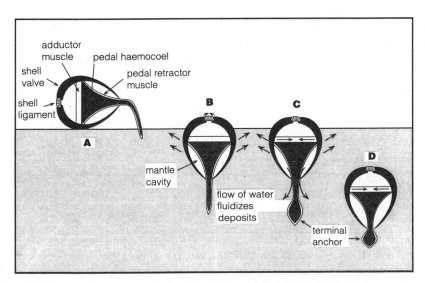

Fig. 5.2 Successive stages of burrowing in the lugworm *Arenicola* (upper) and a generalized bivalve (lower). Based on Newell (1979) from Trueman (1975).

Following initial penetration, *Arenicola* burrows by alternate movements between the terminal anchor formed by the expansion of the head segments, and a penetration anchor of the ridge-like flanges and chaetae. Liquefaction of the sediment is assisted by drawing interstitial water into a cavity which forms when the proboscis is rapidly withdrawn.

In bivalves, the penetration anchor is formed when the shell presses against the substratum when the adductor muscle is relaxed (Fig. 5.2). The foot then extends downwards. The siphons close next, and following contraction of the adductor muscles, water is forced out either side of the foot causing a short-lived liquefaction of the sand. The increase in internal haemocoelic pressure by the contraction of the adductor muscles expands the far end of the foot to form a terminal anchor which presses against and dilates the sediment. The anterior and posterior retractor muscles contract, pulling the shell into the sediment. Penetration is aided by a rocking action of the shell which causes localized thixotropy. The effectiveness of this can be demonstrated by trying to push a dead shell through sediment: much greater forces are required than are needed by the live shell aided by the thixotropic action of squirting water and rocking the shell.

Thus, in most soft-bodied animals, movement through the sediment requires an initial **penetration anchor** forming a solid immovable base from which a protrusible organ, such as a foot or proboscis, can be everted. The sediment becomes liquefied by repeated eversions, allowing movement. A **terminal anchor** is then established, the penetration anchor is released and then longitudinal retractor muscles contract and pull the animals through the sediment. The cycle begins again with the re-establishment of the penetration anchor, and eversions of the digging organ.

Most crustaceans burrow by rapid movements of their limbs which move sediment backwards as they plough through the sand. Again rapid small movements assist in liquefying the sediment, allowing the animals to almost swim through it. Many infaunal crustaceans have spade-like appendages which help in burrowing, but can be used in swimming as well.

5.3.2 RESPIRATION

The infauna of sandy beaches may not suffer the acute desiccation experienced by rocky shore species, but they do have to contend with the problem of low levels of oxygen, which becomes more severe with depth and with finer particles (Chapter 1). Free oxygen is present in significant amounts only on coarse sandy beaches which drain rapidly and thoroughly at low tide, and where oxygen can percolate through the relatively large interstitial spaces. On beaches composed of sediment with more than about 10% of particles smaller than 250 μm, the oxygen concentration of interstitial water is only about 10% of the air-saturation level (Fig. 5.3). On most fine-particle beaches, free oxygen may only occur in the top millimetre or so of the sediment. The lack of oxygen in such beaches is mainly due to poor drainage of water and to high bacterial populations which utilize what little oxygen is present (Chapter 1).

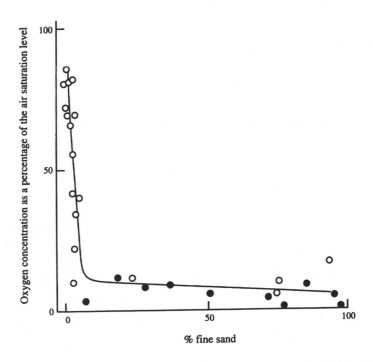

Fig. 5.3 Concentration of oxygen in sediments of different particle size. Solid circles are beaches with a black layer (modified after Brafield, 1978, with permission).

Most species have appropriate physiological mechanisms or behaviours for coping with the problem of low levels of oxygen at low tide. Many have respiratory pigments, such as the haemoglobin of polychaetes and haemocyanin of decapod crustaceans. Other species are able to regulate their oxygen consumption and reduce their activity to minimize demand at low tide. In more stable sediments, many species construct semipermanent burrows which provide access to better-aerated surface water at low tide. Finally, as is the case for many rocky shore species, many species have to resort to anaerobic metabolism for some part of the low tide period. A very readable account of this aspect of respiration of sandy beach organisms is provided by Brafield (1978).

Below the chemocline of the redox potential discontinuity (RPD) layer, not only is it difficult to find oxygen, but sulphide, which is highly toxic to most animals, occurs in high concentrations. Many of the more familiar infauna, such as crustaceans, molluscs and annelids, do not occur within this black layer, unless they possess some kind of burrow or tube which gives them access to the upper sediment and so to oxygenated water. However, many meiofaunal species regularly occur in the black layer and some groups, such as the Gnathostomulida and the turbellarian groups Solenomorphidae and Catenulida, seem restricted to it. The inhabitants of this black zone and the chemocline may represent a self-contained sulphide ecosystem or **thiobios** (Fenchel and Riedl, 1970; Boaden and Platt, 1971; Boaden, 1989). The organisms found in the overlying oxic sediment, many of which are absent from the thiobios, for example Proseriata (Turbellaria), Nemertini, Tardigrada and Archiannelida, have been termed the **oxybios**. The thiobios are assumed to possess physiological adaptations to a lack of oxygen and to the presence of toxic compounds such as hydrogen sulphide and ammonia. However, Reise and Ax (1979) have argued convincingly that the division of sediments into an upper oxic zone and a lower sulphidic region is a gross oversimplification. Many macrofaunal species excavate burrows which cross the chemocline and create oxic microhabitats within the black zone (Fig. 5.4). Whilst all classic thiobiota seem to live where the oxygen concentration is effectively zero, Meyers et al. (1987) found that species were zoned in a continuous fashion along the oxygen and sulphide concentration gradients, so that it is probably not sensible to make a clear distinction between an oxybios and a thiobios. It is also possible that the thiobiotic meiofauna can utilize extremely low concentrations of oxygen that occur in reduced sediments, but which are not possible to measure at present.

5.4 COPING BY MODIFYING BEHAVIOUR

The cyclical nature of tidal, daily, semilunar and seasonal changes produces fluctuating but predictable environmental conditions. Animals function (respire, feed, reproduce) during favourable periods and reduce activity in unfavourable conditions, usually returning to a refuge of some kind or, in the case of sessile animals, retreating inside a closed shell. Mobile animals may return to a home scar, a cryptic habitat or a host organism such as a seaweed clump. The most mobile species leave the intertidal altogether. Which state of the tide is favourable or unfavourable for movement will vary with species and habitat type. For both rocky and sediment shore species, moving to find food or a mate when the tide is out increases the risk of dying from desiccation. Also, many organisms can only effectively move to forage when the tide is in, but at particular states of tide there is a high risk of dislodgement by waves, often leading to death. At night, physiological stress is less when the tide is out. The risk of predation may be higher during night-time immersion than during the day because many predators such as crabs and fish forage into the intertidal at night. Conversely, avian predation is higher when the tide is out and although shorebirds do feed at night, predation will be more intense during the day. Given a predictably fluctuating environment, it is not surprising to find that many organisms exhibit rhythmic behaviour, emerging from a refuge of some kind to exploit available resources at a favourable time (review: Naylor, 1985).

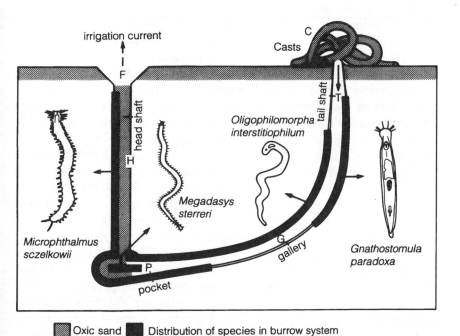

Fig. 5.4 Oxygenated habitats created by a lugworm *Arenicola* burrow. The meiofaunal species are concentrated in areas immediately adjacent to the burrow (adapted with permission from Reise and Ax (1979). A meiofaunal thiobiosis limited to the anaerobic sulphide system does not exist. *Mar. Biol.*, **54**, 229).

Sessile rocky shore animals, such as barnacles and mussels, simply respond to external cues by opening and closing their shells as appropriate after testing the environment by a variety of peripheral sensory structures. No marked endogenous rhythmicity has been shown for these species. Newell (1979) has likened the 'evasive' behaviour of such species to that of burrowing sandy-shore organisms that do not have or need an endogenous component to their activity patterns. Anticipation of change would confer little advantage to a non-mobile species.

In contrast, many mobile intertidal animals have been shown to have endogenous activity rhythms of various kinds. These persist for some time in the absence of environmental cues in constant laboratory conditions. Endogenous rhythms are perhaps most strikingly seen in the hatching and emergence of the rocky shore dipteran fly *Clunio* (Neumann and Heimbach, 1985). In *Clunio*, the larvae and pupae occur low on the shore in red algal turfs. They are only exposed to air on low water of spring tides. Newly emerged winged males copulate with wingless females and this behaviour and the subsequent egg-laying must take place in air. An exact semilunar (14 day) rhythm of hatching is seen and the adaptive advantage of anticipation is obvious in this case: if the fly hatches at the wrong time it drowns! Endogenous rhythms are also well expressed in other rocky shore arthropods, such as *Carcinus maenas* (review: Naylor, 1985). In intertidal gastropods, the endogenous component of the many activity rhythms shown is less clear cut (Table 5.3). Endogenous rhythmicity has been experimentally demonstrated in a few species (e.g. *Littorina nigrolineata*, Petpiroon and Morgan, 1983). It has also been inferred in many other species, particularly in homing species like limpets which return to a home scar or refuge ahead of rising or falling tides (e.g. Hartnoll and Wright, 1977). Recent work has confirmed the endogenous basis of the rhythmic behaviour in limpets (Della Santina and Naylor, 1994). Gastropods are less amenable to laboratory studies on activity, as their behavioural patterns are so easily disturbed (e.g. Petraitis, 1984), and their movements are relatively sluggish, making standard activity-measuring devices difficult to use. This may account for the relatively few studies (compared with those carried out on crustaceans) showing continued rhythmicity under constant laboratory conditions. The flexibility of the activity patterns shown by many gastropods suggests that the endogenous component, although present, can be easily overridden by responses to exogenous factors. Activity patterns can also vary between populations, as for example in *Patella vulgata* (Hawkins and Hartnoll, 1982b, 1983a; Little, 1989; review: Chapman and Underwood, 1992).

Table 5.3 Foraging timing and endogenous rhythms in intertidal molluscs*

Species	Endogenous rhythm?	Timing	Habitat
Molluscs feeding actively when submerged			
Chitons			
Chiton pelliserpentis	?	Night	High-shore rock
		Day and night	Low-shore rock
Prosobranch limpets			
Patella vulgata	Yes	Day	Horizontal rock
Cellana ornata	?	Day and night	Rock
Cellana radiata	?	Day and night	Rock
Neritids			
Nerita plicata	Yes	Day and night	Rock
Nerita atramentosa	Yes	Day and night	Rock
Littorinids			
Littorina nigrolineata	Yes	Day and night	Rock
Littorina littorea	?	Day and night	Rock
Nodilittorina pyramidalis	No	?	Rock
Nodilittorina unifasciata	No	?	Rock
Bembicum nanum	Yes	?	Rock
Trochids			
Melagraphia aethiops	?	Day and night	Rock
Calliostoma zizyphinum	?	?	Rock
Gibbula cineraria	?	?	Rock
Gibbula umbilicalis	?	?	Rock
Monodonta lineata	?	?	Rock
Austrocochlea obtusa	Yes	Day and night	Rock
Pulmonate limpets			
Siphonaria laciniosa	?	Night	Rock
Molluscs feeding actively when emersed			
Chitons			
Acanthopleura brevispinosa	?	Night	Rock
Acanthopleura gemmata	?	Night	Rock
Prosobranch limpets			
Patella vulgata	Yes	Day and night	Vertical rock
		Day	High-shore rock
		Day and night	Low-shore rock
Neritids			
Nerita textilis	?	Day and night	Rock
Nerita polita	Yes	Day	Rock
Nerita albicilla	Yes	Day and night	Rock

Species	Endogenous rhythm?	Timing	Habitat
Pulmonate limpets			
Siphonaria capensis	?	Day and night	Pools
	?	Night	Rock
Siphonaria gigas	?	Night	Rock
Pulmonate onchidiids			
Onchidella binneyi	?	Night	Rock
Molluscs feeding actively			
when awash			
Prosobranch limpets			
Cellana toreuma	?	Day and night	Rock
Collisella limatula	?	?	Rock
Collisella scabra	?	Day	Rock
Scurria stipulata	?	?	Rock
Neritids			
Nerita scabricosta	?	Night (mainly)	Rock
Nerita funiculata	?	?	Rock
Nerita versicolor	?	Night	Rock
Nerita tessellata	?	Night	Rock
Littorinids			
Nodilittorina exigua	Yes	Day and night	Rock
Littorina modesta	?	?	Rock
Pulmonate limpets			
Siphonaria normalis	?	?	Rock
Siphonaria alternata	?	?	Rock
Siphonaria pectinata	?	?	Rock
Siphonaria maura	?	?	Rock

*Data extracted from Little (1989).

Amphipods and isopods inhabiting sediments show highly rhythmic behaviour with a strong endogenous component. The rhythms are such that animals emerge from the sand, feed, and then return to burrow or hide in the strand-line material at the appropriate tidal level. The midshore isopod *Eurydice pulchra* has tidal, daily and semilunar rhythms. It emerges from the sand when the tide comes in, feeds whilst swimming in the water column, ceases activity before the retreating tide and burrows into the sand for the duration of the tide-out period. Activity is greater during the night than in daytime and is also greater on spring, than neap, tides (review: Naylor, 1985).

The semi-terrestrial amphipod *Talitrus* has a circadian rhythm, foraging at night (Williams, 1983). On these excursions it goes down the beach into the intertidal zone before returning to the top of the shore. It also has a semilunar rhythm, being more active on neap tides thus avoiding being swept away on spring tides. In addition to the temporal pattern of activity, the animals orientate themselves to return to the correct position. Transplant experiments to either side of narrow isthmuses have shown that *Talitrus* will hop off in the wrong direction if taken to a beach where the strand-line is in the opposite direction (Ugolini *et al.*, 1986). Visual cues, such as the light/dark boundary on the horizon, have been implicated in this behaviour. Interestingly, the orientation behaviour also has a rhythmic component, with an exploratory and a homing phase. Various cues such as changes in pressure, salinity, agitation, immersion/emersion, or light/dark associated with environmental cycles have been shown to entrain rhythms. The accepted general view is of an approximate endogenous pacemaker which modulates activity patterns and is continually reset by environmental cues to local time.

Plants too can show activity rhythms (Palmer and Round, 1965; Round, 1981). Diatoms, dinoflagellates and euglenoids can all migrate up and down through the sand in response to tides and changes in light intensity. In general, movement to the surface is rapid when the sediments are exposed by the outgoing tide. The algae remain at the surface until an hour or so before the returning tide, when they move back into the sediment. They also migrate down if the conditions become darker. There appears to be an endogenous clock capable of modification by direct responses to the environment. The behaviour maximizes photosynthesis and minimizes the risk of being carried away by water movement when the tide is in.

5.5 DEALING WITH OTHER ORGANISMS

In Chapters 3 and 4 we reviewed the consequences of various trophic interactions for zonation patterns and community structure and composition. In this section, we explore the implications of a variety of predator–prey and grazer–plant interactions for the morphology, life histories and behaviour of the individual protagonists, as well as the development of mutualisms where both partners seem to benefit. Whole books could be written on these interactions and we cannot hope to provide a fully comprehensive account within this chapter. Instead we have selected a few examples covering both animals and plants and rocky and sediment shores.

5.5.1 GASTROPOD SHELL MORPHOLOGY AND DECAPOD PREDATION

Morphological defences, such as thick shells, are one of the most obvious ways by which prey can combat predators. An oft-quoted example is the thicker and more elaborate shells of marine snails in areas where predation from shell-crushing decapod crustaceans appears to be more intense (Kitching *et al.*, 1966; Raffaelli, 1978; Crothers, 1985). One of the best illustrations comes from studies on the dogwhelk *Nucella lapillus* in the British Isles. This species occurs over a large part of the exposure gradient, from exposed bedrock shores to sheltered weed-covered boulder habitats. On exposed shores, the snails have a thin shell and a wide aperture due to the large foot required to cope with the higher risk of wave dislodgement (Seed, 1978; Gibbs, 1993). In sheltered conditions, the snails have thick shells with a narrow aperture and the lip has internal ridges or 'teeth' (Fig. 5.5). Laboratory experiments show that more force is required to break 'sheltered' than 'exposed' whelk shells (Hughes and Elner, 1979; Currey and Hughes, 1982; Crothers, 1983; Appleton and Palmer, 1988) and crabs find it more difficult to chip away at the lip of 'sheltered' snails to get to the flesh. In aquarium trials, crabs presented with both 'exposed' and 'sheltered' snails inflicted much greater mortality on the thinner-shelled forms. Furthermore, when 'exposed' snails were transplanted to sheltered habitats and 'sheltered' snails to exposed shores, snails with the 'wrong' shell type suffered high mortality.

There is compelling evidence therefore that different shell morphologies are selected under different conditions by conflicting pressures of predation and wave action, with a trade-off occurring between shell thickening, aperture shape and foot size. However, there still remains the question why 'exposed' snails have thin shells. Thin-shelled snails do not seem to produce more young with the energy they save by not thickening the shell. In the North American species, *Nucella emarginata*, thick-shelled forms produce the same number of young as those with thinner shells (Geller, 1990), suggesting that the cost involved in making a thick shell is not met by diverting resources from reproduction.

Similar trends can be seen on a much grander scale in gastropods living at different latitudes. At very high latitudes, there are few large decapods and gastropods have thin shells. Conversely, tropical waters support very large crabs and lobsters with powerful chelae, and gastropod shells are often thick, massive and with elaborate spikes and ridges. Such trends are intriguing and suggestive of coevolutionary responses between prey and predators, but in only a few cases has this proposition been adequately tested. For instance, it would be equally easy to argue that the large spines seen on many tropical shells provide stabilizers in a high-energy environment. It should also be remembered that incorporation of seawater calcium into shell material is much more difficult in cold than in warmer waters (Vermeij, 1978).

Many of the above arguments implicitly assume a genetic basis for shell variation in gastropods. There is indeed evidence that some shell characteristics are under genetic control and may therefore be subject to natural selection. In *Nucella* and *Littorina saxatilis*, local population differentiation can occur because these species have direct development

leading to juvenile snails hatching from egg capsules (*Nucella*) or leaving
the parent as 'crawlaways' (*Littorina*). In contrast, *Littorina littorea* has a
pelagic larval stage following release of eggs to the plankton; little
morphological differentiation is seen in this species between exposed
and sheltered shores (Currey and Hughes, 1982). In *Nucella*, however,
there is some evidence of phenotypic responses: individual *Nucella*
respond morphologically to chemical cues released by predators and
damaged conspecifics by growing heavier shells with thicker lips and
better-developed apertural teeth (Palmer, 1991; Gibbs, 1993).

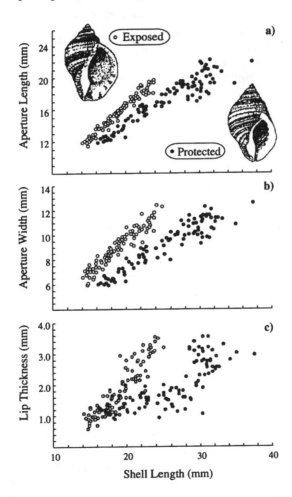

Fig. 5.5 Variation with exposure of shell shape (a and b) and thickness (c) in the
dogwhelk *Nucella* (modified after Palmer, 1991, *Developments in Hydrobiology*, **56**.
Reprinted by permission of Kluwer Academic Publishers).

5.5.2 BEHAVIOURAL ADAPTATIONS TO PREDATION

Sessile or slow-moving animals might seem to have little potential to develop behavioural defences against predators. By definition, they cannot quickly flee and hide, although cockles and trochid gastropods can use violent movements of the foot to escape from predators such as starfish. Several species have developed quite aggressive tactics to cope with would-be predators. Where mussels occur in dense sheets they have been known to tie down and eventually kill predatory dogwhelks by covering them with numerous byssus threads (Petraitis, 1984).

Limpets show a variety of responses to predators (Branch, 1981). Some species deploy nauseous mucus to deter predators, whilst others flee. In several species of *Patella*, individuals that are small relative to the size of predator flee, whereas larger animals will mushroom and then forcibly bring their shell edge down to stamp on the predator's foot (in a whelk) or arm (in the case of a starfish), driving the predator away (Bullock, 1953; Dayton *et al.*, 1977). Branch (1981) has shown that the size threshold of fleeing or fighting in South African limpets is smaller when challenged with a small whelk than when confronted by a large starfish. In the UK, *Patella vulgata* has a fight–flee threshold of around 15–20 mm when confronted with a starfish: small limpets flee, large ones fight. Occasionally a limpet will stamp and then flee. Interestingly, this threshold does not seem to change with size of starfish. Also, the threshold occurs at about the size that limpets first take up a home scar (Hawkins and Fernandes, unpublished).

5.5.3 SEAWEED DEFENCES AGAINST GRAZERS

Preferential grazing on competitively superior seaweeds by sea urchins, limpets, snails and chitons can, under some circumstances, be all-important in structuring intertidal communities. However, if the preferred seaweeds are the inferior competitors, then grazers have a quite different effect, as shown by Lubchenco's experiments on New England shores (Chapter 4). Clearly, the feeding preferences of these herbivores have important consequences for community organization. Choosing the right food is also important for the grazer. In the laboratory, the urchin *Strongylocentrotus droebachiensis* consistently prefers the kelps *Nereocystis luetkeana* and *Laminaria saccharina* to the fleshy *Agarum cribrosum* (Vadas, 1977). Urchins raised on the kelps grew faster and larger than those raised on *Agarum*, and field evidence suggests that reproductive output is also greater in kelp-fed urchins.

What then determines what seaweed grazers will eat ? Many factors are likely to be involved, including chance, but here we consider a few of the more obvious ones: **predictability**, **digestibility** and **edibility**. A casual inspection of a rocky shore shows some seaweeds growing in dense, virtually monospecific stands, whilst others appear to be scattered more haphazardly over the shore in small clumps or patches. For a small, slow-moving herbivore like a limpet or an urchin, these two kinds of seaweed are not equally predictable. Species occurring in dense stands, such as the infralittoral kelps and the intertidal fucoids, are very predictable: they can be relied upon to be in the same place for a long time. On the other hand, ephemeral species like *Ulva* and *Enteromorpha* are less predictable, popping up at different times in different places, usually as a result of disturbance. These latter seaweeds can often escape herbivory in time and space, whilst the predictable species are prime targets for herbivores.

Not all seaweeds can be equally well digested and assimilated by herbivores and the ability to digest seaweeds will also vary between grazer species, making generalizations difficult. Nevertheless, some patterns do emerge within a grazer taxon. For instance, herbivorous fish, which are a common feature of warmer regions (Horn, 1992), possess the enzymes required to cleave the α-linkages of red and green seaweeds, but not the β-linkages of brown algal storage compounds (Lubchenco and Gaines, 1981). Assimilation efficiencies also vary, from 30% to 70% for limpets, snails and urchins (Hawkins and Hartnoll, 1983b), although the values for bluegreen bacteria are much less (< 20%). It is unlikely that a slightly lower digestibility of one seaweed over another would restrict the diet of these herbivores, but where differences are large there could be significant effects on the performance of an individual, as described above for the urchin *Strongylocentrotus*.

One consequence of low digestibility is that some seaweeds survive the passage through a herbivore's gut (Santelices, 1992). If faecal pellets of fish and invertebrate grazers are incubated, incompletely digested material can propagate new plants. Furthermore, field experiments indicate that at least some of the surviving fragments and propagules can reattach on the shore. Interestingly, opportunists such as *Ulva* and *Enteromorpha* survive digestion more frequently and through more species of grazer, than seaweeds that might be considered late successional, such as kelps.

Not surprisingly, those species which are highly predictable to grazers often have a rough texture and large amounts of secondary compounds, so that they are ranked low in feeding trials. Of course, there will always be specialist herbivores with perverse tastes, but in general, thin filamentous or membranous seaweeds, such as *Enteromorpha, Ulva, Ceramium* and *Polysiphonia*, seem more edible than tough, leathery forms like the kelps and fucoids. Other tough species include the encrusting brown *Ralfsia* (but see later) and the red *Apophloea*. Some greens and reds have gone as far as protecting the thallus in an armour of calcium carbonate. These include the familiar red coralline turfs (*Corallinu, Jania*) and coralline paints or crusts (*Lithothamnion, Lithophyllum*) and the warm-water green seaweed *Halimeda*.

With respect to chemical make-up, over 500 secondary compounds (so called because no primary metabolic function is known) have been isolated from marine algae and many of these seem to deter herbivores (reviews: Hay and Fenical, 1988, 1992). In common with other plants, seaweeds produce terpenes, aromatic compounds, acetogenins, amino-derived substances and polyphenolics (Table 5.4). Seaweeds do not possess the alkaloids characteristic of terrestrial plants, although these can be produced by nitrogen-fixing bacteria.

Amongst the reds, the families Bonnemaisoniaceae, Plocamiaceae, Rhizophyllidaceae and Rhodomalaceae are particularly rich in compounds ranging from aliphatic haloketones and brominated phenols to more complex monoterpines, sesquiterpines and diterpines. Concentrations of these substances range from trace amounts to up to 5% of plant dry weight and several common shore plants, such as *Laurencia, Dumontia, Asparagopsis* and *Chondrus*, are unplatable to grazers. The genus *Laurencia* has been particularly well studied and found to contain a wide variety of complex terpenoids (more than 400 compounds, including 16 novel structures only found in *Laurencia*), acetogens and elatol. Some of these compounds kill cells and even whole fish or insects. Others deter feeding by fish and other generalist grazers. Clearly, *Laurencia* earns its common name of 'pepper dulse'.

In the green seaweeds most work has been done on the order Caulerpales, which are common on reefs in tropical areas. Common genera are *Halimeda, Caulerpa, Udotea* and *Penicillus*, which produce terpenoids in amounts ranging from trace to 2% of plant dry weight and which are effective herbivore deterrents. These compounds are often concentrated in areas of new growth. Other greens have been less well studied but *Cladophora fascicularis* is known to produce a brominated diphenyl ether.

Brown seaweeds produce polyphenolics termed phlorotannins derived from phloroglucinol. These have been compared with terrestrial tannins although their chemical origins are quite different. Phlorotannins occur in high concentrations in temperate brown seaweeds, particularly fucoids, at levels between 1% and 15% plant dry weight. Phlorotannins show antibiotic, antifungal and antifouling activities in addition to deterring herbivores. Interestingly, they are absent or in low amounts in tropical browns (< 2% plant dry weight), where species in the order Dictyoales (*Dictyota, Zonaria*) instead produce complex mixtures of terpenoids, acetogenins and terpenoid-aromatic compounds.

Table 5.4 Occurrence of biologically active chemicals in the four major groups of seaweeds*

	Greens	Reds	Browns	Bluegreens
Taxa studied	Families Caulerpaceae, Udoteaceae, Halimedacea	Families Bonnemaisoniaceae, Plocamiaceae, Rhizophyllidaceae, Rhodomelaceae	Order Dictyotales	Family Oscillatoriaceae
	Caulerpa, Halimeda Tidemaria, Penicillus, Udotea, Avrainvillea Chlorodesmis Pseudochlorodesmis,	*Laurencia, Plocamium Asparagopsis, Bonnemaisonia, Ochtodes Chondrococcus, Sphaerococcus,*	*Dictyota Pachydictyon, Glossophora, Dictyoperis, Zonaria, Desmerestria, Sargassum*	*Lyngbya*
Active chemicals	Sesquiterpenoid and diterpenoid compounds	Aliphatic haloketones, brominated phenols, monoterpenes, sesquiterpenes, diterpenes	Phlorotannins, terpenoids, acetogenins, terpenoid-aromatics	Lyngbyatoxin

*Source: Hay and Fenical (1992).

Simple inorganic compounds can also deter grazers. Sulphuric acid is produced by the brown seaweed *Desmerestia* and can account for up to 18% of the dry weight of the plant. Those who make pressed collections of seaweeds would be forgiven for thinking that the sole purpose of this substance is to destroy their entire herbarium, but it is also likely to inhibit feeding by herbivores. For instance, in Chilean kelp beds, the palatable kelp *Macrocystis* only successfully colonizes areas protected by patches of *Desmerestia* which seem to prevent access by urchins (Santelices, 1990).

Many seaweeds show trade-offs between availability, digestibility and edibility in their response to herbivory. Long-lived, predictable kelps are tough and loaded with tannins. Coralline crusts must rank amongst the most predictable of all seaweeds for small grazers and are well defended by their calcareous skeleton. They thrive under intense herbivore pressure as grazers remove potential competitors (see below). Ephemeral forms like *Ulva* are lacking in chemical and physical defences, but can escape in time and space through their rapid growth, early reproduction and high propagule output, to areas where grazing pressure has been reduced by local disturbance.

These different strategies can even be seen within the same species, where alternative life forms seem appropriate to different grazing environments (Slocum, 1980; Littler and Littler, 1980; Lubchenco and Cubit, 1980). Such bet-hedging has confused the systematics of seaweeds by giving rise to grazer-resistant encrusting forms like *Ralfsia* and *Hildenbrandia* that are not species in their own right, but alternative forms of erect frondose seaweeds. For instance, different 'species' of *Ralfsia* crusts have the common foliose seaweeds *Mastocarpus*, *Petalonia* and *Scytosiphon* as alternative life forms. Similarly, the burrowing algae found in gastropod and barnacle shells are in fact the conchocoelis phase of erect seaweeds such as *Porphyra*. The erect forms generally have a high competitive ability but lower grazer resistance, whilst the crustose or burrowing forms have low growth rates and a high resistance under grazing pressure (Table 5.5). All seaweeds pass through a highly vulnerable early stage during which grazing, sweeping by macroalgae and competition from ephemerals can kill the young plant. Many algae have the high growth rate characteristics of ephemerals in early life, allowing them to pass quickly through this critical phase. Only later do they develop the chemical defences and structural attributes which enable them to persist in the face of grazing and competition for canopy space.

Table 5.5 Predictions of the predominant life history for mid- and high-intertidal algae under different grazing regimes*

Grazing pressure	Life history	Examples
Constant and light	Isomorphology – uprights	*Enteromorpha.*
Constant and heavy	Isomorphology – crusts and borers, calcification	Low-shore 'lithothamnia', *Corallina*
Variable and predictable	Heteromorphology: with alteration of production and predominance of morphs on seasonal basis	Burrowing, *Conchocoelis, Porphyra.* Crust '*Ralfsia', Petalonia, Scytosiphon*
Variable and unpredictable	Heteromorphology: continuous production, but not survival of both morphs. Possibly, borers and crusts in heavily grazed midshore areas act as reservoirs from which recruitment can lead to escapes in variably grazed areas at higher levels	Crust '*Ralfsia', Gigartina papillosa.* Burrowing *Conchocoelis, Porphyra*

*Adapted from Lubchenco and Cubit (1980), Slocum (1980), Hawkins and Hartnoll (1983a).

5.5.4 MEIOFAUNA–MACROFAUNA INTERACTIONS IN SEDIMENT

In Chapter 2, we described the major taxonomic groups within the meiofauna and their basic ecology. Here we examine the possibility that predator–prey and competitive interactions between meiofauna and macrofauna have moulded the life-history patterns of some of the larger macrofauna. These ideas were stimulated by the need to explain the conservative nature of the **benthic biomass size spectrum** first reported by Schwinghamer (1981) in intertidal and shallow-water sediments (Fig. 5.6). The spectrum is constructed by measuring the size (volume) of every organism within a sample, from bacteria to macrofauna.

Three modes are always discernable in the spectrum, representing the three major types of sediment organism and their lifestyles: the microbes which cling to sand grains, the meiofauna which move between grains and the macrofauna which push the grains aside. The size troughs separating these three modes are consistently in the range 8–16 μm and 500–1000 μm (Fig. 5.6). This conservatism was originally explained by the way small and large organisms perceive surfaces. Schwinghamer suggested that the larger macrofauna see the sediment on a coarse scale, because the sizes of particles are small relative to the size of the organism. They either live in or plough through what to them is a semi-solid medium. In contrast, meiofauna are of a similar size to the particles themselves so that they must live between the particles. The very smallest metazoans, protozoans and bacteria perceive individual sediment particles as surfaces on which to live. The meio-fauna–macrofauna trough is therefore thought to represent a size of organism that is too large to live interstitially yet too small to live like the macrofauna.

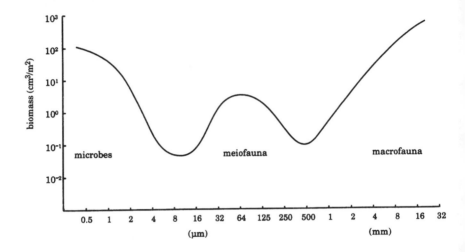

Fig. 5.6 The benthic biomass size spectrum (from various sources). Although termed a biomass spectrum, the amount of fauna in each size class is usually expressed as a volume.

At first sight, this hypothesis makes good sense, but there are alternative explanations. For instance, a dichotomy of life history traits occurs at an organism size of about 45 µg, or 0.5–1 mm, such that smaller organisms tend to have a different suite of traits from larger ones (Warwick, 1989). If intermediate characteristics are not feasible, this may explain why few species occur in the region of the meiofauna–macrofauna trough. A second explanation is the intensity of competitive and predatory interactions between the larger meiofauna and the smaller macrofauna. Some of the larger nematodes can eat capitellid polychaetes their own size, whilst turbellarian flatworms can inflict tremendous mortality on the early settling stages of intertidal invertebrates (Bell and Coull, 1980; Watzin, 1983). Warwick (1989) speculates that such interactions might have played a significant role in the evolution of planktonic larval development in the macrofauna of sediments. The eggs and newly hatched larvae of the macrofauna are similar in size to the meiofauna so that planktonic larval development might be a way in which these species avoid interactions with the meiofauna, the larvae eventually settling at a larger size, in the trough (Fig. 5.7).

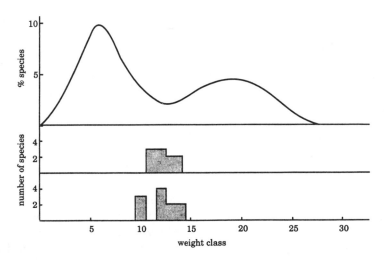

Fig. 5.7 The size (weight) of macrofaunal larvae (shaded, below) compared with the size distribution of meio- and macrofaunal species in sediments (modified after Warwick, 1984, with permission).

5.6 ASSOCIATIONS, MUTUALISMS AND SYMBIOSES

The interactions described above fall into the general class of predator–prey interactions, where it is easy to see that only one participant, the 'predator', benefits. Not all interactions are so clear-cut and it is often possible to see benefits to both parties. These mutualisms probably arise from a long history of grazing, predation or habitat provision, where the joint responses of the protagonists produce intimate associations. These relationships provide much of the fascinating natural history of the shore and constitute the sophisticated mutualisms that may enhance the dynamic stability of intertidal communities as suggested by Paine (1980). Here, we discuss some of the more interesting examples of associations between species and how they may have led to mutualistic interactions. The topic of parasitism has been well dealt with by other authors (e.g. Sousa, 1991) and is not addressed here.

5.6.1 HABITAT PROVISION BY MACROALGAE: UNWANTED VISITORS AND WELCOME GUESTS

Macroalgae provide a diversity of resources for epiphytic organisms: they are a surface for attachment, permanent or temporary shelter; they trap sediment and organic matter; they also provide food, either directly or indirectly via epiphytic plants or animals. Similarly, four types of epiphytic organism can be recognized: primary phytal species, cryptofaunal species, sediment-associated species and algal borers (Williams and Seed, 1992).

Primary phytal species are most common on seaweeds and other macrophytes, but can often be found elsewhere. Many live permanently epiphytically, others are transient. Colonial forms such as bryozoans compete well for space and predominate on many low intertidal or subtidal algae. The fauna of high-intertidal seaweeds tends to be more mobile, enabling animals to move into the humid parts of seaweed clumps or beds in response to desiccating conditions (Williams and Seed, 1992). Sessile species show considerable specificity, not only in the number of species on which they live, but with respect to the particular parts of the plant colonized. Selective settlement of larvae is important in maintaining these distributions. Specificity is particularly pronounced in the tube-dwelling spirorbid polychaetes, which are sometimes named by reference to their common host (e.g. *Spirorbis corallinae* found on *Corallina*). Seaweeds exude considerable amounts of organic carbon, and sessile organisms might also benefit from the nutrition provided by these exudates; *Flustrellidra* has been estimated to take up 20% of the total carbon fixed by *Fucus serratus*, gaining 17% of its energetic requirements in this way (Oswald, 1986).

Cryptofaunal species utilize algal holdfasts (particularly those of kelps) and live amongst turf-forming algae. They generally show little habitat specificity. Algal borers are rarer and more specialized. Algal holdfasts and turfs trap much sediment, promoting the development of an essentially infaunal community, often consisting of many small meiofaunal species.

Epiphytic plants, particularly ephemeral species, are common on large fucoids and kelps. In many cases canopy or other competitive interactions prevent epiphytes from living on the rock itself. Brown filamentous ectocarpaceans, *Enteromorpha* and *Palmaria*, are often found growing on the fronds and blades of seaweeds, but when low-shore fucoid and kelp canopies are cleared, these species are amongst the first colonists of the rock surface (Hawkins and Harkin, 1985). However, abundant growths of sessile epiphytes can lead to decreased performance and even mortality of host plants, unless they are removed by herbivores (review: Williams and Seed, 1992). A good example is the close association between the snail *Littorina obtusata* and the large fucoid *Ascophyllum nodosum* (Williams, 1990). The snail grazes away other epiphytes, thereby reducing the hydrodynamic loadings on plants and decreasing the detachment rate during storms. Although the snail consumes the algal thallus, this does not appear to affect the performance of the plant. In contrast, *L. obtusata* has a severe effect on *Fucus serratus*, a potential competitor of *Ascophyllum*, when the latter occurs in the *Ascophyllum* zone or when snails are experimentally transplanted to *Fucus serratus* plants lower down the shore. *Ascophyllum* produces noxious secondary chemicals (polyphenols) which deter most grazers, but which attract *L. obtusata*. It is one of the few grazers to actually consume the thallus tissue; others feed on epiphytes (Norton and Manley, 1990; Norton *et al.*, 1990). These chemicals may also be used by the snail to ensure it remains in an appropriate zone (Williams, 1990). There are clear mutual benefits from this association.

Most epiphytic grazers appear to be prudent herbivores: their activities rarely excessively damage individual plants. One reason for this is that they crop limited amounts of host tissue and avoid vulnerable areas, such as meristems or reproductive tissue. However, this is not always the case. Hay *et al.* (1987) suggest that grazing by amphipods and polychaetes can remove up to 20% of the blade area of *Dictyota*, whilst occasional population explosions of the snail *Lacuna vincta* feeding on *Laminaria saccharina* in New Hampshire (Fralick *et al.*, 1974) and on *Fucus edentatus* in Nova Scotia (Thomas and Page, 1983) can devastate seaweed beds. Where vulnerable areas of the plant are grazed, the effects of the herbivore can be disproportionate to the amount consumed. For instance, if the blue-rayed limpet *Helcion (Patina) pellucidum* grazes on the holdfast rather than the blade of *Laminaria* spp., this can increase the detachment rate of plants during storms. Generally, studies on amphipods and isopods have shown that severe damage only tends to occur when preferred foods such as epiphytic ephemeral algae or diatoms become scarce.

Thus, the main benefit that accrues to seaweeds by supporting small mobile herbivores is their ability to keep the surfaces clear of other epiphytes. Similarly, predators, such as nudibranchs and pycnogonids, will remove encrusting bryozoans (Todd and Havenhand, 1989). On large, erect seaweeds, sessile epiphytes can increase drag, making them more susceptible to wave dislodgement, and will reduce the frond area available for photosynthesis. Extensive colonization by colonial encrusting invertebrates, such as bryozoans, tunicates and hydroids, can reduce photosynthesis. *Flustrellidra* growing on *Fucus serratus* reduces the rate of photosynthesis on fronds to 5% of that in unfouled controls (Oswald *et al.*, 1984). This can lead to an overall reduction of photosynthesis in plants by around 40% (Oswald, 1986). Lowered photosynthesis in turn reduces growth rate, as shown for *Membranipora* growing on *Macrocystis*. In addition to reducing photosynthesis, encrusting bryozoans reduce the flexibility of algal fronds and the latter are at greater risk of shearing off in turbulent conditions or during storms.

On encrusting algae, such as the crustose corallines, mobile grazers also keep down faster-growing competitors. About 30% of limpets are grazers of crustose corallines (Steneck and Watling, 1982), and the juvenile stages of many limpets, chitons and abalones settle preferentially in pools lined with these crusts, stimulated in many cases by various GABA-like neuroactive compounds (Morse, 1992). These associations have developed into species-specific mutualisms, as illustrated by the interactions between the limpet *Tectura testudinalum* and the encrusting coralline *Clathromorphum circumscriptum* (Steneck, 1992). *Tectura* preferentially settles on this coralline which is also its preferred food. The limpet also survives starfish predation better on this crust. If the limpet is removed, the crust dies due to overgrowth by ephemeral seaweeds. On the other hand, *Clathromorphum* seems well adapted to grazing by this limpet, possessing a thick epithallus which protects a basal meristem. Epithelial cell growth matches the rate of limpet grazing and the reproductive structures are deep inside the crust. *Clathromorphum* is the competitively dominant crust when grazed by this limpet, but thinner, faster-growing corallines become dominant in the absence of grazing. This kind of relationship seems to have evolved independently in several distantly related limpet groups with equally distant coralline groups. Coralline crusts were present in the Precambrian, long before docoglossate limpets first appeared in the late Triassic, so it seems unlikely that calcified crusts evolved in response to limpets. However, it is clear that a generalized relationship now exists between grazers and many species of crustose corallines.

5.6.2 OTHER WAYS OF DEALING WITH FOULING ORGANISMS

Many plants do not rely on cleaning by herbivores, instead making use of their own antifouling mechanisms. These include secretions of mucus-like compounds, various antibiotic defence compounds (Gieselman and McConnell, 1981) and shedding of the epithelium itself (Filion-Mykelbust and Norton, 1981; Russell and Veltkamp, 1982). Many of the defence compounds deployed against grazers (see above) may have a more general function as antifoulants and some have allelopathic effects against competitors. For example, the growth of the seaweeds *Porphyradisius simulus* and *Rhodophysena elegans* is inhibited by the presence of *Ralfsia* (Fletcher, 1975). Laminarian and fucoid polyphenols and tannins are known to have deleterious effects on the settlement of bacteria (Al-Ogily, 1985) and certain invertebrates (Gieselman and McConnell, 1981), whilst *Chondrus crispus* has been shown to secrete chemicals, including hydrogen peroxide, capable of inhibiting diatom growth (Kahfaji and Boney, 1979; Huang and Boney, 1984). However, the density of epiphytic plants and animals on many seaweed species is a testimony to the failure of these defence mechanisms in some individual plants.

Animals are also prone to fouling and show various antifouling and cleaning behaviours. These can include the use of the foot in *Mytilus* (Hawkins, pers. obs.) and *Calliostoma zyzyphinium* (Jones, 1984) to wipe the shell clean and perhaps lay down a protective mucus coating. Animals with limbs, such as amphipods and decapods, can groom themselves, and epizooites will also be lost when they moult. Echinoderms use pincer-like pedicellariae to remove settling propagules. Conversely, animals such as spider crabs actively encourage a growth of living seaweeds which provide camouflage. Sea urchins, such as *Paracentrotus*, trap loose materials such as seaweeds and shells to cover their upper surface. This may be both to increase crypsis and, in the case of algal coverings, to provide food.

5.6.3 GARDENING BY MICROPHAGOUS GRAZERS

The final example of a close association between herbivores and plants is gardening. This can be formally defined as the modification of plant assemblages, caused by the activities of an individual grazer within a fixed site, which selectively enhances a particular plant species and increases the food value of the plants for the grazer. The most obvious examples are highly territorial species in which individuals defend patches of algae (or garden) against interlopers (Reviews: Branch, 1981; Branch *et al.*, 1992). These areas are usually quite small and, in the case of the southern African *Patella cochlear*, just surround the periphery of the axis of rotation of the individual limpet. Usually if the territorial species is removed, various interlopers of other species move in and the gardened alga disappears. In *Patella cochlear*, small limpets will maintain gardens on the backs of shells of larger individuals. The exact mechanism whereby the garden develops is unknown. Mucus may be involved because gardening limpets of the genus *Lottia* in California produce more sophisticated mucus than homing species, non-homing generalist grazers or predators. Also, the mucus of territorial *Lottia* enhances algal growth (Connor and Quinn, 1984).

Recent work (Davies *et al.*, 1992a) on UK shores has shown that the homing limpet *Patella vulgata* produces a more costly mucus than the non-homing snail *Littorina littorea*. Microalgal adhesion and growth performance is enhanced by *Patella vulgata* mucus. Although this is not gardening in the same sense as in *Lottia* and *P. cochlear*, this behaviour of *P. vulgata* does represent an enhancement of grazing resources. Trail-following behaviour will ensure that expensive mucus and its admixture of microbial food is re-ingested, reducing the considerable cost of mucus production, which can represent 20–30% of the energy consumed (Horn, 1989; Davies *et al.*, 1990). Similar gardening activities have been suggested for nematodes and other small sediment metazoans (Riemann and Schrage 1978; Chapter 6).

OVERVIEW

Shore plants and animals have evolved a variety of morphological, physiological and in the case of animals behavioural mechanisms to withstand the rigours of the shore, particularly aerial exposure and wave action. Those with larvae must regain the shore and minimize the risk of mortality by selective settlement. On rocky shores they must attach and resist dislodgement as they grow. On depositing shores, most animals and some microscopic plants avoid the rigours of emersion and waves by burrowing. Alternation of aerial and aquatic exposure also brings respiratory problems to both rocky and sediment-shore animals. Some rocky shore animals close themselves off by shutting shells or closing operculae and respire anaerobically. Similarly, burrowing organisms have to deal with low oxygen levels in sediments. Meiobenthic animals living in the anoxic black layer have to deal with hydrogen sulphide and ammonia.

In many cases animals opt to cope with environmental changes by behavioural rather than physiological mechanisms, although some do both. Many animals and some sandy-beach diatoms have rhythmic behaviour which allows anticipation of predictable environmental change associated with the tides and day/night cycles.

Plants and animals deploy a variety of defences against grazing and predatory species and fouling organisms. Some interactions are one-sided, but quite sophisticated mutualisms have coevolved, including algal gardening by grazers.

The shore as a system 6

The seashore is part of a larger coastal ecosystem where production from the shore is exported offshore and other material is imported to the shore from adjacent waters. There is also transfer of material between the shore and terrestrial coastal ecosystems. Because of the transfer of material the shore can be considered an **open** ecosystem, although some coastal systems can be **semi-enclosed** (e.g. lagoons, some estuaries). This material can be in the form of dissolved inorganic and organic nutrients, tiny organic particles, larger organic particles and chunks of debris and detritus, invertebrate larvae and juvenile fish. Adult fish, crustaceans, birds or even reptiles actively migrate into the intertidal and nearshore zone to feed and reproduce. The purpose of this chapter is to examine in broad terms how the shore functions as a system and interacts with the adjacent open sea and terrestrial communities. We begin by describing how nearshore oceanic processes affect shore communities. Then we take several examples of marine systems and outline the flows of energy and material through them. Finally, we consider the ways in which the different life-history stages of various species exploit the shore during their circuits of migration and link the shore with offshore and terrestrial systems.

6.1 THE OCEANOGRAPHIC CONTEXT

Oceanographic current systems are a major factor affecting the climate experienced by the land. Not surprisingly, they are also a major influence on shore ecosystems as they largely determine the temperature and nutrient status of the adjacent waters. Their importance is well illustrated by comparing shores that occur at the same latitude but on different sides of the Atlantic. South-west Britain supports a warm-temperate biota, whereas shore communities in Labrador, at the same latitude, would be considered to be boreal Arctic.

The difference of course is due to the effects of the cold Labrador Current and the warmer Gulf Stream which bathes south-west Britain. On a more local scale, the south and west coasts of the British Isles and the western coasts of France have a much more diverse, and essentially southern, fauna compared with North Sea coasts.

One of the earliest and best-documented examples of the influence of offshore conditions on shore ecology was the pioneering work done by the Stephensons in South Africa, summarized and extended by Branch and Branch (1981) (Fig. 6.1). The west coast of southern Africa is typified by cold, nutrient-rich water influenced by the Benguela Current, with upwelling occurring for much of the year. Just around the Cape there is water of intermediate character due to the mixing of the cold Benguela waters with those of the warmer Agulhas Current. Further east, an essentially warm-water, subtropical community is more typically found and there are lower levels of nutrients. The western Cape has large and extensive kelp beds, very high intertidal production and large populations of seabirds, which feed offshore and deposit nutrients in the form of guano onto the shore. The rocky shore zonation patterns and community composition change with region (Fig. 6.1). On the west coast, algae are far more luxuriant and browns and greens are more common. Whilst production is lower on the south and east coasts, diversity, particularly of animals, is higher.

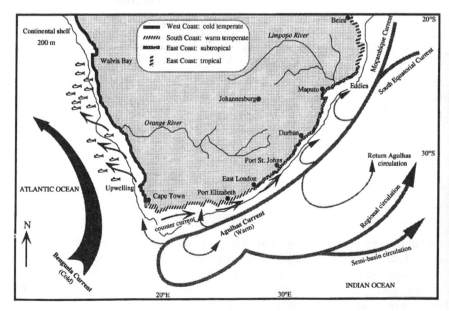

Fig. 6.1 Major oceanographic features around the coast of southern Africa and associated shore communities (pages 187, 188 and 189) (adapted with permission from Branch and Branch, 1981, *The Living Shores of Southern Africa*).

WEST COAST ZONATION

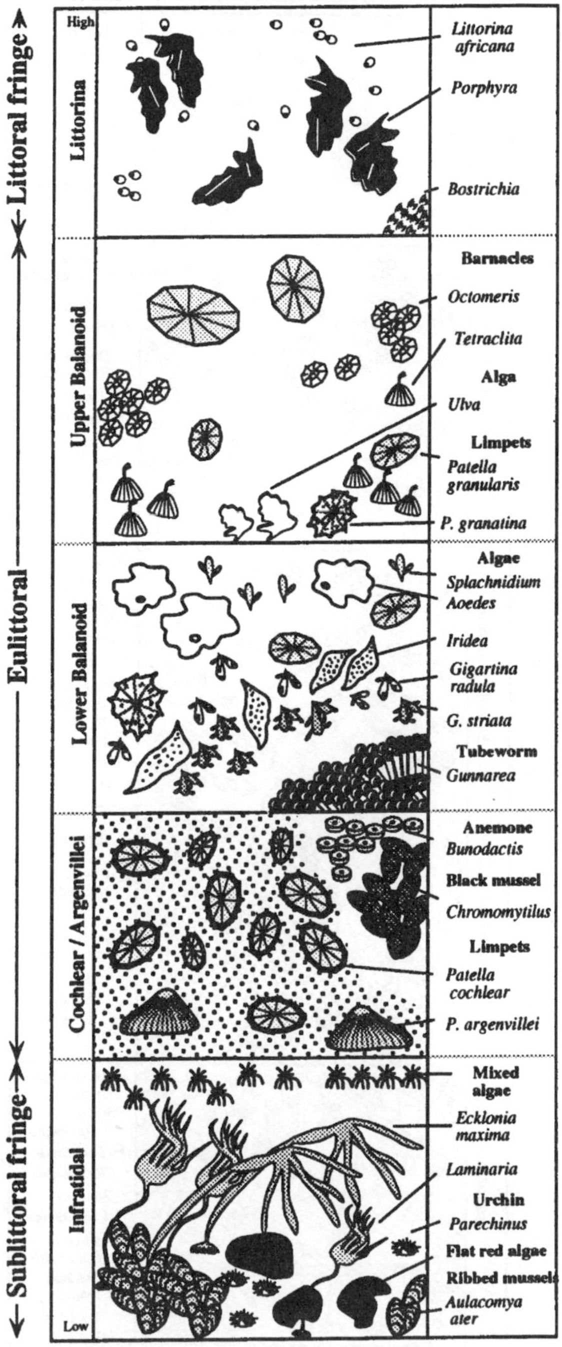

Littoral fringe →

← Littoral fringe

High

Littorina

- Littorina africana
- Porphyra
- Bostrichia

← Eulittoral →

Upper Balanoid

- **Barnacles**
- Octomeris
- Tetraclita
- **Alga**
- Ulva
- **Limpets**
- Patella granularis
- P. granatina

Lower Balanoid

- **Algae**
- Splachnidium
- Aoedes
- Iridea
- Gigartina radula
- G. striata
- **Tubeworm**
- Gunnarea

Cochlear / Argenvillei

- **Anemone**
- Bunodactis
- **Black mussel**
- Chromomytilus
- **Limpets**
- Patella cochlear
- P. argenvillei

← Sublittoral fringe →

Infratidal

- **Mixed algae**
- Ecklonia maxima
- Laminaria
- **Urchin**
- Parechinus
- **Flat red algae**
- **Ribbed mussel**
- Aulacomya ater

Low

SOUTH COAST ZONATION

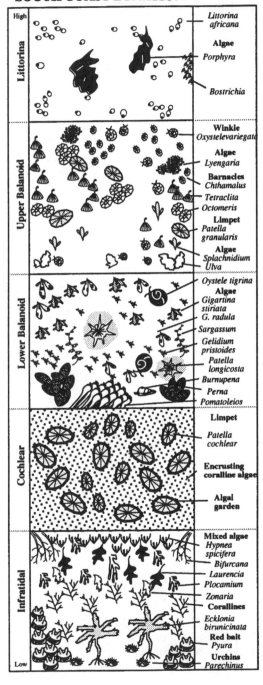

Littorina
- Littorina africana
- **Algae**
- Porphyra
- Bostrichia

Upper Balanoid
- **Winkle** Oxystelevariegata
- **Algae** Lyengaria
- **Barnacles** Chthamalus
- Tetraclita
- Octomeris
- **Limpet** Patella granularis
- **Algae** Splachnidium Ulva

Lower Balanoid
- Oystele tigrina
- **Algae** Gigartina stiriata G. radula
- Sargassum
- Gelidium pristoides
- Patella longicosta
- Burnupena
- Perna
- Pomatoleios

Cochlear
- **Limpet** Patella cochlear
- **Encrusting coralline algae**
- **Algal garden**

Infratidal
- **Mixed algae** Hypnea spicifera
- Bifurcana
- Laurencia
- Plocamium
- Zonaria
- **Corallines**
- Ecklonia biruncinata
- **Red bait** Pyura
- **Urchins** Parechinus

EAST COAST ZONATION

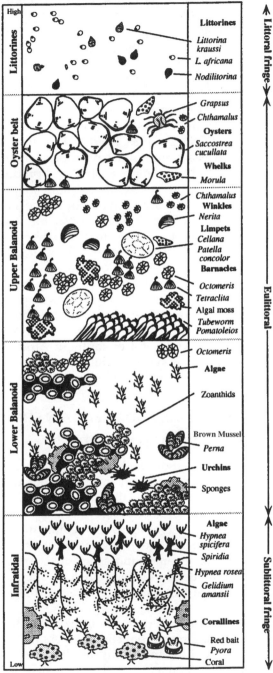

High

Littorines

Littorines
Littorina kraussi
L. africana
Nodilitorina

Oyster belt

Grapsus
Chthamalus
Oysters
Saccostrea cucullata
Whelks
Morula

Upper Balanoid

Chthamalus
Winkles
Nerita
Limpets
Cellana
Patella concolor
Barnacles
Octomeris
Tetraclita
Algal moss
Tubeworm Pomatoleios

Lower Balanoid

Octomeris
Algae
Zoanthids
Brown Mussel
Perna
Urchins
Sponges

Infratidal

Algae
Hypnea spicifera
Spiridia
Hypnea rosea
Gelidium amansii
Corallines
Red bait
Pyora
Coral

Low

Littoral fringe

Eulittoral

Sublittoral fringe

Upwelling areas can have a particularly marked influence on shore communities (Fig. 6.2). Algal biomass and primary production tend to be high and in areas of seasonal upwelling, such as Oman (Elliott and Savidge, 1990; Savidge *et al.*, 1990, 1992; Brock *et al.*, 1991; Currie, 1992), seaweeds become more prolific as the upwelling proceeds. Upwelling areas are also rich in plankton, providing considerable resources for sessile suspension feeders, whilst detrital inputs from enhanced seaweed production fuel microbial production that can be utilized by detritivores and suspension feeders alike (Velimirov *et al.*, 1977; Field *et al.*, 1980a,b; Branch and Griffiths, 1988; Field and Griffiths, 1991).

In contrast, the seaweed communities of tropical shores with waters of low nutrient status tend to be impoverished and production is low. A great variety of grazers and severe desiccation stress control algal biomass on these shores. Most seaweeds occur in the sublittoral or are associated with coral reefs (see Brosnan, 1992; for a comparison of temperate and tropical shores). Here, most of the production is by symbiotic algae within the coral itself (review: Spencer Davies, 1992).

It is likely that under nutrient-rich conditions, macroalgae have a better chance of escaping grazing by passing rapidly through their early stages and continuing to grow quickly, thereby swamping or outpacing grazer activity. Thus, under some circumstances 'bottom-up' forces predominate over 'top-down' control by consumers (Menge, 1992). It could be argued that the west coast of southern Africa is essentially a 'bottom-up' upwelling-driven system, while the south and particularly east coasts are perhaps controlled from the top down by consumers.

6.2 THE FLOW OF MATERIAL AND ENERGY IN SHORE SYSTEMS

6.2.1 PRIMARY PRODUCTION: RATES AND FATES

Primary production on the shore and in the shallow sublittoral can be impressive when compared with that in other kinds of system (Table 6.1). Large kelps often produce annually well in excess of a kilogram of carbon per square metre of shore (1000 gC m^{-2} year^{-1}). Similarly, rocky intertidal seaweeds and macrophytes of sediment shores, such as seagrasses, salt-marsh grasses and mangrove vegetation, have annual primary productivities of several hundred gC m^{-2}. Annual microphyte production within sandy beaches is typically low, often only a few tens of gC m^{-2} (Table 6.1, page 193), but is higher on surf beaches and some mudflats (see below).

What happens to this material? About 90% of the macrophyte primary production is not grazed directly. It enters the local food chain partly as particulate detritus and partly as dissolved organic matter (DOM). Much of the detrital material is then exported from the locality.

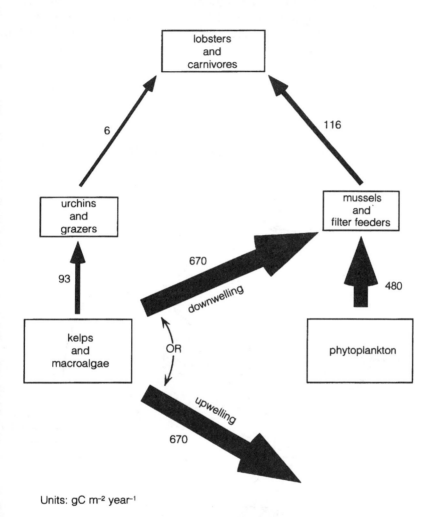

SOUTH AFRICAN KELP FOREST

Units: gC m⁻² year⁻¹

Fig. 6.2 Major flows of energy in South African kelp forests (data from Field and Griffiths, 1991.

This can be seen by comparing the energy flow pathways between the major shore components shown in Figs 6.2 and 6.4–6.9. These webs have been greatly simplified to allow an appreciation of the relative magnitudes of direct grazing and detritus-based components. As can be seen, very little of the macrophyte production is grazed directly, most passing into detritus-based food chains. Only a fraction of the fresh detritus is ingested directly by larger multicellular organisms, much of it being colonized and broken down by microbial organisms. These are in turn grazed by invertebrates, which therefore feed on macrophyte production indirectly at one stage removed. Odum and Heald (1975) recognize at least four ways by which the organic material in freshly fallen mangrove leaves can be utilized by other organisms and these routes probably apply equally well to other sources of macrophyte litter (Fig. 6.3).

ROUTES FOR MACROPHYTE LITTER

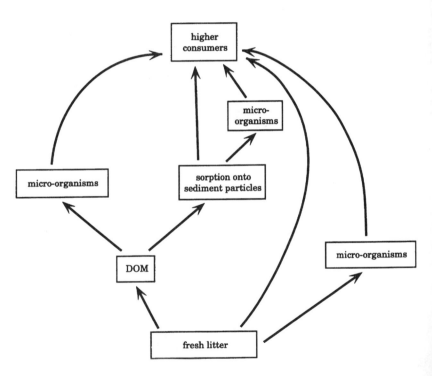

Fig. 6.3 Routes by which marine macrophyte litter is utilized by other organisms (data from Odum and Heald, 1975).

On hard surfaces, microalgal films can be extremely productive, more so than macroalgae on more exposed shores (Fig. 6.4). The film itself consists of organic molecules, heterotrophic bacteria, cyanobacteria,

DOM
dissolved organic matter

diatoms, fungi, protistans and the early stages of macroalgae. Much of this microbial production is utilized directly by grazers. Mucus from diatoms and the trails of foraging gastropods is also an important element of these films and may enhance the recruitment of seaweed propagules and adhesion of diatoms (Connor and Quinn, 1984; Davies *et al.*, 1992a,b). Much of the mucus produced by herbivorous gastropods will be reingested and hence recycled. Some will be exported to coastal waters. It has been suggested that territorial limpets 'garden' algae using mucus (Connor and Quinn, 1984).

Table 6.1 Estimates of primary production in intertidal and shallow water systems. All values gC m^{-2} year^{-1}, based on the following conversion from the original literature: 1gC = 12 Cal or 50 kJ. Asterisked values are extrapolated from the short-term estimates originally expressed as amounts per day

Species and site	Production (gC m^{-2} year^{-1})	Source
KELPS		
Kelp (*Ecklonia maxima, Laminaria, Macrocystis*) community (South Africa)	760	Field and Griffiths (1991)
Laminaria pallida (South Africa)	1330	Mann (1982)
Laminaria hyperborea (UK)	1225	Mann (1982)
Macrocystis pyrifera (California)	800–1000	Mann (1982)
Ecklonia radiata (Western Australia)	1500	Mann (1982)
ROCKY INTERTIDAL ALGAE		
Sargassum platycaspum (Caribbean)	500*	Mann (1982)
Encrusting corallines (Caribbean)	1000*	Dawes *et al.* (1991)
Distyopteris justii (Caribbean)	200*	Dawes *et al.* (1991)
Dictyota dentata (Caribbean)	900*	Dawes *et al.* (1991)
Homosira banksii (New South Wales)	200	King *et al.* (1991)
Fucoids (Atlantic Canada)	700	Mann (1982)
Fucus vesiculosus (Baltic)	350	Wallentinus (1991)
Fucus spiralis (Spain)	120	Niell (1977)
Fucoids (UK)		
Exposed	16	Hawkins *et al.* (1992a)
Semi-exposed	160	Hawkins *et al.* (1992a)
Sheltered	1250	Hawkins *et al.* (1992a)
Microalgal films (UK)		
Semi-exposed	100	Hawkins *et al.* (1992a)
High shore, exposed	60	Workman (1983)
Low shore, exposed	100	Workman (1983)
SEAGRASSES		
Zostera marina (Northern Carolina)	350	Mann (1982)

Zostera marina (Australia)	190–800	Hillman et al. (1989)
Posidonia australis (Australia)	140–190	Hillman et al. (1989)
Amphibolis spp. (Australia)	340–430	Hillman et al. (1989)
Cymodocea spp. (Australia)	365–657	Hillman et al. (1989)
Halodule spp. (Australia)	280	Hillman et al. (1989)
Thalassia testudinum (Florida)	900	Mann (1982)

SALT-MARSH

Spartina (southern USA)	500	Mann (1982)
Spartina, Limonium and Salicornia (UK)	400	Mann (1982)

MANGROVE

Rhizophora mangal (Florida)	350–500	Mann (1982)
Mangrove (New Zealand)	310	Knox (1986)

SAND AND MUDFLATS

Sand microalgae (New South Wales)	50	King et al. (1991)
Microphytobenthos (Netherlands)	25–57	Wolff (1977)
Benthic microalgae (UK)	143	Mann (1982)
Benthic microalgae (UK)	31	Baird and Milne (1981)
Benthic microalgae (UK)	5	Steele and Baird (1968)
Benthic microalgae (general)	20–200	Reise (1992)
Benthic microalgae (New Zealand)	170	Knox (1986)

On sediment shores *in situ* microphyte production seems relatively unimportant compared with imported suspended material. This is in the form of phytoplankton and detritus, the breakdown of which is the basis of production by heterotrophic bacteria. On exposed sandy (surf) beaches characteristic of South Africa, Australia and the west coast of the USA, phytoplankton production in the surf zone can be extremely high and is the major input of energy into the system (Brown and McLachlan, 1990; Fig. 6.5). Diatoms such as *Anaulus* are concentrated into patches by surf circulation and accumulate in areas of reduced water movement close to the beach. In the afternoon they sink and attach, by a thick mucus coat, to sand grains at the sediment surface, where they spend the night. In the morning they lose their mucus coat, are resuspended and concentrated towards the beach by waves (McLachlan and Lewin, 1981; reviews: Brown and McLachlan, 1990; Talbot *et al.*, 1990). As on rocky shores, mucus production contributes significantly to both particulate and dissolved organic carbon, about 40% in the example shown in Fig. 6.5. Whether this is considered *in situ* primary production or imported depends on where the seaward boundary of the beach is drawn and hence whether the surf zone is included.

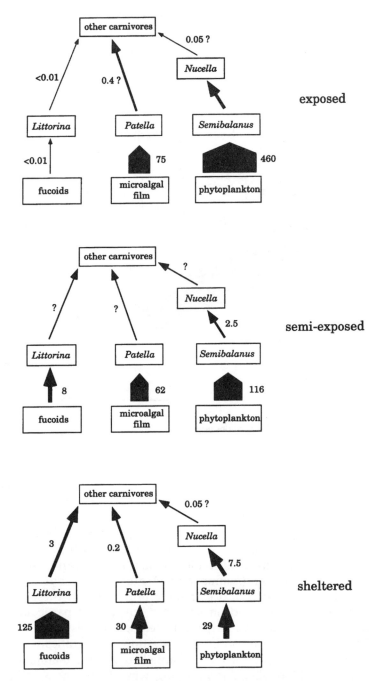

Fig. 6.4 Major flows of energy (gC m^{-2} year^{-1}) on exposed, semi-exposed and sheltered rocky shores on the Isle of Man, UK (data from Hawkins *et al.*, 1992a).

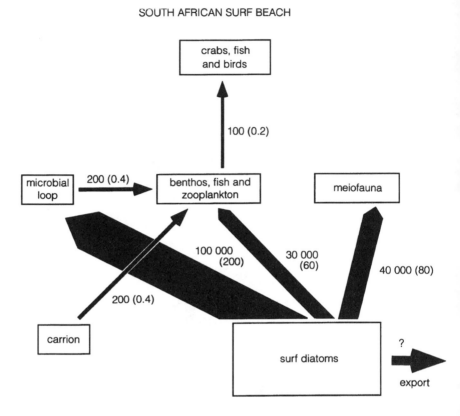

SOUTH AFRICAN SURF BEACH

gC per linear metre of shore year⁻¹

(approx. gC m⁻² year⁻¹)

Fig. 6.5 Major flows of energy as gC per linear metre of shore⁻¹ and, in brackets, rough correction to area (gC m⁻² year⁻¹) in a South African surf beach (data from Brown and McLachlan, 1990).

Clearly, small primary producers can play a significant role in the trophodynamics of both rocky shore and surf-beach systems. On more sheltered mudflats the contribution of microphytes can be substantial, as in the Wadden Sea (Asmus and Asmus, 1985), but heterotrophic bacterial production is also extremely important on these shores (Fig. 6.6). These microbes feed on imported suspended particulate material and are in turn grazed by larger multicellular organisms.

Depending on the shore in question, a significant amount of sediment bacterial production is channelled into the meiofauna (Figs 6.5 and 6.6). Traditionally, this grouping was viewed as a self-contained system (predators, scavengers, herbivores and detritivores are all present), interacting little with the larger macrofauna but perhaps having importance for nutrient cycling and mineralization processes. However, it is now clear that small epibenthic predators, such as juvenile fish, crabs and shrimps, prey on meiofauna, especially copepods and nematodes (Warwick *et al.*, 1978; Gee, 1989).

YTHAN ESTUARY, SCOTLAND

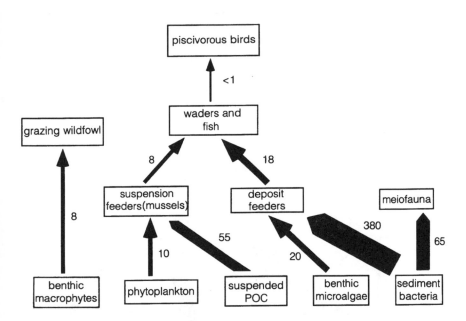

Fig. 6.6 Major flows of energy (gC m⁻² year⁻¹) in the Ythan estuary, north-east Scotland (data from Baird and Milne, 1981).

Reimann and Schrage (1978) have suggested that the mucus trails left by nematodes and other meiofauna as they move through the sediment traps organic matter which in turn supports a dense microbial growth. This is reingested, mucus and all, by the animals returning along the same trail. The mucus-trap hypothesis is similar to that suggested for rocky shore gastropods (Chapter 5 and above). Nematodes may garden ecto-symbiotic microalgae and bacteria on the outside of their cuticle (Gerlach, 1978), but it has yet to be demonstrated that they obtain significant amounts of their food in this way (Jensen, 1988).

6.2.2 EXPORT AND IMPORT OF PRIMARY PRODUCTION

A key feature of most shore ecosystems is the high degree of water movement due to tides, wave action and currents. Even in the most sheltered estuarine habitats, moving water drains the flats and the associated salt-marsh or mangrove vegetation. Because so much of the primary production by macrophytes is not grazed whilst alive, but instead enters the system as detrital particulate carbon (Figs 6.7–6.9) or as dissolved organic carbon, much of the carbon fixed at a particular site will inevitably be transported away by water movements. On rocky shores, the relative amounts of material imported and exported will vary with exposure. Studies on the Isle of Man, UK, suggest that sheltered shores are net exporters of energy and exposed shores are net importers, mainly in the form of phytoplankton (Hawkins *et al.*, 1992a) (Fig. 6.4).

NORTH AMERICAN SEAGRASS BED

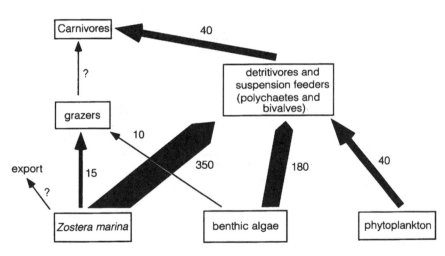

Fig. 6.7 Major flows of energy (gC m^{-2} year^{-1}) in a North American seagrass bed (data from Thayer *et al.*, 1975 and Mann, 1982).

Estimates of the amount of primary production entering the water column through dissolution to seawater varies. Values of this **dissolved organic matter** (DOM) can be very high, as much as 1–10 mgC l^{-1} and extreme values of 100 mgC l^{-1} have been documented over algal mats (Schramm, 1991). These figures must be viewed against estimates of only a few mgC l^{-1} for particulate carbon in coastal waters. Up to 40% of the carbon fixed by seaweeds can be lost as DOM, and salt-marsh plants and seagrasses are thought to contribute significantly to the amount of DOM in coastal waters. This process also affects phytoplankton, of course, and in the surf zone of South African sandy beaches (Fig. 6.5) about 25% of the primary production is available as dissolved organic material.

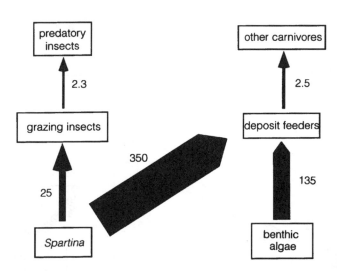

GEORGIA, USA SALT-MARSH

Fig. 6.8 Major flows of energy (gC m^{-2} year^{-1}) in a Georgia, USA, salt-marsh (data from Teal, 1962).

Most of this DOM is in the form of low-molecular-weight carbohydrates or free sugars and free or combined amino acids. These compounds are readily taken up by bacteria, which are grazed by flagellates, which are eaten in turn by ciliates, thus forming a microbial loop, which is now recognized as an important feature of all marine food chains, both inshore and offshore. For instance, in a study of a sandy beach on the south-east coast of Africa, McLachlan and Romer (1990) estimated that in terms of carbon consumption, the microbial loop was quantitatively the most important part of the system (Fig. 6.10). Some of the DOM may also be taken up directly by shore invertebrates *in situ*. For example, some large nematodes in the family Oncholaimidae, which because of their massive buccal armature have always been viewed as predators, are now known to take up significant amounts of DOM (Chia and Warwick, 1969; Lopez *et al.*, 1979; Jensen, 1987). One of the puzzles that first made ecologists aware of the significance of DOM in the trophic ecology of these nematodes was the scarcity of largish prey in their guts.

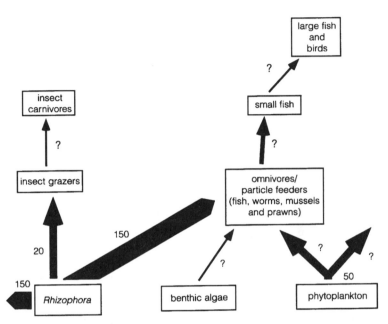

Fig. 6.9 Major flows of energy (gC m⁻² year⁻¹) in a New World mangrove system (data from Odum *et al.*, 1982).

Of course not all DOM remains dissolved. As discussed above, much of it will be directly taken up by microbes and metazoans and hence transformed to **particulate organic matter** (POM). Another transformation route is adsorption onto the surfaces of inorganic particles or air bubbles; this may be particularly important in intertidal systems where air bubbles are continually being created and where particles are resuspended by water turbulence (Schramm, 1991).

There is significant loss, transport or export of particulate carbon from the macrophyte-dominated systems shown in Figs 6.7–6.9, but the actual amounts will vary from place to place according to local hydrography. North American east coast salt-marshes appear to be net exporters of both dissolved and particulate organic carbon (Table 6.2), a process known as outwelling. Estimates of outwelling from mangroves are scant and contradictory (Woodroffe, 1985; Hutchings and Saenger, 1987), but like those for kelp forests and rocky shores, will probably depend on local conditions, such as whether upwelling or downwelling persists (Fig. 6.2) or whether nearshore hydrographic barriers occur, such as salinity fronts. Nevertheless, most of these systems are probably net exporters of DOM and POM, much of which must then enter their fringing sandy beaches and mudflats. Indeed, it is not possible to account for secondary production in sandy beaches and mudflats without significant imports of organic material from elsewhere. Thus, in the western Wadden Sea the annual consumption of energy by micro- and meiobenthos is about 260 gC m^{-2} and for macrobenthos about 90 gC m^{-2}. Benthic microalgal production is of the order of 100 gC m^{-2}, which falls far short of the total requirements of the benthos (350 gC m^{-2}) and has to be supplemented by a further 250 gC m^{-2}, representing about 70% of the total, in the form of imported detrital carbon (Knox, 1986). Similarly, in shallow sandy areas in north-west Scotland, the annual demands of the benthos amounts to 50 gC m^{-2}, but only 5 gC m^{-2} is produced *in situ* by microalgae, leaving a further 45 gC m^{-2} (90% of the total demand) to be imported into the beach as detritus (Steele and Baird, 1968).

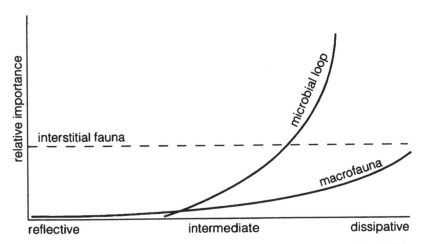

Fig. 6.10 Relative importance of three trophic groups over a range of South African sandy beaches (adapted with permission from McLachlan and Romer, 1990).

The input of POM from macrophyte-dominated systems can often be dramatic, such as the stranding of enormous quantities of kelp, shed or ripped off during storms to leave a deposit up to 1 m or so deep at the top of the shore. For instance, it is estimated that Kommetjie Beach in South Africa receives annually about 2000 kg of kelp (wet weight) per linear metre of shore (Griffiths *et al.*, 1983; Koop and Lucas, 1983). This material supports a rich and diverse macrofauna often dominated by insects, oligochaetes and high-shore amphipods (sand hoppers) and isopods. The larvae of insects can be particularly important in strandlines (Stenton-Dozey and Griffiths, 1983). Strandlines are in turn fed upon by a variety of birds (Puttick, 1979).

Stranded kelp detritus looks impressive, but it is often difficult to assess how much of this material is actually utilized by consumers. Similarly, where does the detritus which fuels deposit-feeders in other systems come from? To tackle this problem, we have to examine the biochemical make-up of the imported material itself and the tissues of the invertebrates presumed to be feeding on it. Fortunately, different kinds of primary producers fix and metabolize carbon in different ways and in the process change the ratio of two stable isotopes of carbon, ^{12}C and ^{13}C, from the ratio in which they occur naturally as inorganic carbon. By measuring the relative proportions of these two isotopes in detrital POM we can get a good idea of what kind of primary producer (bacteria, diatoms (benthic or planktonic), kelps, seagrasses, salt-marsh or mangrove plants) the material originally came from. Furthermore, because this material will be transformed into invertebrate consumers, the tissues of the latter will also tend to have a $^{12}C/^{13}C$ ratio similar to that of the original source of fixed carbon. Using this technique, we can trace trophic pathways through food webs (Peterson and Fry, 1987). The approach can be extended to the stable isotopes of other elements and has been a powerful tool for revealing the extent to which many invertebrates are fuelled by microbes living symbiotically within their tissues (e.g. Conway and Capuzzo, 1990).

Table 6.2 Annual flux of organic carbon between salt-marshes and coastal waters*. Minus signs indicate export

Site	Carbon flux (gC m^{-2}year^{-1})	
	DOC	POC
Great Sippewissett, Cape Cod, Massachusetts		−76
Flax Pond, Long Island, New York	−8.4	+61
Canary Creek, Lewes, Delaware	−38	−62
Gott's Marsh, Patuxant River, Maryland		−7.3
Ware Creek, York River, Virginia	−80	−35
Carter Creek, York River, Virginia	−25	−116
Dill Creek, Charleston, South Carolina		−303
Barataria Bay, Louisiana	−140	−25

*Source: Knox (1986).

Strandings of secondary consumers can also result in significant inputs of organic material. Jellyfish in particular are often stranded along with other neuston (e.g. the bubble raft snail *Ianthina*) living in the surface layer of the ocean waters, and mass strandings can occur (e.g. the by-the-wind-sailor *Vellela vellela*, Kemp, 1986). These stranded animals are eaten by sandy shore scavengers such as crabs. On the beaches of southern Africa the whelk *Bullia* is a scavenger *par excellence*, feeding on stranded material (A.C. Brown, 1982).

Rocky shores can naturally switch from being net exporters of primary production to net importers with time. For example, moderately exposed shores on the Isle of Man can sometimes be dominated by fucoid algae and sometimes by barnacles and limpets (see series of photographs in Hawkins and Jones, 1992). The interactions between consumers and producers on these shores not only determine the structure of the community, but the way in which energy flows (see Hawkins *et al.*, 1992a for further details). In this regard, escapes from consumers by good recruitment and rapid growth not only have repercussions for community structure but also for the pathways of energy flow.

6.2.3 THE IMPORTANCE OF SMALL ORGANISMS FOR ENERGY FLOW

One of the reasons why meiofauna are overlooked, and hence not particularly well understood, is their small size. The same could be said for microalgae. Yet for both these groups it is precisely because of their small size that it is imperative to include them in studies of energy flow. Except on very exposed sandy beaches (and the deep sea), the biomass of the meiofauna is considerably less than that of the macrofauna, even though their numerical abundance may be several orders of magnitude higher (Table 6.3). One might think therefore that energy flow in these systems would be completely dominated by the macrofauna, but this is not the case. For instance, Fenchel (1969) has suggested that at some sites, ciliate protozoans alone could account for eight times the metabolism of the macrofauna on a sandy beach. The standing stock or biomass of meiofauna is therefore a poor indicator of their importance. More relevant is how that biomass is distributed amongst the individual organisms. One gram of biomass may represent 5000 individual nematodes, whereas 1 g of macrofauna might be a single polychaete. Other meio- and microfauna are even smaller (Fig. 6.11). This statistic becomes important when the metabolic rates of small poikilotherms like marine invertebrates are considered. The weight-specific metabolic rate (equivalent to oxygen consumption) increases dramatically as organisms become smaller (Fig. 6.12), so that 1 g of meiofauna utilizes much more oxygen, and hence energy, than 1 g of macrofauna.

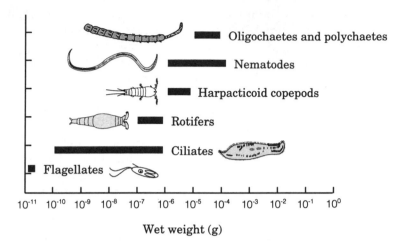

Fig. 6.11 Range of sizes of the main meiofaunal groups (adapted with permission from Mann, 1982).

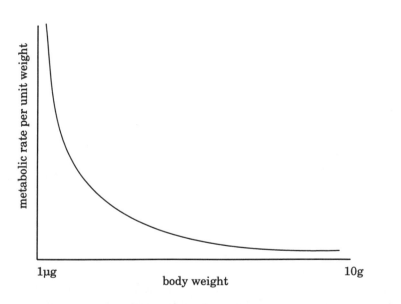

Fig. 6.12 Relationship between organism size and metabolic rate.

Body size is also related to generation time. Most meiofaunal species have more than one generation a year, some as many as ten (Gerlach, 1971). If an average of three generations a year is assumed, and for every 1 g of meiofauna 3 g are produced each generation (Gerlach, 1971), then a 1 g standing stock of meiofauna would produce 9 g over the year. Most of this biomass is dissipated by predators or other natural factors. In contrast, the longer generation times of the macrofauna mean that their turnover rates are 10–50% those of the meiofauna (Gerlach, 1971). It follows that in terms of using up energy and producing more of themselves, the meiofauna are likely to be much more important than estimates of their standing stock suggest. In fact, the ratio of **production to biomass**, the **P/B** ratio, is about 10 for meiofauna and often only about 2.5 for macrofauna (Table 6.3; McLachlan and Romer, 1990).

Table 6.3 Biomass and production of meio- and macrofauna in estuarine intertidal sediments*

Site	Meiofauna			Macrofauna		
	Biomass (gC m^{-2})	Production (gC m^{-2}year^{-1})	P/B ratios	Biomass (gC m^{-2})	Production (gC m^{-2}year^{-1})	P/B ratios
Ythan estuary	2.1	20.0	10.0	15.1	37.0	2.4
Lynher estuary	1.2	13.5	11.3	5.4	5.4	1.0

*Based on data from Baird and Milne (1981) and Warwick et al. (1979).

Meiofauna are also reasonably abundant on rocky shores (review: Hicks, 1985) yet there have been few efforts to include them in energy flow studies. A notable exception is the work of Gibbons and Griffiths (1986), who estimated densities, biomasses and productivities of macrofauna and meiofauna at five shore levels spanning the entire intertidal gradient at False Bay, South Africa (Fig. 6.13). Whilst the macrofauna were twelve times as important on this shore in terms of biomass, they were only three times as important in terms of productivity. Clearly, rocky shore ecologists need to follow the direction taken by sediment ecologists and include meiofauna in their energy flow webs. Of course, a similar argument can be made for the tiny algal propagules and other components of the microbial film on rocky shores. Because of their fast turnover rates maintained at least in part by the grazing activities of limpets (Chapter 5), their productivity can be equivalent to that of the macrophytes on some shores (Field, 1983). This is an important but neglected area of rocky shore ecology. The compositions of algal films have only recently been described (e.g., MacLulich, 1986; Hill and Hawkins, 1990, 1991) and much work remains to be done in the technically difficult area of measuring their rates of production and consumption by grazers.

6.3 LINKS BETWEEN THE SHORE AND OTHER SYSTEMS

6.3.1 PLANKTONIC LARVAE OF SHORE ORGANISMS

Clearly there is significant exchange of organic material between the different kinds of shore depicted in Figs 6.2 and 6.4–6.9. To this flux must be added the large numbers of larvae and algal propagules that are exported by macrobenthic invertebrates and plants into coastal waters, only a proportion of which return and settle. Many larvae utilize energy produced in the plankton and provide a food source for offshore pelagic and benthic systems, establishing an important link between the shore and offshore ecosystems.

The links are well illustrated by barnacles, the nauplii of which can spend up to 4–6 weeks feeding in the plankton, where they are in turn preyed upon by a variety of other zooplankton, including larval fish. In Boreal–Arctic species, such as *Semibalanus balanoides*, larval release is synchronized with the diatom-dominated spring phytoplankton bloom and is triggered by the presence of a particular diatom, *Skeletonema*. If the release of larvae is well timed and matches the bloom, then settlement can be high (Barnes, 1956; Connell, 1961a; Hawkins and Hartnoll, 1982a). On strongly indented coastlines or shores adjacent to islands, the incidence of onshore winds during the settlement season increases the likelihood of transport of the larvae onshore (Hawkins and Hartnoll, 1982a; Shanks, 1986; Bertness *et al.*, 1991) thereby influencing recruitment success (Barnes and Barnes, 1977). Embayed areas which are fed by rivers will have flushing times depending on the riverine input. During high flow conditions, larvae will be swept offshore (Gaines and Bertness, 1992). On more linear coastlines this effect is less important (e.g. Kendall *et al.*, 1985).

In upwelling areas of the Californian coast, recruitment success has been strongly linked to upwelling events which bring larvae onshore (Gaines *et al.*, 1985). Variation in the supply of recruits subsequently influences the whole community structure of the shore, because massive settlements can completely swamp predators in some years (Sebens and Lewis, 1985). This effect is more marked in species that release a single brood. Where larval dispersal is restricted or water is funnelled through narrow channels, larvae are retained and concentrated and there is consistently high settlement, as in sea lochs, inlets and narrow bays. This also occurs in areas with offshore fronts which trap larvae, such as the Gulf of Maine. It is not surprising that settlement of key animals, such as barnacles and mussels, is high and that biological interactions are intense on these shores (Chapter 4). Similarly, settlement densities of barnacles in the Firth of Clyde are high and this may account for the strong interactions seen in Connell's work on competition in barnacles

(Chapter 3), but which are not generally seen elsewhere in the British Isles (Hawkins and Hartnoll, 1982a).

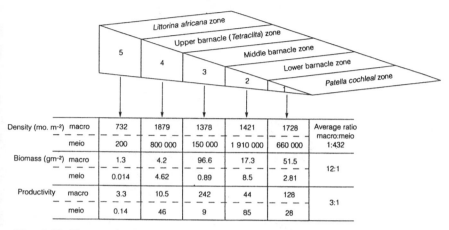

Density (mo. m⁻²)	macro	732	1879	1378	1421	1728	Average ratio macro:meio
	meio	200	800 000	150 000	1 910 000	660 000	1:432
Biomass (gm⁻²)	macro	1.3	4.2	96.6	17.3	51.5	12:1
	meio	0.014	4.62	0.89	8.5	2.81	
Productivity	macro	3.3	10.5	242	44	128	3:1
	meio	0.14	46	9	85	28	

Fig. 6.13 The relative importance of meio- and macrofauna at different levels on a rocky shore in South Africa (modified with permission after Gibbons and Griffiths, 1986).

In warm-temperate, subtropical and tropical areas, multiple brooding is much more common in barnacles. These species tend to reproduce for longer periods of the year and feed primarily on phytoplankton dominated by flagellates. Such trickle recruitment is highly susceptible to stochastic events. Recent work has indicated that the northern limits of some shore barnacles in Europe are likely to be governed by events in the plankton: the adults reproduce, but once the larvae are released, early broods fail (Burrows *et al.*, 1992).

Barnacle larvae effectively 'sample' the offshore ecosystem before settling, and in areas where northern and southern pairs of species coexist, settlement of these species can be used as bio-indicators of conditions offshore. In south-west Britain, Southward (1967, 1980, 1991) and co-workers (Southward *et al.*, 1995) have shown that changes in the balance of the northern *Semibalanus balanoides* and southern *Chthamalus* spp. over the last 80 years reflect other changes offshore. Dominance by *Semibalanus* is associated with a pelagic community characterized by offshore northern species including the indicator chaetognath *Sagitta elegans* with herring or mackerel being the main pelagic fish. Conversely *Chthamalus* dominance reflects a system with a more southern and inshore zooplankton assemblage (including *Sagitta setosa*) and the pilchard, *Sardina pilchardus*. Thus, the relative abundance of *Semibalanus* and *Chthamalus* tells us a great deal about the distribution of major water

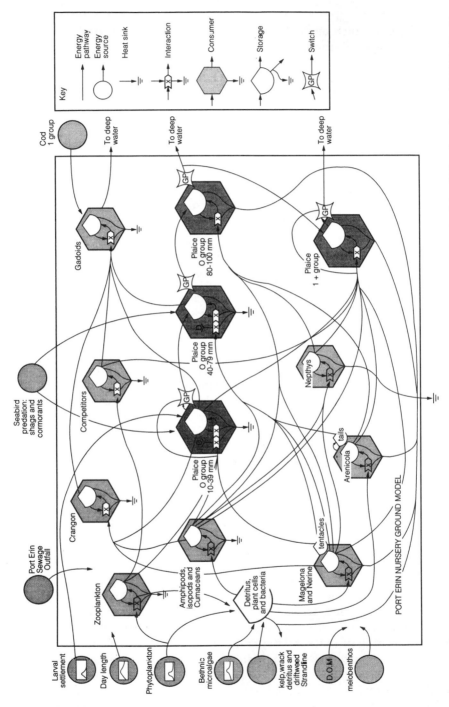

Fig. 6.14 An energy flow model of a flatfish nursery ground on a shallow sandy bay on the Isle of Man (courtesy of R.D.M. Nash). Energy circuit symbols follow Odum (1971).

masses and this has practical applications: it is far cheaper to count a few barnacles than go on expensive oceanographic cruises.

Interestingly, some coastal system simulation models are more stable when they include the role of pelagic macrobenthic larvae than when they exclude such larvae (Warwick, 1989; Chapter 5), implying that the coupling of benthic and pelagic systems by benthic larvae may have significance for the dynamics of the entire inshore system.

In the previous sections we have discussed how events in the plankton affect the benthos, but benthic communities can also significantly affect local plankton assemblages, mainly because of the high filtration rates of bivalves. Adult individual cockles, mussels, oysters and clams can process 1–3 l of water per hour (Jørgensen, 1990); where bivalve populations are dense they may filter daily a volume equivalent to the tidal volume of the estuary or salt-marsh. This is the case for the ribbed mussel *Geukensia demissa* in salt-marshes on the east coast of the USA (Jordan and Valiela, 1982), for the mussel *Mytilus edulis* in the Ythan estuary (Raffaelli, unpublished) and probably for many other semi-enclosed regions. Rates of water processing by *Mytilus edulis* are typically 10 m^3 m^{-2} $hour^{-1}$ and could remove up to half the phytoplankton and 20% of the dissolved amino acids in the water passing over the bed. Little of the bacterioplankton is removed unless flow velocities over the mussel bed are much lower (two orders of magnitude) than the typical speeds of 10–20 m min^{-1} (Jørgensen, 1990).

6.3.2 LIFE HISTORY AND MIGRATION PATTERNS

Other organisms utilize shores temporarily as nursery grounds (Fig 6.14), or episodically as in migratory fish and birds. Several species of migratory fish use estuaries as a corridor. They pass from fresh water to spawn in the sea – **catadromous** species like the eel *Anguilla* – or from the sea to spawn in fresh water – **anadromous** species like the Atlantic salmon *Salmo salar* or the lamprey *Petromyzon*. Some species actually spawn in intertidal areas. For example, Pacific herring *Clupea pallidus* lay their eggs in sticky masses on seaweeds in shallow water and capelin go one stage further, by laying their eggs on sandy shores above the surf zone. One of the most spectacular examples of using sandy beaches purely for breeding are the various species of turtles which lay eggs on the shore. Although the juvenile hatchlings migrate straight back out to sea, both the eggs and the vulnerable juveniles provide food for a variety of intertidal predators.

In tropical and subtropical areas, the juveniles of many species of commercially important crustaceans, particularly penaeid prawns, use estuaries, lagoons and mangroves for feeding grounds before migrating offshore to spawn. Many species of fish also spawn away from these intertidal areas. On hatching, the larvae move into inshore areas, having metamorphosed into juveniles by the time they reach these habitats. The importance of kelp beds, estuaries, mangroves, seagrass beds and other sheltered bays as nursery areas lies in the high biomass of small invertebrate prey organisms and the refuge from predators provided by the high turbidity and the habitat heterogeneity created by the vegetation and the shallow water. Also, the water is shallow and hence warms up more quickly, providing good conditions for growth. Movement into estuaries might also reduce the physiological stress due to osmotic differences between juvenile marine fish and their environment. Seawater fish have to drink water to make up losses due to a more concentrated external environment (Rankin and Davenport, 1981).

In view of the large number of juvenile fish found in these habitats, including coral reef species in mangroves, they could be extremely important in the maintenance of commercial offshore stocks of fishes (Gunter, 1967; Haedrich, 1983). For example, it has been argued that five of the six most important commercial fishery species in the USA may be dependent on estuaries (Smith *et al.*, 1966). On the Atlantic coast of the USA, there are extensive estuaries, especially in the south, and these habitats may be responsible for at least half the commercial landings each year (Day *et al.*, 1989). In New South Wales, Australia, the estuary-dependent part of the total fish catch is estimated at 66% by weight and 70% of the value of A\$27 million (Knox, 1986). In the North Sea, the Forth Estuary, Scotland, supports large numbers of juvenile and overwintering fish and it has been estimated that this estuary supports 0.54%, 0.45% and 0.05% of the stocks of similar-sized plaice, cod and herring respectively (Elliott *et al.*, 1990). Although these figures seem small, if similar values apply to other North Sea estuaries then together they could account for most of the juvenile phase of the major commercial fisheries. The larger estuarine systems such as the Wash and the Wadden Sea seem particularly important (Beek *et al.*, 1989; Hovenkamp, 1991; Kerstan, 1991).

Whilst the migrations of a variety of vertebrate consumers, including grey whales and walruses, regularly take in inshore waters for feeding, the best data on consumption rates of shore visitors come from studies on shorebirds. Every year enormous numbers of waders and wildfowl visit sheltered sandy beaches and mudflats during their autumn and spring migrations. Not all areas are used for the same purpose. Some estuaries are used as overwintering areas, others as staging posts during migration, others for breeding and yet others as cold-weather refuges. Furthermore, different species or populations of shorebirds may use the same estuary for different purposes. The impact of these birds on intertidal invertebrates and macrophytes varies greatly from site to site (Chapter 4). For instance, in salt-marsh and seagrass beds, estimates of the amount of plant material consumed by wildfowl range from 1% to 50% of the standing crop (Table 6.4). Similar estimates of the amount of the invertebrate standing crop taken by waders vary from 6% to 44% (Table 6.5), whilst seaducks may be major predators of rocky shore invertebrates over the winter (Mathieson *et al.*, 1991). When shorebirds leave these areas, for other feeding grounds or for terrestrial breeding areas, they take with them the material and energy consumed on the shore.

Table 6.4 Impact of grazing wildfowl on macrophyte standing crop*

Macrophyte	% Consumed
Zostera marina	1
Ruppia maritima and *Zostera marina*	1–25
Ruppia cirrhosa	20
Zostera spp.	30–75
Potamogeton spp.	13
'Submerged aquatic plants'	50

*After Thayer *et al.* (1984)

Table 6.5 Consumption by shorebirds of invertebrate production*

Site	% Production consumed
Gravelingen estuary	6
Waddensee intertidal	17
Langebaan lagoon	20
Ythan estuary	36
Tees estuary	44

*After Baird *et al.* (1985)

6.4 ENERGY FLOW AND FUNCTIONAL INTERACTIONS ON SHORES

The main energy flows shown in Figs 6.2 and 6.4–6.9 show that some pathways are more important than others in the transfer of energy between different trophic compartments. Thus, in kelp beds and the rocky intertidal, POM in the form of suspended detritus and phytoplankton is the major route of material flowing up the food chain to fuel higher trophic levels on the shore. In contrast, the amount of kelp or seaweed directly cropped by herbivores is probably less than 10% of the primary production. Yet experimental manipulations of grazers often reveal that these relatively minor consumers play a keystone role in controlling the dynamics of these systems (Chapter 4). Disruption of the trophic links between sea urchins (or limpets) and their macroalgal food can result in dramatic effects which cascade through the entire system. The scale of these effects is out of all proportion to the amount of energy flowing along the urchin–kelp link (Fig. 6.15).

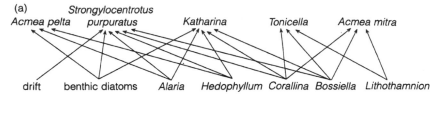

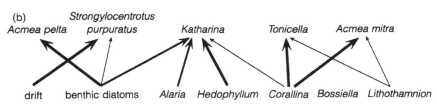

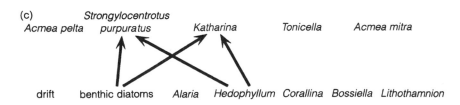

Fig. 6.15 The importance of trophic interactions in a North American Pacific coast rocky shore food web as gauged by (a) energy flow, (b) binary linkage and (c) derived from experimental manipulation (adapted with permission from Paine (1980) *J. Anim. Ecol.*, **49**.)

Energy flow studies and manipulative field experiments should be seen as complementary approaches to the study of the shore as a system, but they may not always provide similar answers about the functional importance of a particular organism! Finally, a note of caution. A lot of effort can go into constructing energy budgets that might only apply to a particular phase in a dynamic shore community, or with a particular population structure of a key organism that may change in space or time. Construction of the budgets becomes an end in itself and all sorts of assumptions and short cuts are taken to pursue this aim. Constructing budgets can sometimes be at the expense of real understanding of the system and methodological constraints may be so great as to render any comparisons between systems sketchy at best.

OVERVIEW

The ecology of shores is strongly influenced by events offshore. Like many interface zones, coasts are highly productive. Much of this production passes through the detrital pathway, paralleling many terrestrial and freshwater systems. This contrasts with open sea systems where production is channelled through herbivorous copepods and by the microbial loop which largely processes dissolved organics. In coastal waters, microbial processing of detritus is particularly important in making fragments available to higher trophic levels, particularly suspension feeders. The shore also receives considerable material from the shallow subtidal in the form of stranded detritus originating from kelp beds on both rocky and sediment shores. Macrophyte-dominated communities (seaweed beds, mangroves, salt-marshes) also export material to nearshore waters. Intertidal organisms export larvae offshore and the recruitment patterns upon their return are strongly influenced by nearshore coastal hydrography.

The intertidal is used as a rich feeding ground by crustaceans and fish and by adult shorebirds. The shore plays an important role in the migrations of some of these species and, in the case of nursery grounds, a crucial phase of their life history is spent there. The pathways of energy in intertidal communities can be influenced by key consumers, the removal of which can lead to shores switching from being net producers to net consumers.

Human impact on the shore

7

Humans have a long history of exploiting the coast for food, transport and as a place to live. Clear evidence of occupation and usage is provided by large prehistoric middens of shellfish (Meehan, 1982) and other marine remains, both at remote sites and in the heart of waterfront developments of modern cities (Cunliffe and Hawkins, 1988). A typical example is at the mouths of the rivers Plym and Tamar in Devon (UK). Here, there is extensive evidence of seminomadic occupancy in the Stone Age, followed by Bronze Age and Iron Age settlements (Cunliffe and Hawkins, 1988). There is also a Roman port which after the Dark Ages became one of the leading British trading and naval ports of the late medieval and Tudor period. Plymouth has remained an important port-city to the present day: coastal fishing boats, dockyards, the merchant navy and the Royal Navy, tourists, recreationalists and marine biologists all use the complex of creeks, soft shores, rocky shores and estuaries entering Plymouth Sound. This pattern of progressive exploitation is seen in many coastal areas throughout the world.

In this chapter, the various human impacts on shores and estuaries are broadly covered, and selected case studies are considered in more detail. First, we discuss the age-old collecting of shore plants and animals. Next we consider the chronic effects of pollutants, focusing on organotin leachates from antifouling paints and domestic sewage pollution. This is followed by an account of more acute incidents including detailed assessment of the consequences of catastrophic oil spills. We then go on to briefly discuss introductions of alien species. Finally, we consider the broader implications for coastal ecology of coastal development and land reclamation in relation to predicted sea level rise in response to global warming.

7.1 COLLECTING MARINE PLANTS AND ANIMALS FOR FOOD, BAIT AND CURIOS

7.1.1 TARGET SPECIES

In prehistoric times vast numbers of shellfish were gathered by people, at first without tools, then using simple devices such as suitably shaped stones. Large numbers of nearshore and migrating fish in estuaries and lagoons have also been caught since the earliest times (Meehan, 1982), using harpoons, bows and arrows, hooks and lines, traps and nets. Both plants and animals have been pressed into service to capture fish. South American Indians use various plant extracts to stun or kill fish, such as rotenone which clogs the gills. In Japan, cormorants have been traditionally used to capture coastal fish, their throats constricted by binding to prevent them swallowing.

A heavy dependence on foraging for shellfish on the shore was typical of many of the semi-nomadic indigenous peoples encountered by Europeans for the first time in the 16th–18th centuries, for example the bushmen, called 'strandlopers' by the Dutch at the Cape, Native Americans and Native Australians (Meehan, 1982; Alfredson, 1984). Collecting of shellfish was, and often still is, an important element of the activities of peoples with permanent settlements who also fished and farmed, such as the Polynesian cultures and Native Americans. In many cases sea shells or their products (e.g. wampum in north-east America, cowries in Polynesia) were used as currency. In Europe, widespread subsistence collecting of shellfish was common from prehistoric times until the late Middle Ages.

In northern Europe relatively few species are collected today, mainly winkles (*Littorina*) and cockles (*Cerastoderma*). Most mussels and oysters are cultured in some way. In France, Spain, Portugal and the Atlantic Islands (Canaries, Azores, Madeira), many species are avidly collected for food, including sea urchins, stalked barnacles (*Pollicipes*), large acorn barnacles (*Megabalanus* in the Atlantic Islands), various bivalves, sea cucumbers, octopus and a variety of crabs. Limpets, one of the major constituents of the archaeological middens throughout Europe, are no longer eaten much, except in the Atlantic islands, where they are highly prized. They are also heavily exploited on many of the Mediterranean islands and the Latin passion for 'mariscos' has been exported to South America. 'Opihi' are prized in the Hawaiian Archipelago. In North America, European settlers acquired their taste for clams from the native Indians.

Elsewhere in the developed world, many shore animals are now gourmet items on the menu, especially in Japan and Korea, and shellfish collection has switched from being a subsistence activity to a highly profitable commercial operation purveying luxury items. For instance, in Victorian times in the UK, oysters were the food of the poor; recipes for steak and kidney puddings included oysters as a padding to save on the more expensive meat and offal. Now oysters are only included in puddings in the most up-market restaurants. Meanwhile in poor countries the relentless pressure on food resources by a growing population has increased the exploitation of the shore considerably.

Collection of seaweed for food is widely practised in the Far East, where many species are also now cultivated. In Europe certain seaweeds have traditionally been eaten, such as *Palmaria*, *Chondrus*, and especially *Porphyra* (laver). This market is now being developed with the emphasis on health foods, but most seaweed is collected or cultured for industrial processes. Red algae, such as *Gelidium*, *Gracilaria* and *Pterocladia* are used in the manufacture of agar (Armisen and Galatas, 1987), and *Chondrus* and *Euchema* are used for carrageenans (Stanley, 1987). Kelps and wracks are harvested to make a variety of products (alginates) where their long-chain sugars have useful properties. Alginates end up in a huge range of cosmetics, pharmaceuticals and processed foods: beer, ice cream, face cream and shampoo all contain some seaweed. In western countries these seaweeds are not yet cultivated on a scale large enough for industrial uses, although commercial-scale cultivation is under way in China, Japan, Korea and elsewhere in the Far East. Cultivation of *Laminaria* in this region is almost always for food, although in China a portion of the crop which is not of high enough quality is used in the manufacture of alginates (Tseng, 1987). Hand harvesting and collection of drift seaweeds deposited on beaches is being replaced by mechanized harvesting, allowing large areas of the shallow subtidal and intertidal to be industrially exploited; this is the case for *Laminaria* and *Ascophyllum* in Norway, *Ascophyllum* in Maine and *Macrocystis* in California.

In addition to the direct collection of food, marine invertebrates are extensively collected as bait for commercial and recreational fishing. Sheets of mussels are removed for the worms underneath them, boulders are overturned for soft crabs, and extensive areas of sand and mudflats are dug for worms.

Collection of ornamental shells and other shore animals as curios has become an important source of income in many parts of the world. In addition to local sale to tourists (Newton *et al.*, 1993), this trade has developed into a major export business in some regions, such as East

Africa, the Philippines, Indonesia, Thailand and India (Wells, 1981; Wood and Wells, 1988).

7.1.2 COMMUNITY CONSEQUENCES

The harvesting of shore plants and animals, whether for food, bait, curios or industrial use, not only affects the target species, but may have unforeseen consequences for the rest of the community (Wells and Alcala, 1987). Under low-to-medium-intensity harvesting regimes the classical signs of 'growth overfishing' are apparent, seen as a reduction in size as target species are removed faster than they can grow. Under very heavy exploitation, the population can be reduced to the point where it cannot reproduce enough to sustain itself – so-called 'recruitment overfishing'. Populations of externally fertilizing sedentary species are probably reduced below the density threshold at which successful fertilization can occur. In some protandrous species, such as limpets, removal of the larger females will seriously distort the sex ratio and reduce the output of eggs by the population. Isolated islands, like the Azores and Hawaii, seem particularly prone to recruitment overfishing because much of the reproductive output is probably dispersed away from the islands.

Individual species can be so severely exploited that they become extinct in parts of their range. For example, the sea urchin, *Paracentrotus lividus*, was formerly common along much of the north French coast, but more recent surveys in the 1970s and early 1980s suggest that it has become rare in this region. This could be partly due to climatic changes because this species is near its northern limit and colder weather during the 1970s and 1980s may have reduced recruitment. However, a likely contributing factor is severe exploitation of urchins for their gonads, which are a delicacy in France. In East Africa, cowries which have ornamental value are scarce on shores where shell collecting is intensive, but are abundant at locations that are more remote and less populated (Newton *et al.*, 1993). In the Canaries the large intertidal limpet *Patella candei candei* (also called *Patella crenata*) has been overexploited to the point of localized extinction. Now, it is only abundant on one of the less populated islands (Fuerteventura) and the uninhabited Salvage Islands. It is, however, found in middens left by the indigenous Guanche people throughout the archipelago, prior to Spanish colonization in the 14th century. *Patella ferruginea* is also locally extinct in parts of the Mediterranean and rare where it does occur. This fate nearly befell the limpet *Patella aspera* on certain islands in the Azores, which showed signs of severe recruitment failure. Fortunately, a recent ban on collecting in parts of the archipelago has probably saved this species.

Overfishing is not a recent phenomenon. In ancient middens the average size of specimens tends to get smaller with increasing length of occupation of the site. Additionally, there is usually a shift from large, favoured species to smaller, less favoured ones as localized extinction or overexploitation occurs. Sometimes there is a shift in the species assemblage as people forage further away from base. Usually this shift is from species typical of sheltered areas to those of more exposed coasts.

Removal of a particular species, or set of species, can have knock-on effects for the whole community. Removal of space-occupying species such as stalked barnacles, acorn barnacles, mussels, or large tunicates such as *Pyura* (taken for bait) will open up free space and start successional sequences (Chapter 4). Removal of sedentary grazers such as limpets will also start small-scale successional events under light exploitation, but quite widespread colonization of algae will occur after heavy exploitation. Removal of predatory gastropods (for food or curios) or starfish (for curios) will increase the frequency of their prey items such as barnacles or mussels. Competitive interactions will also be distorted: less favoured species such as the gastropod *Siphonaria* will flourish due to the reduction of preferred human food items, like *Fissularia*, which are competitive dominants (Oliva and Castilla, 1986).

Removal of dominant canopy algae, such as fucoids and kelps, from the middle of high-latitude shores and around low water at lower latitudes will have marked effects on the community (Benedetti-Cecchi and Cinelli, 1992a,b). In the absence of the canopy, successional events will be initiated and it may take several years for recovery to occur. The rate of recovery will partly depend on the size of the patch and the type of harvesting regime (Foster and Barilloti, 1990; Schiel and Nelson, 1990). If the plant is trimmed back and still has the capability of vegetative proliferation, as in *Ascophyllum*, regrowth will be rapid within a year or two (Sharp, 1987; Sharp and Pringle, 1990). If the basal clumps are disrupted, recolonization by new propagules will be extremely slow. The giant kelp, *Macrocystis*, which is mechanically harvested in California, has several meristems throughout its length because of its branching growth pattern. This allows regrowth of existing plants after harvesting. The space opened up in the canopy will also allow the rapid growth of juvenile plants the growth of which is retarded due to shading by adults (North, 1968). There seem to be few ecological effects of harvesting *Macrocystis* at present (Barilloti and Zertuche-Gonzales, 1990). Harvesting of laminarians, such as *Laminaria hyperborea*, removes the blade and part of the stipe including the meristem so that recovery of the population can only occur by settlement of propagules. These come from the tiny gametophyte phase scattered throughout the kelp forest. Recolonization to a mature community will take about 2 years, on the basis of trial, small-scale clearings (Kain, 1975), but will take longer in larger plots.

Removal of individual species for bait, whether by stripping off sheets of mussels or colonial animals on rocky shores, or by digging on beaches, causes severe physical disruption of the habitat. Similar effects occur with clam-digging for food, but where animals are raked from the sediment, as has been traditionally done for cockles, the effects will be less severe. Recently, concern has been expressed about the use of suction-dredging for the collection of shellfish and to a lesser extent calcified seaweeds (maerl), where the sediment is liquefied before sucking out the plants and animals (Cook, 1991).

In mud and sandflats much of the oxygenation of the sediment occurs through bioturbation, that is, the burrows and burrowing activity of the larger infauna (Chapter 4). Bait digging and suction dredging will destroy burrows and many of the larger burrowers. Overall the effect is to reduce the abundance of the target species. Because these are usually large and long lived, recolonization can be slow. However, the rest of the community may recover quite quickly, especially if sediment transport is high (Hall *et al.*, 1990a).

Exploitation of top predators may have had long-lasting consequences for shore assemblages. The removal of sea otters in North America could well have significantly affected sublittoral kelp forests through their interactions with sea urchins (Chapter 4). The expansion of sea otters into California seems to have had a dramatic impact on populations of the Pismo clam, *Tivela stultorum*, which otters excavate by digging pits (M.J. Foster *et al.*, 1991). On Chilean shores, the predatory whelk *Concholepas concholepas* has long been subjected to strong fishing pressure. Establishment of a marine reserve has allowed the survival of larger size classes, with dramatic effects on the mussel *Perumytilus purpuratus*, which is now restricted to a narrow refuge zone at the top of the shore (Castilla and Duran, 1985). Overfishing for lobsters and fish may have played a role in the overgrazing (by urchins) of kelp forests in the North Atlantic (Chapter 4).

Surprisingly few quantitative or experimental studies have been made of the effects of human exploitation on shore communities. This is largely because in areas with heavy exploitation of shore populations, there are few unexploited marine nature reserves to allow comparison (but see Chilean example above).

Recent work in New South Wales (Kingsford *et al.*, 1991) has quantified the extent of human usage of shores for largely recreational purposes, which include collecting invertebrates for food or bait. This work showed considerable spatial and temporal variation of exploitation when examined on various scales. More people affected the shore during school holidays than during term time, but surprisingly, there were no differences between weekends and weekdays. There was greater usage of the shore on sunny than on overcast days. This would have implications for organisms affected by activities such as boulder turning, as they would be exposed to even greater desiccation stress.

Work in southern Africa has shown a generalized convergence in the characteristics of exploited communities at various sites, whilst unexploited communities have more divergent characteristics (Hockey and Bosman, 1986). In every case, the population structure of exploited species is shifted markedly to smaller size classes. The numbers of grazers and the cover of exploited space-occupying species such as oysters decreased in exploited compared with unexploited sites. Non-exploited species, such as barnacles and ephemeral algae, increased in abundance.

Trampling during recreational activities has effects similar to human predation but is unselective (Hockey and Bosman, 1986; Bally and Griffiths, 1989; Povey and Keough, 1991). Damage is usually slight and is mostly restricted to the animals such as the more fragile barnacles (Bally and Griffiths, 1989). In Australia the brown alga *Hormosira banksii* was easily damaged, with over 20% of the biomass of an individual being removed by a footstep (Povey and Keough, 1991). In an experimental study near Cape Town, short-term effects on barnacles, amphipods and isopods were seen, but in the long term the shores were indistinguishable from untrampled control areas (Bally and Griffiths, 1989). In Australia, Povey and Keough (1991) showed that algal mats, whether of *Hormosira* or coralline algae, were damaged; this indirectly increased the abundance of molluscs. Povey and Keough (1991) suggested that disturbance by trampling could tilt the balance from an assemblage of *Hormosira* and other algal mat species to barer patches dominated by molluscan grazers. More work needs to be done on the effects of trampling as leisure and tourism increases and more people use the shoreline for recreation.

7.2 CHRONIC POLLUTION

Chronic pollution is usually associated with industrial or domestic effluents and, as will be seen in the more detailed case studies, antifouling leachates. Changes in run-off and water catchment characteristics due to urbanization and intensive agriculture result in elevated levels of nutrients (leading to eutrophication) and of pesticides in estuaries and coastal waters. Suspended solids due to storm water discharges, dredging for navigation, and industrial activities such as unloading bulk carriers, can also be a major problem in some locations (review: Clarke, 1992).

7.2.1 HEAVY METALS

Although considerable attention has been given to body burdens of contaminants and sublethal effects, very few studies have examined the consequences of heavy metal pollution at the population or community level. In areas affected by industrial discharges and particularly run-off from past and present mining activities, very high levels of a variety of toxic metals can be found. Some, such as zinc and copper, can be partitioned off from the active metabolism by storing in various granules, as in barnacles (Pullen and Rainbow, 1991) and littorinids (Nott and Langston, 1989). Another method of detoxification is by the formation of metallothioneins. These are low-molecular-weight proteins which form a complex with the metal, thus binding it and reducing its toxicity. The formation of metallothioneins can be induced by exposure to low concentrations, and has been found in many intertidal animals (Bebianno and Langston, 1991, 1992a,b).

There is much evidence for the tolerance of heavy metals in populations at contaminated sites (Bryan and Hummerstone, 1971). Tolerance can be acquired by prior exposure, but localized selection can also occur in species with a restricted dispersal, leading to tolerant strains. The acquisition of genetically based tolerance is unlikely to occur in many animals with broadly dispersed larvae because there is little opportunity for localized selection in populations. Of the shore species investigated, *Fucus vesiculosus* and various algae in highly contaminated environments are capable of developing resistant strains. *Hediste (Nereis) diversicolor* has also been shown to have inherited tolerance (Hately *et al.*, 1989), and this may also occur in *Littorina saxatilis*. In species having widely dispersed larvae, such as the shore crab *Carcinus maenas*, and the polychaetes *Nephtys* spp. and *Glycera convulata*, tolerance in contaminated areas is thought to be a phenotypic response.

One of the few kills of shore animals that can be attributed with certainty to heavy metal poisoning was the death of over 2000 birds, mainly dunlin *Calidris alpina*, in the Mersey estuary, UK, in 1979, and again in smaller numbers the following year (Bull *et al.*, 1983). The dead birds had more than 10 ppm lead by dry weight in their livers, with 30–70% in the form of tri-alkyl lead. The source of this lead was factory effluent, which reached the birds through *Macoma balthica* (1 ppm) and *Nereis diversicolor* (0.2 ppm), on which the birds had fed.

7.2.2 COOLING WATER

Another chronic impact which may influence shore communities is the cooling water effluent from power stations (review: Clarke, 1992). This is usually several degrees warmer than ambient water temperatures and may contain chlorine used to prevent fouling organisms clogging water channels. Effects are usually restricted to the immediate vicinity of the outfall. Brown seaweeds such as *Ascophyllum* and *Fucus* were eliminated from a rocky shore heated to 27–30 °C by a power station in Maine, whilst the ephemeral species *Enteromorpha intestinalis* increased significantly near the outfall. On another thermally affected shore in Maine, however, few overall changes were found, except near the discharge where changes in *Ascophyllum* and *Fucus* were detected. Alteration to the discharge to enhance dispersion of the heated water resulted in the full recovery of *Ascophyllum*, but this may have in turn displaced *Fucus vesiculosus*, which disappeared (Vadas *et al.*, 1976).

Impacts in tropical waters tend to be greater, perhaps because the species are more stenothermic and an increase in temperature can push some species above their lethal limit. Large areas of seagrass beds have been killed in this way by discharges from nuclear power plants at the Turkey Point installation in Florida, and the smaller areas of coral present were similarly affected (Thorhaug *et al.*, 1979; review: Clarke, 1992). Puerto Rican mangroves have also been shown to be adversely affected by thermal pollution (Banus, 1983). However, the importance of complicating factors such as chlorine in the water is not clear, although Jokiel and Coles (1990) showed that corals could be killed by elevated temperatures alone.

7.2.3 HYDROCARBONS

Discharges containing hydrocarbons from oil refineries have considerable effects on intertidal communities. One of the best-documented examples is the effects of the Fawley oil terminal near Southampton on the adjacent salt-marsh (Dicks and Hartley, 1982; review: Clarke, 1992). Between 1953 and 1970, oil and chemicals in the refinery effluent, plus small accidental discharges from the onloading terminal and the plant, caused a loss of vegetation from a large area of the salt-marsh. In 1970, steps were taken to improve the effluent. By 1975 the hydrocarbon content had been reduced to one-third (31 ppm to 10 ppm), and the discharge rate by one-quarter (28 000 to 21 000 m³ h⁻¹). Vegetation had recolonized much of the area by 1980, but the sediments remained contaminated. An impoverished fauna, mainly oligochaetes, was present in the lightly contaminated areas, but even these were rare. The common estuarine polychaete *Hediste (Nereis) diversicolor* was also much reduced in numbers (Dicks and Hartley, 1982).

7.2.4 SEWAGE AND EUTROPHICATION

During the 19th century, improvements in sewage disposal helped eliminate many of the great epidemic diseases, such as cholera and typhus, which were common in the industrial cities of Europe and North America. Most of the new sewers ran untreated into nearby rivers, lakes or onto the shore. Sewage treatment methods have been developed since then and are widely used on plants discharging into inland waters. Unfortunately, it is only in recent years that the practice of discharging raw sewage directly into the sea has been questioned. In many countries, including the UK, raw sewage may be discharged onto the shore or to just below the low water mark. Large amounts of sewage solids can increase sedimentation and water turbidity, nutrients can be locally elevated, and there can be considerable **biological oxygen demand** (BOD). Slowly biodegradable and nonbiodegradable solids, such as tampons and contraceptives, litter the foreshore and end up in the strandline. Industrial effluents are often mixed with domestic wastes so that elevated levels of heavy metals and other persistent contaminants can occur. The most severe effects of domestic pollution are in semi-enclosed water bodies such as estuaries and sheltered bays. Usually sewage is only one contributor, with nutrients also coming from agricultural run-off (Raffaelli *et al.*, 1989). High discharges can lead to bottom waters becoming anoxic, or an anoxic 'plug' of water oscillating with tides up and down the estuary.

The organic material and inorganic nutrients in domestic sewage usually increase the secondary production of the shore (Niemi and Warheit, 1989; Brown *et al.*, 1990). On many sand flats and mudflats, this may be revealed by a shift in species composition towards polychaetes and oligochaetes. These species have appropriate feeding mechanisms and life-history characteristics similar to those documented from stressed sublittoral communities (Pearson and Rosenberg, 1978). However, the effects of enrichment are usually seen earlier and more obviously in meiofaunal populations (Warwick *et al.*, 1990b; Raffaelli,

1992). This enhanced production may support large numbers of migrating shorebirds and compensate for the loss in feeding habitat from land-claim and barrages. As a consequence, improvements in sewage treatment and disposal may lead to a decline in shorebird numbers (Van Impe, 1985; Bryant, 1987; Raffaelli, 1992).

Whilst moderate organic enrichment might be seen as having positive effects on higher trophic levels, nutrient enrichment by sewage wastes can also stimulate blooms of opportunistic benthic macroalgae, especially the greens *Enteromorpha, Cladophora* and *Ulva* (Knox, 1986). Moderate growths of these algae provide an additional habitat dimension, a food resource and can stabilize sediments, but dense algal mats have catastrophic effects on the underlying invertebrate assemblages through deoxygenation of the sediment. Extreme blooms of benthic algae are likely to lead to a decline in numbers of birds and fish able to utilize sand and mudflats.

On rocky shores with greater water movement, the effects of even the largest sewage outfalls are localized to a few hundred metres of the outfall at most. In many of the smaller discharges, ecological effects are extremely limited, usually no more than 10 m or so either side of the pipe. However, unsightly sewage-derived plastics will disfigure the shore for quite a large area. Pollution effects are most pronounced in enclosed bays, harbours and estuaries. Here, residence time of the water is high and pollutant dispersal is dependent mainly upon tidal cycles to carry the effluent offshore. In some areas, there are problems with the discharge of effluent from fish farms leading to localized eutrophication of coastal waters (Wallin and Haakenson, 1991). Because of this it is sometimes difficult to distinguish between the effects of sewage outfalls, river-borne discharge such as fertilizers, and natural nutrient run-off, although the combined effect is often a flush of benthic macroalgae in areas such as the Brittany coast (mainly *Ulva* sp.), and Venice lagoon (*Ulva, Gracilaria*).

Although the ecological impact of sewage is probably limited on most shores, there can be significant public health problems. Filter feeders, such as mussels and clams, accumulate and concentrate sewage-derived pathogens and many beds of shellfish cannot be exploited, or can only be eaten after purification for 48 hours or more in sterilized sea water (depuration). Until recently, it was thought that bathing in sewage-contaminated waters presented few risks unless the contamination was gross and visible. Human pathogens were thought to die rapidly under the combined action of high-salinity sea water and ultraviolet radiation. However, many coastal areas have a high turbidity which provides protection from UV radiation, and recent epidemiological studies have shown that the incidence of irritating complaints such as various ear, nose and throat infections and gastro-enteritis are greater if bathing in contaminated water has occurred. These complaints may cause only a few days' illness, but they can ruin a holiday. Debate continues about whether life-threatening diseases can be contracted by bathing in sewage-contaminated sea water. In particular, the persistence of viral pathogens in the environment is not well understood and is the cause of legitimate concern.

One strategy to clean up beaches is to lengthen outfalls and discharge sewage at a considerable distance offshore. Gross solids are usually screened out first (primary treatment). If dispersal is good and winds and currents do not bring material back onshore, this can be a viable option, although there will be some localized effect on the benthic communities around the outfall. Problems have arisen when pipes are of inadequate length or are inappropriately sited, allowing transport of sewage back onshore. Increasingly, the option of treatment to reduce suspended solids and the BOD of the discharge is being adopted (secondary treatment). Tertiary treatment to further remove nutrients and reduce BOD is the ideal solution, although this is a costly process.

7.2.5 ANTIFOULING PAINTS

Scientists have long known that leachates, such as copper, from antifouling paints cause contamination in port areas, but public concern only mounted with growing evidence about the deleterious effects of paints containing organotins. Tributyltin (TBT) and its breakdown products are toxic to a wide variety of marine life (Bryan and Gibbs, 1991), and sublethal effects have been shown to occur at extremely low concentrations, down to nanograms (10^{-9} g) per litre. Unfortunately, concentrations in inshore waters can be as high as 40 ng l^{-1}, and in marinas and harbours may reach 880 ng l^{-1} (Cleary and Stebbing, 1985). Concentrations are particularly high in the water surface microlayer due to the lipophilic nature of TBT and the high concentrations of rich organics and lipids found there. Concentrations can be 27 times as high as in immediate subsurface waters, and even 200 km offshore in the North Sea, levels of up to 1.2 ng l^{-1} have been recorded in this microlayer (Coghalan, 1987). TBT also becomes concentrated in sediments, where it binds to suspended particles and levels in excess of 1000 μg l^{-1} have been found in estuaries in South West England (Langston et al., 1990). In the mid and early 1980s, much of the TBT in inshore waters came from paints applied to yachts and other recreational craft. Earlier versions of TBT-based paints were of the contact-leaching variety, with an exponentially decaying release of biocide into the water. Large pulses of TBT in the water column were typical of spring and early summer when boats were first placed in the water. Second-generation TBT paints are self-polishing co-polymers, where the paint constantly sheds its outer surface, presenting a smooth new surface to the water. Although there is a rapid first release of biocide this is followed by a more constant release over the next 5–7 years. TBT-based paints were widely used because they were very effective: the reduction in drag by use of these paints was estimated to save the US Navy alone nearly 2 million barrels (about 90 million US gallons) of fuel a year.

It is not surprising that a biocide designed to kill marine fouling organisms will have an ecological impact. Formerly, however, most such effects would be restricted to industrialized estuaries and ports, where they would be masked by effects arising from the plethora of other pollutants, including an extremely high BOD. The expansion of leisure boating and marinas in recent years has widened the effect to previously clean estuaries, inlets and bays. On indented coastlines, whole stretches of inshore water can be affected, for example the English Channel. When boats were present in high numbers in enclosed waters, TBT levels in the late 1980s were high enough to kill both commercial and noncommercial species, as well as to affect the growth and reproduction of others.

Early attention focused on death and deformities in oysters (Alzieu *et al.*, 1986), which had disastrous consequences for the aquaculture industry. Subsequently it emerged that stenoglossan molluscs, including various whelks that are important predators on rocky shores, were particularly susceptible to TBT (Bryan *et al.*, 1986; Gibbs and Bryan, 1986). Very low concentrations cause the expression of the phenomenon called 'imposex', the super-imposition of male characteristics on the female snail. This involves the development of a penis-like outgrowth in the female and the formation of a vas deferens which proliferates to block the female genital duct and effectively sterilize the female. Blockage of egg release eventually leads to the death of the female. The degree of imposex in dogwhelks is significantly correlated with TBT concentrations. Imposex is initiated at levels of < 1 ng l^{-1}; blockage of the female duct and effective sterilization of some females occurs at 1–2 ng l^{-1} and is found in all females in areas with concentrations averaging 6–8 ng l^{-1}. Females eventually disappear from these shores. Imposex has been recorded in over 45 species of neogastropod molluscs worldwide (Ellis and Pattisina, 1990), although sterilization has been shown in only a few of these.

In the UK, dogwhelks are now absent or very rare in many ports and harbours. Even a small number of boats can lead to their virtual disappearance or a population dominated by old males. *Nucella* lays egg capsules directly on to the shore and has no larval dispersal stage so that recruitment into affected areas from remote sources by rafting of juveniles is low. Because colonization can only come from exposed populations with different genetic characteristics (including chromosome complement), any juveniles reaching more sheltered, affected shores may have a shell shape and growth strategy unsuited to sheltered conditions where crab predation is high (Chapter 5). This is likely to reinforce the effect of failed reproduction at affected sites. Thus, even where TBT paints are now banned, recovery will be slow.

Many studies have shown the importance of whelks as intertidal predators (Chapter 4); they can have a profound effect on the population structure and dynamics of barnacles and mussels, their main prey items. A reduction in their abundance could lead to denser barnacles, which in turn would be expected to increase the likelihood of escapes of algae from grazers. There is no firm evidence yet for this supposition, but research is under way.

These responses to very low contaminant concentrations make dogwhelks a superb indicator species for TBT. Examination of whelks is also cheaper and easier than chemical monitoring methods and dogwhelks are now being used to monitor recovery in many countries which have banned TBT paints on small boats. In coastal waters of Europe and America, TBT levels still give cause for concern. In the UK, TBT paints can still be used on ocean-going ships over 25 m in length, which spend little time in port, and many countries have yet to ban their use. Nevertheless, the fate of the sex-life of the humble dogwhelk caught the imagination of the general public and politicians alike, and helped in putting sensible legislation in place in many countries in the late 1980s.

7.3 ACUTE EFFECTS ON SHORES

7.3.1 BLOOMS OF TOXIC ALGAE

Red tides of toxic dinoflagellates and other algae are natural events that can cause widespread mortality of animals on both hard and soft shores. Plants seem less affected and the overall effect is usually to diminish the abundance of space-occupying animals, such as barnacles and mussels, and grazers, leading to greater algal domination on affected shores (Southgate *et al.*, 1984). There is, however, increasing evidence that human-caused eutrophication of coastal waters may be responsible for some 'red tides', including the severe blooms of *Gyrodinium aureolum*,

Ceratium spp. and *Chrysochromulina polylepis* in the North Sea and Skagerrak in 1988 which devastated inshore fish, shore invertebrates and aquaculture operations (Smayda, 1989, 1990; Lindahl, 1993).

7.3.2 DUMPING

Dumping of waste on the shore has an acute and often irreversible effect. Various industrial and mining wastes have been dumped in the past including colliery waste (Limpenny *et al.*, 1992). These coarse particles tend to scour the shore, killing hard substrate organisms, but they are too large and unstable to be colonized by sediment dwellers. In many cases, waste has been used to infill the shore for reclamation behind sea walls. An unusual form of dumping, which can be seen along the Cumbrian coast in the UK, is of blast furnace waste (Perkins, 1977). This was dumped hot and molten straight onto the shore in the early days of the Industrial Revolution. It solidifies as an interesting conglomerate resembling concrete but containing lumps of iron, firebricks, and industrial debris. Its properties vary from place to place, from a friable material which is easily eroded to a very hard material with many small depressions. A reasonably normal rocky shore flora and fauna has developed on this hard 'slagcrete' which in places stabilizes the impoverished boulder shores typical of most of this stretch of coast. It is, however, the only rusty rocky shore that we have ever seen!

7.3.3 OIL SPILLS

Whether from wrecked ships, or from accidents on offshore or land-based installations, oil spills are probably the most dramatic marine pollution incidents. The consequent high mortalities of sea birds and other wildlife and the devastation to littoral and nearshore areas give rise to grave public concern and have prompted considerable scientific attention.

Oil spills are the most frequent catastrophic type of disaster. Table 7.1 lists various spills to give an idea of the amounts involved and the proportion that comes ashore. The toxicity varies with the type of oil and the degree of breakdown at sea before it arrives on the shore. Physical smothering is often a more important cause of mortality than chemical poisoning, and clean-up operations often kill more marine life than the oil itself. Different habitats have different sensitivities to oil spills: exposed rocky shores are least sensitive, then come sandy beaches and gravelly or cobbly shores, followed by seagrass beds and finally by sheltered rocky shores, mudflats and salt-marshes, which are the most sensitive.

Table 7.1 Selected major oil spills*

Installation or ship	Date	Oil spilt (tons)[†]	Oil beached (tons)[†]
Persian Gulf	Jan. 1991	>1 million	?
Ixtoc 1	03 Jun. 1979	500 000	12 000
Amoco Cadiz	16 Mar. 1978	233 000	80 000
Torrey Canyon	18 Mar. 1967	100 000	35 000
Khark 5	19 Dec. 1989	70 000	?
Metula	09 Aug. 1974	51 000	42 000
Exxon Valdez	24 Mar. 1989	38 000	4 500
Urquiola	12 May. 1976	30 000	25 000
Aragon	30 Dec. 1989	25 000	?

*From Hawkins and Southward (1992), after various sources.
[†]One ton = 2240 pounds = 1016 kg.

On sediment shores, stranded oil can percolate down through the sediments and persist for a long time because low oxygen levels will inhibit its aerobic breakdown. Application of dispersants can make matters worse, driving hydrocarbon residues deep into the sediments. The retention of oil spreads toxic effects over a much longer time scale. For example, oil persisted in some sandy beaches for up to 5 years after the *Torrey Canyon* oil spill, and leaching from sediments a year after the *Amoco Cadiz* oil spill caused renewed contamination. After the *Torrey Canyon* and *Amoco Cadiz* oil spills, infaunal bivalves (particularly razor shells: *Solen, Ensis, Pharus,* and the cockle *Cerastoderma edule*) and heart urchins (*Echinocardium*) were badly affected. On the other hand, bivalves such as *Tellina* spp. and *Mya arenia* survived the *Amoco Cadiz* spill well. After the *Braer* oil spill in Shetland in 1993, subtidal razor shells were similarly badly affected. Strandline sand hoppers (*Talitrus saltator*) and the eulittoral amphipod *Ampelisca brevicornis* suffered very high mortality after the *Amoco Cadiz* spill whilst other amphipods and isopods were less affected. Polychaetes were remarkably resilient to the *Amoco Cadiz* spill, only *Lanice conchilega* was slightly affected. In the longer term, the biomass of opportunistic species, particularly capitellid worms, increased rapidly after the incident, peaking after one year before declining as resident species began their slower recolonization (Conan, 1982). A similar dominance by capitellid worms also occurred after an oil spill off Massachusetts, USA (Sanders, 1977), and at oil-polluted sites near oil terminals in the Gulfs of Elat and Suez (Amoureux *et al.*, 1978).

In salt-marshes, the light fractions of oil are the most toxic, but the heavier fractions and crude oil can completely smother the vegetation (Baker, 1979; reviews: Long and Mason, 1983; Clarke, 1992). Oil readily becomes trapped by both the vegetation and the sediments, and may persist for a long time. Annual species are most sensitive; either they are killed or their reproduction is severely impaired. Reseeding is required from remote sources and it can take 2–3 years before the populations recover. Shallow-rooted annuals, such as *Suaeda maritima*, and *Salicornia*, have little or no food reserves and are particularly susceptible. Other species (e.g. *Juncus maritimus*) will survive single light exposures, but frequent light oilings will kill them. Certain plants with underground storage organs, such as *Armeria maritima* and *Plantago maritima* are very resistant. Experimental applications of oil show that more than four successive oilings appeared to be the critical level at which most species succumbed. Eight to twelve oilings result in severe die-back and exposure of the underlying mud. As oil decomposes, nutrients are released and renewed luxuriant growth can occur in any surviving species. Rapidly growing mat-forming species, such as the grasses *Agrostis stolonifera* and *Agropyron pungens*, can proliferate, quickly out-competing more susceptible species. After the *Amoco Cadiz* (Conan, 1982), oily footprints were still apparent in salt-marshes 5 years after the spill and *Juncus* plants were still yellowish and weakened. It seems that recovery can take more than a decade in *Puccinellia maritima* and *Juncus maritimus* communities.

The *Torrey Canyon*: a case study of oil pollution

The *Torrey Canyon* spill was one of the first major oil pollution incidents to be documented by shore ecologists. Although its effects may have been less sensational than those of the *Exxon Valdez* spill in an ecologically sensitive and beautiful wilderness area, the detailed history of the recovery process on the shores affected by the *Torrey Canyon* spill is unparalleled (Smith, 1968; Southward and Southward, 1978; Hawkins and Southward, 1992).

The *Torrey Canyon* was carrying 119 000 tons of Kuwait crude oil to the refinery at Milford Haven, South Wales. At 0850 h on Saturday 18 March 1967, the ship hit the Pollard Rock, part of the Seven Stones off the south-west coast of the UK, while steaming on autopilot at a full speed of 16 knots. The ship's momentum was such that the bottom was damaged for more than half its length. Jettisoning of oil in an attempt to gain buoyancy, and breaching of the forward fuel and cargo tanks, resulted in 30 000 tons of oil being lost in the first few hours. A further 20 000 tons were lost over the next seven days during salvage operations, but these efforts were abandoned when the ship's back broke, releasing a further 50 000 tons. The remaining 20 000 tons on board the tanker were destroyed after ignition by aerial bombing.

The 30 000 tons spilt at the time of stranding and the subsequent 20 000 tons released during the week of abortive salvage operations reached the shores of Cornwall a few days later. The oil was deposited on very high spring tides (the highest for 50 years), and therefore was not washed away on subsequent high waters.

The bulk of the oil (50 000 tons), released when the ship broke up, began to approach the Cornish coast, but was swept by a northerly wind southwards towards France. Despite spraying at sea with dispersants, the oil reached Brittany a few weeks later where it was virtually continuous along 90 km of coast, forming deposits up to 30 cm deep. In contrast to the UK's preferred option of spraying with dispersants, in France treatment was mainly by removal (often by hand), although dispersant spraying was carried out in some localities at the urgent request of the region's tourist industry.

What clearly distinguished the *Torrey Canyon* disaster from previous and subsequent oil spills was not the quantity of oil that came ashore, but the volume and indiscriminate application of dispersants to remove oil from sandy beaches and rocky shores. Mortality caused directly by oil on the shore was limited. The toxicity of the oil was relatively low initially, and evaporation of the lighter fractions while the oil was at sea reduced it further. Thus, the main ecological consequences of the *Torrey Canyon* oil spill in Cornwall were due to the intense use of dispersants during the spring of 1967. Over 10 000 tons of dispersants were applied to the 14 000 tons of oil that came ashore in Cornwall.

The first-generation dispersants available in 1967 proved highly toxic to marine life, but at the time of application this was not well known. Subsequent laboratory studies showed a 24-hour LC-50 (the concentration at which 50% of the test organisms die) varying from 0.5 to 5 ppm on sublittoral organisms, and from 5 to 100 ppm on intertidal organisms, with the limpet *Patella* being highly susceptible. These lethal concentrations were much lower than those needed to disperse the oil, and consequently all animals and many algae were killed in areas of the shore close to dispersant spraying.

In addition to their toxicity, dispersants are now known to alter the physical properties of sandy beaches and this is thought to account for the majority of observations made on beaches at the time (Webb, 1991). Surfactants alter the cohesiveness of sand through a decrease in interstitial water surface tension. Large sand movements were noted following the application of detergent, as well as the formation of quicksand. Mixtures of oil and detergent also increased anaerobic bacterial biomass, bringing the black RPD layer to the sediment surface, whilst the oil–detergent emulsion penetrated far deeper into the sand (down to 50 cm) than would have been the case for oil alone. Surfactants also allow water to drain much more effectively from sand, thereby increasing the proportion of air in the beach at low tide. On the flood tide, interstitial air bubbles strengthened by surfactant molecules may be more difficult to dislodge, and these would block the pores and curtail water circulation in the beach (Webb, 1991).

The physical effects documented for sandy beaches were reasonably short lived. In contrast, the effects on rocky shores of the removal of much of the biota were profound and long-lasting. The loss of the limpet *Patella vulgata* had a special significance in this respect, because this grazer is a 'keystone' species in the north-east Atlantic, responsible for structuring midshore communities on moderately exposed and exposed rocky shores (Chapter 4).

Table 7.2 summarizes information on the patterns of recolonization of various rocky shores between 1967 and 1977. The similarity of the overall pattern allows a generalized account of the course of recolonization in the midshore region. Following the death of grazing animals, a dense flush of ephemeral green algae (*Enteromorpha, Blidingia, Ulva*) appeared which lasted up to a year (Fig. 7.1). After about 6 months, large brown fucoid algae (mainly *Fucus vesiculosus* and *F. serratus*) began to colonize the shores. Very few animals were present under these dense growths of algae. Any surviving barnacles were overgrown and eventually died, whilst the dense canopy prevented subsequent recruitment of barnacles by the sweeping action of the *Fucus* fronds and by presenting a barrier to larval settlement. This was probably reinforced by the dense population of the predatory dogwhelk *Nucella* which built up under the canopy so that barnacles declined to a minimum on most shores between 1969 and 1971 (Table 7.2 and Fig. 7.2 for Porthleven).

Patella first recolonized the shores during the early winter of 1967–8 and survived well in the damp conditions under the extensive *Fucus* canopy, preventing subsequent recruitment of *Fucus* by grazing. As the plants aged, grazing of the holdfasts reduced *Fucus* further and between 1971 and 1975 the shore became very bare with even less algae than before the spill.

With the disappearance of *Fucus*, the abnormally dense population of limpets abandoned their normal homing habit and migrated in lemming-like fronts across the shore. The pronounced peak shown in 1972 (Fig. 7.1), occurred when this front crossed the area sampled at Porthleven. Barnacle numbers increased on all shores once the fucoids declined. Following the bare period between 1974 and 1978, the shore went through a phase of increased *Fucus* cover, although overall cover never exceeded 40%. Limpet density increased with some fluctuations

Table 7.2 The time course of recolonization of rocky shores in Cornwall, UK, expressed in years from the date of the *Torrey Canyon* disaster in March 1967 (from Southward and Southward, 1978 and Hawkins and Southward, 1992)

	Lizard Point exposed, Vellan Drang	Lizard Point sheltered, Polpeor Cove	Porthleven west of harbour	Maen Du Pt Perranuthnoe	Sennen Cove exposed, 300 m east	Sennen Cove near pier	Cape Cornwall	Godrevy	Trevone exposed, sewer rocks	Trevone sheltered, MTL reefs
Relative exposure to waves	+++	++	++	++	+++	+-	+++	++	+++	+
Amount of oil stranded	+	++	+++	+	++	+-+	++	++	++	++
Dispersant treatment	+	+++	+++	++	++	+-+	++	0	++	+++
Persistence of oil/oil–dispersant mix*	< 1	< 1	< 1	< 1	< 1	< 1	< 1	< 1	< 1	< 1
Years until:										
Maximum *Enteromorpha* cover	1	1	1	1	0–1	1	1	None	0–1	1
Maximum *Fucus* cover	2–3	1–3	1–3	2–3	1–3	1–3	1–2	None	2	1–3
Minimum barnacle cover	2	2	3	4	3	3	3	< 0.5	2	2–6
Maximum number *Patella*	?†	6	5	5	?	3	3	N/A‡	3	5
Start of *Fucus vesiculosus* decline	4	4	4	4	4	5	3	N/A	3	4
All *Fucus vesiculosus* gone	5	6–7	6–7	6	5	6	6	N/A	5	8
Increase in barnacles	4	6	6	5	4	6	4	1	3	7
Patella remain reduced	?	6	8	7	6–7	8	7	N/A	6	N/A
Species richness regained	5	9	> 10	8–10	9	9	8–9	2	5–6	> 10

*Time (years) taken for 90% to disappear.
†?, no quantitative data.
‡N/A, not applicable (see text).

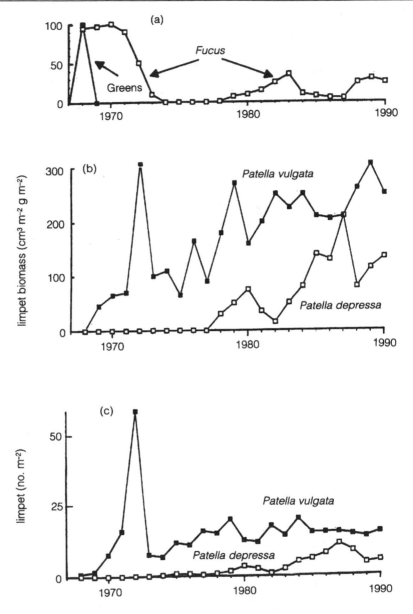

Fig. 7.1 Long-term changes in the major components of the midshore community on a rocky shore of south-west England following the *Torrey Canyon* incident (adapted with permission from Hawkins and Southward, 1992).

from 1975 before dropping in the early 1980s and then rising again, probably reflecting normal levels of spatial and temporal variation typical of limpet populations. During the period of dominance by the initial colonizing cohorts recruitment of limpets was low (e.g. 1969–1972), presumably due to intense competition between age classes. Subsequently, recruitment improved, apart from the occasional dip (1977, 1981), and after 1978 the population generally had 60–70% juvenile limpets under 15 mm in length (Fig. 7.2(c)).

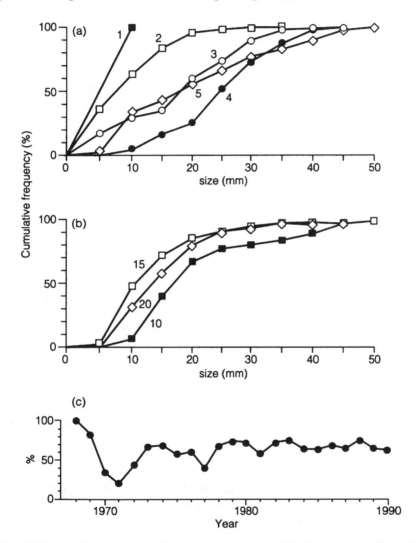

Fig. 7.2 Population structure of limpets on Porthleven MTL flat, a shore affected by the *Torrey Canyon* spill (adapted with permission from Hawkins and Southward, 1992). (a,b) Cumulative size frequency, 1, 2, 3, 4, 5, 10, 15 and 20 years after the spill. (c) Percentage of juvenile *Patella* smaller than 15 mm each year after the spill.

Barnacles have been monitored since the 1950s as part of a long-term assessment of the effects of climate on the English Channel ecosystem (Southward, 1967, 1980; Chapter 6). The mean and standard deviation for counts in the 12 years prior to the spill give some idea of natural temporal variation. After the initial decline following the spill, the barnacle population slowly increased at all shore levels. At a shore level equivalent to high water of neap tides, all counts from 1979 onwards were within one standard deviation of the pre-spill mean, although only one value (in 1990) actually exceeded the mean. At mid-tide level, a similarly irregular pattern was observed, with exceptionally high counts just after the bare phase in 1976. Lower on the shore, there was much greater fluctuation which probably reflected the greater influence of biological interactions, including predation and competition with spasmodic settlements of *Mytilus*, and this may account for the decline observed in the late 1980s.

How long was recovery?

Before recovery can be assessed, normality must be defined. The normal condition in the eulittoral of exposed shores of Devon and Cornwall is one of small-scale spatial and temporal fluctuations in the major components of fucoids, barnacles and limpets. Isolated patches of *Fucus* occur, but they are never more than clumps of a few plants, and total cover rarely exceeds 20%. The patchiness and fluctuations are partly generated by variation in recruitment and small-scale differences in microhabitat, predation and physical disturbance (Chapter 4). 'Recovery' is defined here as a return to these normal levels of spatial and temporal variation.

After the initial massive increase in *Fucus*, and a similar but aphasic increase in the key herbivore *Patella*, subsequent fluctuations have been much smaller. *Fucus* cover was clearly abnormal for the first 11 years, and was perhaps slightly elevated in the early 1980s, before fluctuating around normal levels after 15 years or so. The abundance and population structure of *Patella vulgata* were clearly abnormal for at least 10 years, probably 13.

Thus, the time scale for recovery at these sites seems to be at least 10 years. If limpet population structure and barnacle densities are used as criteria, then 15 years may be more realistic. These time scales are not surprising when the long life spans of the main organisms are considered: *Fucus* 4–5 years, *Patella* up to 20 years but usually < 10 years, and *Chthamalus* at least 5 years and possibly 20. If we estimate that the average life span of a limpet is 7–10 years, it is highly likely that population structure will take 15 years or so to stabilize.

The time scale for recovery is clearly much longer than was thought by many ecologists in the early 1970s. Dense growths of seaweeds were seen as a sign of recovery, rather than of a highly disturbed community, and there were suggestions that the system had returned to normal within 2 years. Even the more pessimistic considered that only a few more years were needed for a complete return to normal, but Southward and Southward (1978) rightly dismissed optimistic forecasts and the myth of rapid recovery. At that time, they could only assert that some shores heavily treated with dispersants had not returned to normal after 10 years, whilst many had taken at least 5–8 years. It is now clear that it may take 15 years or so for the worst-affected shores to recover. In contrast, recovery at the only shore where oil was substantially untreated because of its proximity to a seal breeding area (Godrevy) was rapid and almost complete within 3 years. Similar work in Panama in a deep mud associated with fringing mangroves (Burns *et al.*, 1993) has suggested that 20 years or more is required due to the long-term persistence of oil trapped in anoxic sediments and subsequent release into the water column.

Lessons learned

It was very quickly learnt from this incident that large-scale use of dispersants causes acute toxic effects. In the few weeks taken for the oil to cross the English Channel, a very different approach was adopted by the French and the Channel Islanders – manual removal or the use of suction devices with dispersants applied sparingly. These lessons were absorbed by those in charge of responding to the *Santa Barbara* blow-out in 1969 and subsequent spills such as the *Amoco Cadiz* in 1978. Impetus was also given to the development of less toxic dispersants (NAS, 1989) and to physical dispersal and mechanical collection. In most instances, manual methods (whether removal of oiled plants or use of absorbent materials) seem to cause less disturbance than that associated with the trampling and movement of equipment, vehicles and vessels during mechanized operations.

Considerable hope has been raised by the possibility of enhancing natural biodegradation 'bioremediation' by adding nutrients which limit bacterial growth to the oil, particularly in oleophilic media. Laboratory tests and field trials have been encouraging and we should be in a better position to judge once their effectiveness during the *Exxon Valdez* and their limited use in the 1991 Gulf War clean-ups has been fully evaluated (Hoff, 1993). Current opinion on the effectiveness of bioremediation varies from overoptimism to extreme scepticism. The results that are emerging from applications in Alaska after the *Exxon Valdez* spill are often conflicting. Concerns have also been expressed about eutrophication caused by bioremediation and toxic side effects of some preparations. However, with further field trials and experience of appropriate use, bioremediation still holds considerable promise.

Clearly, the *Torrey Canyon* incident was an early example of a major ecological disaster made much worse by an inappropriate response. A more considered approach to spills has emerged and more sensible procedures have evolved which were implemented during the *Exxon Valdez* spill. The least expensive and most ecologically sound option for restoring exposed and moderately exposed shores covered by oil is probably to do nothing, but, as Foster *et al.*, (1990) point out, during an environmental crisis social pressure to 'do something', and the political need to be seen to be doing something, often outweigh ecological considerations.

7.4 INTRODUCTION OF NEW SPECIES

The biogeography of shores has been changed over the last 200 years or so by introductions of species from one part of the world to another. Most introductions have been an accidental by-product of sea trade. Others have been deliberate attempts to improve fisheries or to provide suitable aquaculture species. Occasionally, other species are accidentally introduced along with the target species.

Introductions, both deliberate and accidental, have been successful to various degrees. Most dramatically they lead to full-scale invasions with rapid expansion of range, displacement of native species and severe disruption of community dynamics. More usually, however, a steady but gradual expansion follows initial invasion, with the range and abundance eventually stabilizing. Sometimes, introductions persist only in small isolated pockets, neither expanding nor contracting. Many introductions just fade away and most accidental ones are probably never even noticed. Sometimes an introduction will only persist if the environment is changed by humans as well. Table 7.3 provides a summary of various types of introductions throughout the world. In this section, we consider some of the better-documented cases of invasions and briefly consider their consequences for shore communities.

Most accidental introductions are generally either from fouling organisms growing on the undersides of ships or from organisms living in the ballast water. Not surprisingly, many species of barnacles have invaded new areas in this way, one of the best documented being the Australian species *Elminius modestus* (review: Lewis, 1964; Fig. 7.3). This

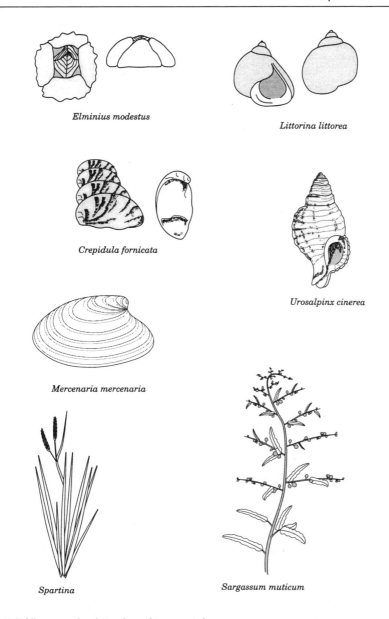

Elminius modestus

Littorina littorea

Crepidula fornicata

Urosalpinx cinerea

Mercenaria mercenaria

Spartina

Sargassum muticum

Fig. 7.3 Alien species introduced to coastal areas.

species was first detected on British shores just after the Second World War, almost certainly having entered UK waters with wartime convoys. The amount of shipping at this time must have provided large concentrations of potential colonists and antifouling was the least of the

Table 7.3 Introductions of alien species around the world*

Mechanism of introduction	Taxa introduced	From	To	References
Lessepsian migrations (via the Suez Canal)	Fishes			
	Sardinella aurita	Mediterranean	Arabian Sea (rare)	Dutt and Raju (1983)
	> 30 spp.	Red Sea/Indo–Pacific	Mediterranean	Ben-Tuvia (1973)
	Crustacea			
	Neptunus pelagicus	Red Sea	Mediterranean	Calman (1927)
	Processa aequimona	Red Sea	Mediterranean	Rees and Cattley (1949)
	Panaeus japonicus	Red Sea	Mediterranean	Ben-Tuvia (1973)
	P. semisulcatus	Red Sea	Mediterranean	Ben-Tuvia (1973)
	Metapenaeus monoceros	Red Sea	Mediterranean	Ben-Tuvia (1973)
Migration via the Panama canal (rare)	Fishes			
	Tarpon atlanticus	Western Atlantic	Eastern Pacific	Hildebrand (1939)
Deliberate introduction through aquaculture	Mollusca			
	Ostrea edulis	Europe	Eastern USA	Loosanoff (1955)
	Crassostrea gigas	Japan	Tasmania	Thompson (1952)
	Crassostrea virginica	USA	Hawaii	Edmondson and Wilson (1940)
	Fishes			
	Oncorhynchus spp. (Pacific salmon)	USA	Chile, Tasmania, New Zealand	Elton (1958)
	Morone saxatilis (striped bass)	Eastern USA	Western USA	Scofield and Bryant (1926)

Table 7.3 contd.

Mechanism of introduction	Taxa introduced	From	To	References
Accidental introduction through aquaculture	Macrophytes			
	Sargassum muticum	Japan	Europe, NE Pacific	Druel (1973), Farnham et al. (1973)
	Mollusca			
	Urosalpinx cinerea	USA	UK	Cole (1942)
	Crepidula fornicata	USA	UK	Walne (1956)
	Mya arenaria	Eastern USA	Western USA	Elton (1958)
Transport on ships as fouling organisms or in ballast water	Phytoplankton			
	Biddulphia sinensis	Asia	Northern Europe	Hardy (1956)
	Macrophytes			
	Falkenbergia rufulanosa	Australia	Europe	Walker et al. (1954)
	Asparagopsis armata	Australia	Europe	Walker et al. (1954)
	Mollusca			
	Littorina littorea	Europe	Eastern USA	Carlton (1989)
	Crustacea			
	Elminius modestus	Australia	Europe, South Africa	Crisp and Chipperfield (1948), Sandison (1950)
	Balanus improvisus	N. Hemisphere	UK	Bishop (1951)
	B. eboneus	Eastern USA	UK	Bishop (1951)
	Eriocheir sinensis	Asia	UK	Elton (1958), Barnes (1980)

*Further reading: Elton (1958), Por (1978), Carlton et al. (1982), Carlton (1989), Godeaux (1990).

many worries of ship owners. Furthermore, once docked, ships might not leave port for many weeks. *Elminius* rapidly spread around the British Isles and mainland Europe, partly by larval dispersal and partly by shipping movements. The species does particularly well in estuaries and bays and can displace native barnacles such as *Semibalanus balanoides* and *Chthamalus montagui*. Complete displacement has not occurred because the native species do better on more exposed coasts, which provide a reservoir from which recruitment onto more sheltered shores can occur. *Elminius* is now an established part of the north European shore fauna, occurring from Scandinavia to Portugal. Its overall effect on the dynamics of rocky shore communities has been small as it has simply replaced some individuals of a group or guild of co-occurring barnacles the coexistence of which is maintained by fluctuations in recruitment (Southward, 1991). Another barnacle immigrant to the UK is the warm-water *Balanus improvisus*. This species is restricted to areas such as Swansea Docks, where the water temperature was raised by warm-water discharges (Naylor, 1965).

An accidental immigrant that has attracted much attention on both the Pacific coast of North America and northern Europe is the brown macroalga 'Jap-weed' *Sargassum muticum* (Fig. 7.3). This species invaded both regions as a result of oyster transplantation. The spread of this seaweed was rapid and aided by floating fragments which remain reproductively active. It has spread along the south coast of the British Isles eastwards and westwards from the Isle of Wight, and has expanded its range throughout much of Denmark, western Sweden, Holland, Belgium and France. In the UK, it dominates a low-shore habitat of mixed stones and sand or boulders. In areas with tidal movement, it grows extremely fast, clogging approaches to small ports and harbours. It also grows profusely in deep rock pools. No native macroalgal species really dominates the same habitat as *Sargassum*, although it may have displaced some species such as *Chorda filum* from unstable habitats, and may reduce the abundance of *Cystoseira* and *Halidrys* growing in low-shore pools. Its complex growth form (Fig. 7.3) provides an ideal microhabitat for small animals and its presence can actually enhance species richness (Withers *et al.*, 1975; Critchley *et al.*, 1990; but see Norton and Benson, 1983). *Sargassum* performs very differently in its native environment compared with its newly invaded territory. In its native Japan, *Sargassum* rarely attains a large size and cannot be considered an ecological dominant, whereas in Europe it can grow very long, reaching an extremely high biomass, and can be the most abundant species in suitable habitats. Although *Sargassum* has also spread in the United States, it is never as luxuriant as in Europe. Perhaps an assemblage of competitors and grazers similar to that in the Indo–Pacific keeps it in check in North America. Alternatively, some aspect of the physical environment may favour *Sargassum* in Europe. More recently concern has focused on *Undaria*. This Japanese seaweed has spread all over the world (e.g. New Zealand) and after deliberate introduction to the Mediterranean and Atlantic coasts of France was very recently found in the UK in the Hamble (Fletcher and Manfredi, 1995). This vigorously growing kelp will have a major influence on the British flora.

Other species that have been introduced along with cultivated shellfish include the slipper limpet *Crepidula fornicata* and the oyster drill *Urosalpinx cinerea* (Fig. 7.3) from the USA. Both species were introduced into UK waters when oyster beds were relaid with the Virginian oyster *Crassostrea virginica*. They became a pest on both introduced beds and amongst native oysters *Ostrea edulis*, the natural predators of which (e.g. *Ocenebra erinacea*) were not as voracious as *Urosalpinx*. The filter-feeding *Crepidula* competed for space and food with native oysters and cluttered up the beds. Interestingly, TBT antifouling paints are thought to have led to a decline in the numbers of *Urosalpinx* in the UK (Gibbs *et al.*, 1991).

In the eastern United States, the edible periwinkle *Littorina littorea* (Fig. 7.3) was introduced into Newfoundland in the middle of the 19th century, probably in ballast. It subsequently spread down much of the eastern seaboard and a succession of papers have shown how important *L. littorea* now is for structuring rocky shore communities in north-east America (Chapter 4). It has also reduced the abundance of, and in some cases displaced, the native mud snail *Ilyanassa obsoleta* in creeks and salt-marshes (Carlton, 1982, 1989) and outcompetes the native congeneric *L. saxatilis* (Behrens-Yamada,1987; Behrens-Yamada and Mansour, 1987), with which it co-occurs naturally in Europe. Matthieson *et al.*, (1991) review invasions and range expansions of several other rocky shore species along the eastern seaboard of the USA.

Introductions of alien sediment-dwelling invertebrates appear to have had less dramatic consequences. A small population of *Mercenaria mercenaria*, the American hard shell clam, was probably deliberately introduced into Southampton in the 1920s. This clam now breeds successfully in the Southampton, UK, area. Studies on its reproductive biology (Mitchell, 1992) showed that spawning occurred as water temperature rose rapidly to 18—19 °C, some 3–4 °C lower than in the USA. This indicates that some physiological acclimation has occurred in the UK. Increased water temperatures in Upper Southampton Water with the commissioning of Marchwood power station in the late 1950s led to spawning earlier in the year. The hard shell clam may have displaced the soft shell clam *Mya arenaria*, itself an earlier introduction, but the main consequence of its introduction was the initiation of a fishery which changed from hand-digging to suction dredging in the 1970s, causing considerable habitat disruption and damage to non-target species.

Sediment habitats have been significantly affected by the introduction of a salt-marsh plant, the North American cord grass *Spartina alterniflora*. This species was introduced towards the end of the 19th century to stabilize and reclaim high-intertidal mudflats. Unfortunately, *S. alterniflora* crossed with the native *S. maritima* to produce an infertile hybrid (*S. townsendii*). This then gave rise to fertile *S. anglica* by diploidy. The result was an aggressive, fast-growing plant that has now colonized extensive areas of intertidal mudflats in Britain. As a result, the successional development of salt-marsh towards a vegetation characterized by terrestrial plants has accelerated, leading to substantial loss of intertidal feeding grounds for shorebirds (Goss-Custard and Moser, 1988) as well as the virtual ousting of indigenous *Spartina* species (review: Gray and Benham, 1990).

The Suez Canal, completed in 1869, provides an artificial link between the species-rich Red Sea, with its Indo–Pacific biota, and the rather impoverished version of the Atlantic biota present in the Mediterranean. Many Red Sea species have used this corridor to invade the Mediterranean, taking advantage of the predominantly northerly flow of water, with far fewer species migrating in the other direction. During the operation of the Canal, the passage of species has become easier as the high salinity (50‰) Bitter Lakes, once a formidable physiological barrier to invasion, have become less saline. Populations of several Red Sea species are now established in the eastern Mediterranean. These migrations have been termed Lessepsian, after the builder of the Suez Canal, Ferdinand de Lesseps (review: Por, 1978).

Invasions are continually occurring: Korean sea squirts *Styela clava* are common in British docks, and North Atlantic shore crabs *Carcinus maenas* have been found at Cape Town. The Japanese oyster *Crassostrea gigas* has been found in 1993 successfully breeding on the North Wales coast. Few of these invasions have been as dramatic or catastrophic as those in freshwater systems. For example, the freshwater mussel *Dreissena* from eastern Europe now clogs up vast amounts of the water supply system in the USA and has drastically affected the ecology of the Great Lakes, whilst the Nile perch has had dramatic effects on the ecology of African tropical lakes. However, the potential for ecological disaster in coastal waters remains, especially if pathogens are transferred along with other introduced species and these then find novel hosts. One of the world's most important scallop fisheries, that for *Argopecten gibbus* off the east coast of the USA, has recently been devastated in this way by the protozoan parasite *Marteilia*, thought to have been introduced along with oysters.

7.5 COASTAL ZONE CHANGE

Coastal zones face a variety of pressures. In many parts of the world, the intertidal zone is being progressively squeezed between the encroaching onshore development and what is probably the most pervasive threat facing the world's coastline: rising sea levels (e.g. Reid and Trexler, 1992). Sea levels have been rising in many parts of the world since the last glaciation, particularly in the heavily industrialized North Atlantic region and the developing and densely populated Indo–Pacific rim countries, but the process is now thought to be accelerated by global warming, the result of anthropogenic greenhouse gases (see below). Coastal development will also increase as the world population grows. On a more local scale, there will be greater pressure for the use of estuaries with large tidal ranges for generation of renewable sources of energy as fossil fuels become more expensive. The consequences of these changes for the shape and nature of the coastline and the ecology of intertidal areas are therefore worth considering in some detail.

Coastal development results in the infilling of lagoons, reclamation of marshes and foreshore and an overall linearization and hardening of the coastline. Historically, the Dutch have fought a long battle with the North Sea and by an extensive system of dykes have extended their landmass and turned the semi-enclosed marine Zuider Zee into the freshwater IJsselmeer. However, most coastal development is piecemeal and insidious and hence difficult to effectively regulate. The cumulative effects of such developments within a particular estuary can be depressing. For instance, about 2500 ha of mudflats in the Firth of Forth, Scotland, have disappeared over the last 200 years through a series of individually unspectacular schemes, leading to a reduction in the estuary's fish biomass by 50% (McLusky *et al.*, 1991). This scale of loss of intertidal zone has occurred or is forecast for many UK areas (Prater, 1981) and must apply to most of the world's industrialized estuaries.

Development, whether by dykes or coastal defences, seaside promenades, residential 'marinas' or dock complexes, tends to shorten the foreshore and reduces the extent of mudflats, upper-shore creek and salt-marsh systems. It can also prevent interchange of material between the sea and sand dune. The export of strandline seaweed into sand dunes can be a valuable source of nutrients and stabilizer of blown sand, and foraging land animals move from the dunes onto the foreshore.

Most importantly, the profile of the mid and upper shore becomes steeper, and particles become coarser as erosion increases. This will lead to greater instability of the sediments, lower organic matter content and hence lower productivity of deposit-feeding invertebrates. Inevitably this will affect the productivity of the shore as a nursery ground for fish and as a feeding ground for birds – effects compounded by the loss of areas of mudflat (Goss-Custard and dit Durrel, 1990).

Sea defences generally linearize the coastline, reducing heterogeneity and habitat diversity. In addition, there will be a switch at upper shore levels to typical rocky shore species on more open coasts and an impoverished estuarine hard substrate fauna in brackish water. Eroded sediments will also scour the bottom of any sea walls, reducing the productivity and diversity of hard substrate communities. These foreshortening effects will be exacerbated if sea level rise accelerates as predicted.

7.5.1 SEA LEVEL RISE

Perhaps the greatest potential threat facing shorelines worldwide is the rapid rise in sea level predicted over the 21st century as a result of global warming (Houghton *et al.*, 1990). A general warming of the Earth's climate is likely to occur through the persistent accumulation of so-called **greenhouse gases** in the atmosphere. As their name suggests, these gases act like the glass in a greenhouse by preventing the re-radiation of heat from the Earth's surface back up through the atmosphere. The most abundant of these gases is carbon dioxide, concentrations of which have increased consistently since the Industrial Revolution in Northern Europe in the mid-18th century. Data from ice cores indicate that pre-industrial (AD1750) levels of carbon dioxide in the atmosphere were about 280 ppm, whilst records from the Mauna Loa Laboratory in Hawaii show a continuing and convincing upward trend from 310 ppm in the late 1950s to 340 ppm in the late 1980s (Ince, 1990). In other words, CO_2 levels have increased by 25% since 1750.

Other gases, namely methane, chlorofluorocarbons (CFCs), nitrous oxide and ozone, also contribute to global warming (Hekstra, 1990) and, although CO_2 predominates, it is by no means the most potent. Methane levels in the atmosphere are only about 1.7 ppm, yet this gas may account for 30% of the greenhouse effect (CO_2 is responsible for 50%). Similarly, one molecule of CFC is estimated to be 20 000 times as potent a greenhouse gas as one molecule of CO_2. There are uncertainties as to how the levels of these gases might increase in the future, but clearly we should be concerned about even modest increases, given their marked effects and long residence times in the atmosphere.

Defining the relationship between atmospheric levels of these greenhouse gases and the amount of global warming is difficult, and there is considerable debate about the expected rise in temperature and how this might affect sea levels (Hekstra, 1989; Schneider and Rosenberg, 1989). Estimates range from 0.5 °C to 2.5 °C by the year 2030, with most in the range of 1–2 °C. Such increases in temperature might lead to sea levels rising by a few centimetres to several metres by the end of the 21st century, depending on actual temperature increase. Sea levels will rise by four main routes: thermal expansion of the upper ocean, melting of glaciers and small ice caps, disintegration of the massive Greenland ice sheet and, most catastrophically, the break-up of the western Antarctic ice sheet.

The concept of thermal expansion is relatively straightforward, and this is likely to be the most immediate cause of a sea level rise of about 10–12 cm by 2030 (Fig. 7.4).

Understanding how valley glaciers will respond to warming is more difficult (Oerlemans, 1991). Most will appear to advance, as the meltwater lubricates the downhill-sliding ice mass, but this will be followed by severe retreats. The question of what will happen to the polar ice sheets is far more problematic. These ice sheets are maintained in a dynamic state by snowfall, melting and disintegration into icebergs. Greenland is warmer than Antarctica and a slight rise in temperature would lead to a sea level rise of several centimetres, through the melting of the Greenland ice sheet. Because of the lower temperatures in Antarctica, this area will experience heavier snowfalls from an overall increase in water vapour in the atmosphere. Thus, the scenario for much of the Antarctic is one of increasingly icy conditions, perhaps even lowering sea levels by many centimetres. However, this rosy picture does not apply to the massive ice sheets of the western Antarctic. These are many kilometres thick and attached to the sea floor in places, thereby holding back the mainland sheets. If the points of contact between the sheets and the sea floor are eroded by warming, then the ice would be pushed out to sea, to be quickly followed by mainland ice. The effects on the world's coastlines would be catastrophic, sea levels could rise by 1–2 m over a 10 year period. Fortunately, this bleak picture is probably a worst-case scenario which may never occur unless warming is particularly severe.

Sea level rise will also vary from area to area, depending on regional geological movements. Areas that are subsiding, such as the Mediterranean, will experience more severe effects than those areas

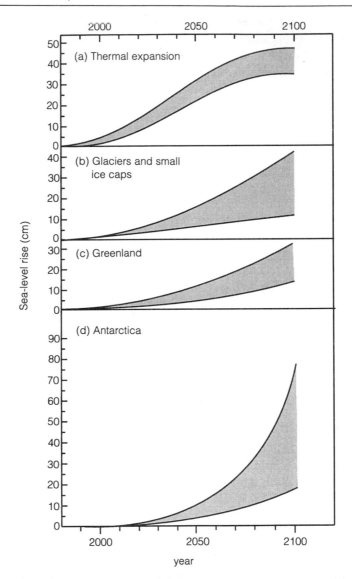

Fig. 7.4 The possible contribution of different factors to sea-level rise over the next 100 years (redrawn with permission from *The Rising Seas* by Martin Ince. Published by Earthscan Publications, Ltd, London, 1990). Shaded areas represent the range of predicted rises in sea level.

being uplifted, such as Scandinavia. This disparity is well illustrated by the scenario for the UK, which has areas of both types (Fig. 7.5). Sea level around the UK is currently rising at a rate of about 1–1.5 mm per year, but over the next 50 years this is expected to increase to at least 5 mm a year (Boorman *et al.*, 1989). Estimates of total rise vary from 0.5 m to 3.5 m over the next 100 years and 5 m over the next 200 years according to Boorman *et al.* (1989), whilst Houghton *et al.* (1990) suggest rises of 0.2 m by 2030 and 0.65 m by 2100. The consequences of this will vary throughout the UK (Fig. 7.5). The north and west (Scotland) is still rising because of isostatic rebound following the loss of its glacial ice mass, but this will not be sufficiently fast to entirely escape the predicted rises in sea level. Conversely, the south and east of the UK is subsiding, so that sea level rise will be much greater here.

The effects of sea level rise on the ecology of coastal systems will vary with habitat. In general, harder steeper shorelines, such as cliffs, may well experience only minimal effects, with species simply shifting their zones upwards. However, cliffs susceptible to erosion will be under greater attack because a sea level rise is likely to be accompanied by a more severe wave climate and increased frequency of storms (Ince, 1990).

Sediment shores, with their shallow profiles, will be dramatically affected (Goss-Custard *et al.*, 1990; Siefert, 1990). Sediment supply from rivers will be slowed down and alongshore drift will be disrupted, increasing the shore's susceptibility to erosion. In low-lying regions, such as the east coast of the USA, this could lead to massive coastline retreat, with dramatic salinity effects on any extensive wetland areas immediately behind the coast (Fig. 7.6).

In more sheltered habitats, especially estuaries, there is likely to be an accumulation of sediment, as river-borne material is deposited earlier and open coast sediment is imported on the tide (Goss-Custard *et al.*, 1990). In theory, it might be possible for these habitats to migrate inland, establishing a new estuary or marsh several hundreds of metres back from the original setting. However, in many areas this is unlikely to be permitted. In wealthy industrialized regions like North America and Europe, much commerce, industry, agriculture and settlement is around estuarine or low-lying areas. There will undoubtedly be public pressure to prevent inundation of these interests and this will be manifested by creating new sea defences and raising existing ones. The economic costs of this would be enormous (Crosson, 1989). For the UK, the cost would be several thousand pounds sterling per linear metre of coastline (Boorman *et al.*, 1989), and similar figures probably apply to other countries.

However, the ecological costs could be even more staggering (Beukema *et al.*, 1990). There could be a marked loss of mudflat and salt-marsh as these habitats are prevented from migrating inshore by sea

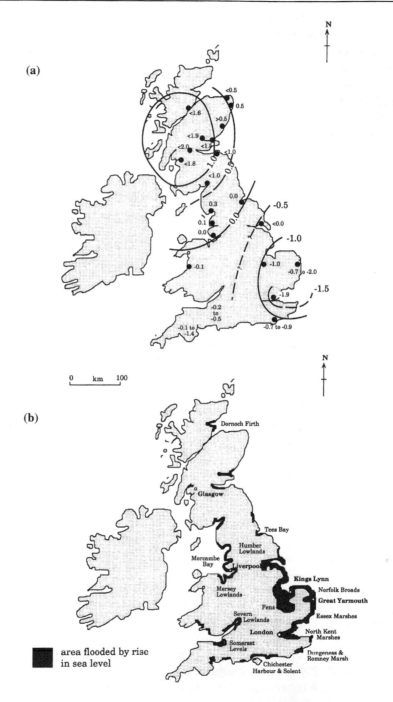

Fig. 7.5 Changes in sea level around the UK. (a) Estimated current rates (mm year⁻¹) of crustal movement in the UK. Isolines join areas with similar rates of movement (reproduced with permission from Sheenan, 1989). (b) Areas of Britain vulnerable to sea-level rise (reproduced from Boorman, L. A. *et al.* (1989).

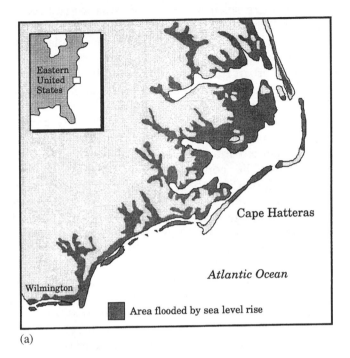

(a)

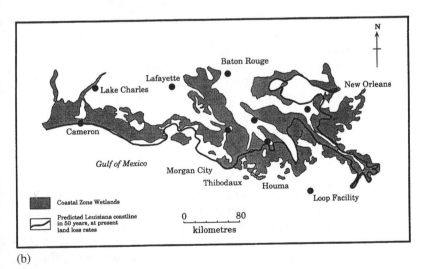

(b)

Fig. 7.6 Predicted transgressions (shaded) of the sea into coastal areas of the USA. (a) The North Carolina coast, assuming a 1.5 m rise in sea-level. (b) The Louisiana shoreline in the year 2030. (Redrawn with permission from *The Rising Seas* by Martin Ince. Published by Earthscan Publications, Ltd, London, 1990.)

walls. Erosion will increase the turbidity of estuaries and coastal waters and this could interfere with primary productivity and reduce the feeding efficiency of filter-feeding animals. The displaced sediments may choke or block existing channels. The loss of fringing vegetation and the increase in urban run-off will increase the amount of nutrients entering coastal waters and could promote coastal eutrophication. One possible benefit is that displaced sediment may be transported elsewhere to create new spits and possibly even new barrier islands. However, intertidal mudflats, already under pressure in many estuaries, will be further squeezed into a steep-sided, narrow band supporting fewer invertebrates. The consequences of this for the millions of migrating shorebirds that use these flats every winter as staging posts, refuelling stops and cold-weather refugia are likely to be grave. Similarly, the status of these areas as nursery grounds and migration corridors for fish would be severely affected.

In addition to cliffs, beaches and mudflats, tropical shores support mangroves and coral reefs. The effects of sea level rise are difficult to evaluate for either habitat, but the prognosis is not good (Wafar, 1990; Reid and Trexler, 1992). Because of the relatively stable conditions under which mangrove trees fix sediment around their roots, they may find it hard to keep pace with a rapidly rising sea level, because the latter is likely to be accompanied by increased rates of erosion. Whilst trees might persist for some time, the mangrove ecosystem proper may never develop. In areas now supporting extensive mangrove systems, like Tonga in the Pacific, few mangrove areas existed during historical periods of rapid sea level rise.

Coral reefs are essentially sublittoral systems and extend 30 m or more below the sea surface. At first sight, therefore, coral reefs might be expected to cope well with the predicted rise in sea level. However, estimates of the annual increase in height of coral reefs vary between 1–3 mm and 7–10 mm (Wells and Edwards, 1989), whilst sea level might rise at 11 mm per year according to some models for tropical waters. Also, the faster-growing coral tends to comprise branching species, not those species which make up the massive physical structure of the reef. Branching species are more susceptible to storm damage, predicted to increase as sea levels rise. Whether reefs will keep up or give up in the race with sea level is far from certain (Wells and Edwards, 1989). A more worrying problem is that of coral bleaching caused by higher temperatures killing the zooxanthellae (symbiotic algae) within the coral polyps (Brown, 1990). Many corals seem to require a steady all-year-round temperature of about 29 °C and increases by only a degree or so can cause the release of the polyp symbionts. Corals are thus a good example of how the different components of climate change might interact to affect coastal species.

OVERVIEW

Of the various impacts on shores, human food gathering, recreation and chronic discharges are reversible if the collecting/effluent/leachate is stopped or environmental standards improved. The recovery of the salt-marsh at Fawley and recovery under way by dogwhelks from TBT leachates are good examples, although full recovery may take some time. After environmental disasters such as oil spills, recovery is possible through natural recolonization. This can take longer than some optimistic interpretations: at least 15 years in the case of the rocky shores devastated by the *Torrey Canyon* spill or 20 years in mangroves. Inappropriate responses to oil spills make matters worse, although with experience the authorities are slowly getting it right.

In most cases, pollution, whether chronic or acute, tends to favour short-lived opportunistic species. They can quickly colonize after acute impacts and their short generation time allows development of tolerance to chronic discharges. An exception is salt-marshes, where annual angiosperms are highly vulnerable. Removal of long-lived space-occupying species as well as grazers and predators also favours ephemeral species. Furthermore, eutrophication stimulates the growth of ephemeral green algae such as *Ulva* and *Enteromorpha*.

In recent years, restoration techniques have been pursued in aquatic habitats. Attempts have been made with some success to restore kelp beds, seagrass beds and salt-marshes. Even disused docks on degraded estuaries in inner cities can be managed to support thriving marine communities (Hawkins *et al.*, 1992). Recovery need not be a passive process depending on natural recolonization and succession, it can be actively accelerated (see various papers in Thayer, 1992).

Introductions are probably irreversible because eradication is usually impossible once a toe-hold occurs, but the effects of introductions, both deliberate and accidental, are being combated by greater awareness of the risks posed by alien species. Quarantine measures are tighter and there are stricter controls on the movements of marine plants and animals.

The largest threat to shores is also irreversible. Coastal development protected by sea walls is likely to lead to an impoverished, linear habitat of coarser-sediment beaches backed by sea walls with a restricted rocky shore fauna – a situation made worse by accelerated sea level rise through global warming. Sea defences can, however, be designed to minimize ecological impact and can increase habitat diversity. Offshore island breakwaters, designed using sediment transport models, can result in beaches being naturally replenished. They also lead to a diversity of sediment types with consequent increases in the variety of the infauna. The breakwaters can also provide new rocky shore habitats in areas previously without hard substrata. Overtopping of sea defences can also be combated by the development of artificial saline lagoons behind the defences as a safety valve – one which at the same time helps restore one of the most threatened coastal habitats. Also, the politically controversial decision of retreating from the coast to permit some inland transgression of at least some estuaries must be given serious consideration. Without this protection and management, the highly productive and accessible interface between land and sea will degenerate in the more populated parts of the world.

Only by the zoning of natural areas to control development, maintaining chains of nature reserves and worldwide tightening of emission standards will we be able to safeguard the future of our shores and estuaries. Fortunately the idea of integrated coastal zone management is catching on, but action is always better than trendy buzz-words.

Studying shores **8**

Shores are easy to study compared with the systems they divide: the land and the open sea. They have sharp environmental gradients and a suite of usually well-described and easily identifiable species. Although plant life is restricted to algae and a few higher plants, examples of most of the major animal groups can be found on the shore. There are good identification guides or keys to most groups. Access to the shore is usually easy and, in many parts of the world, shores are publicly owned or common property. Because of these features they provide an ideal outdoor laboratory for learning about systematics and taxonomy, ecophysiology, population and community ecology. It is not surprising that many educational courses include shore ecology within the curriculum. Also, because shores are more accessible and easier to investigate than offshore marine systems, they are often the focus of biological monitoring programmes aimed at detecting the potential effects of changes in coastal activities and contamination. In this chapter we consider some of the general principles underlying the study of shore communities.

Rocky shores are particularly easy to study: they are two dimensional; most species are sessile and even the mobile animals are generally slow moving; the larger, most conspicuous fauna and flora can often be easily and accurately identified from illustrated guides; plants and animals are of a similar size (in contrast, say, to the disparity between trees and insects) and relatively short lived; abundance can be easily estimated as percentage cover or density in a non-destructive way, allowing the dynamics of populations or communities to be studied without too much interference. Most importantly, they are amenable to an experimental approach. This allows manipulation in a variety of ways: individuals can be transplanted to different parts of the vertical or horizontal environmental gradients to test their abilities to survive environmental stress or biological interactions; species thought to be important in structuring the community by occupying space, by canopy shading or sweeping effects, grazing or predation, can be experimentally removed or their densities can be altered.

Sandy beaches and mudflats are not as amenable as rocky shores. The habitat usually has to be physically destroyed and sieved to quantify the infauna, and even then, faster-burrowing polychaetes and bivalves may escape detection altogether. The experimental approach is trickier on sediment shores because it is difficult to non-destructively sample an area and cages may alter the sediment environment. However, because the fauna is intimately associated with the physical structure of the habitat, sandy beaches and mudflats provide opportunities for investigating these associations in detail.

The chapter begins by describing the basic approaches needed for surveying the shore, followed by a closer look at methods for studying rocky, sandy and muddy shores. We then consider the general principles for undertaking manipulative experiments on rocky and sediment shores. More advanced students about to embark on their first attempts at quantitative descriptions or experimental manipulations should also consult appropriate guides: Baker and Wolff (1987) is a particularly good starting point for investigations of coasts and estuaries, along with Hawkins and Jones (1992) for British rocky shores and Holme and McIntyre (1984) for sediment shores.

8.1 DESCRIBING THE SHORE

8.1.1 BROAD-SCALE SURVEYS

Shores can be described at a variety of scales, from very broad biogeographic surveys, through regional surveys, to between-shore comparisons of a few sites, to detailed surveys of a particular shore or part of a shore. In many instances, describing one or a series of shores is like making a painting. At first a rough sketch is drawn and then the detail is painted in later. Going straight for detail is not a good idea – it is quite easy to recognize a face from a rough sketch, but not from a beautifully detailed painting of a nose. Before you start any kind of descriptive survey the aims of the study must be absolutely clear. Make sure you know what question you are attempting to answer! Once the aims are defined, the methodology can be decided upon to undertake the study with the required accuracy, precision and appropriate attention to detail. We advocate a hierarchical approach, with more accurate and precise methods being used as more detail is required. To use a further analogy, when using a microscope it would not be sensible to go straight to oil immersion. It is better to do low-magnification work first and then, when the overall structure is known, to look in more detail.

Qualitative descriptions

Broad-scale descriptions may require only a non-quantitative approach. This is particularly true for rocky shores, where the broad patterns can easily be seen. The classic qualitative accounts of rocky shores, by Stephenson and Stephenson and others outlined in Chapter 2, set the scene and provided the big picture, allowing more detailed studies to be carried out within this framework. If a series of quantitative transects had been done at intervals around the coasts described by the Stephensons, very little coverage would have been possible with the resources available. The broad-brush approach they adopted is much more difficult on sandy and muddy shores, where limited destructive sampling and classification of the sediment types are required.

When carrying out such broad-scale assessments, it is best to describe the major characteristics and habitats first. For instance, is the shore rocky, sediment or mixed? Is it exposed or sheltered? If rocky, is it bedrock or predominantly boulders? What are the sizes of the boulders or the approximate grades of the sediment? Is the shore evenly sloping or highly broken? What is the rock type? At this stage major communities can also be described. For example, is the midshore covered by large seaweeds or barnacles or mussels? The best way to gather this kind of information is to use a standardized form, as done by the Joint Nature Conservancy Committee in the UK in their broad-scale surveys. The checklist allows other factors relevant to the nature conservation value of the shore to be noted, such as coastal developments, recreational activity, pollution and sea defences. Extensive use of photography or video is essential for such broad-scale surveys.

Semi-quantitative approaches: abundance scales

The initial broad-scale approach can be taken further by estimating the abundance of the major species found on the shore. To undertake broad-scale biogeographic studies of a suite of ecologically important rocky shore species, Crisp and Southward (1958) devised semi-quantitative abundance scales. The abundance of organisms at a particular locality could be ascribed to one of a limited number of categories, such as abundant, common, frequent, occasional, rare, not found, this number was usually five or six, but it was later expanded by some authors to seven or eight). These scales have the advantage of allowing a quick assessment of abundance for a particular point of a shore or even for a whole shore. They were usually based on a logarithmic or semi-logarithmic progression (1–10, 10–100, 100–1000 etc.). A good example is the work on biogeographic limits in the English Channel (Fig. 8.1). Such scales can be successfully adapted to sediment shores as well. They are, however, limited by between-observer variability. It is best to make a few quick counts in quadrats of various size on rocky shores, or do a limited number of sievings of set areas of sandy shores to establish the right point on the scale. When comparing a number of shores, the maximal abundance of species is usually estimated.

Abundance scales are also useful for making rapid semi-quantitative comparisons of zonation patterns on several shores within a region. Each shore is levelled (i.e. surveyed) to provide a series of vertical stations at appropriate shore heights. The abundance of the species is assessed using an appropriate scale, usually within an area of about 20 m² at each station. This method is probably better on rocky shores where the environmental gradients are sharper and zonation patterns more distinct. The advantages of this approach are its speed, allowing an integration of abundance over a wide area by avoiding problems of patchiness. It is also suitable for heterogeneous habitats such as boulder fields or rock pools, where quadrats are always difficult and sometimes impossible to use. The major difficulties with abundance scales are the differences between operators and problems of statistical analysis, because there is no measure of variability. These problems make it very difficult to detect trends over long time periods using this approach.

8.1.2 THE QUANTITATIVE APPROACH

Quantitative estimates of abundance using quadrats are used for detailed surveys. On sediment shores, destructively removing all the material for subsequent sorting is the only approach possible. On rocky shores, destructive methods are used if estimates of biomass are required for studies of standing crop or energy flow, but non-destructive methods should be used wherever possible. This permits the same area to be studied over time without disrupting community composition (but statistical problems may arise; see later) and is clearly preferable for ethical and conservation reasons. Unfortunately, destructive sampling is unavoidable for sampling the fauna associated with algae or sheets of sessile animals. It is also sometimes the only effective way of dealing with algal turfs, where many small seaweed species are interwoven.

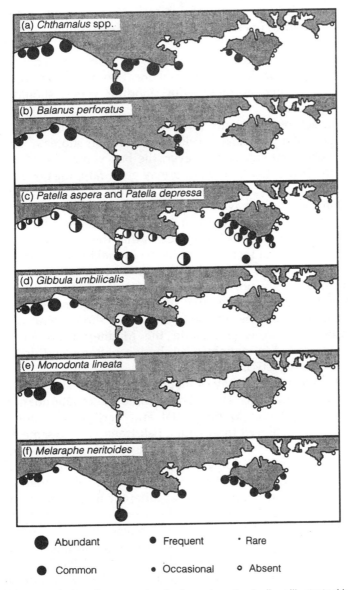

Fig. 8.1 The use of abundance scales for broad-scale studies, illustrated by work on the biogeographical limits of rocky shore species in the English Channel (modified with permission after Crisp and Southward, 1958). The half black symbols in (c) refer to *Patella depressa*.

Sampling using quadrats and corers

Various sampling strategies have been used for quantifying patterns of abundance on shores, and in particular describing zonation patterns. On rocky shores it is best to minimize microhabitat variation, for example, by only considering open, seaward-facing, free-draining bedrock. Rock pools, crevices, and the undersides of boulders all have different environments and species and they should be treated separately. Sediment shores at least appear more homogeneous, but clearly different areas, such as those with standing water, should also be sampled separately.

Belt transects of contiguous quadrats are useful for detecting absolute shore heights of a particular species, especially on shores of small tidal range or horizontal extent. However, the belt should be of a sensible width. If the belt is too narrow, you may get a false idea of abundance and the vertical range of zonation. If the belt is too wide, there is a danger that conditions may vary across it due to the horizontal gradient of wave exposure. Also, wide belt transects are labour intensive. The major disadvantage with belt transects is that they are difficult to treat statistically and hence there are problems in making comparisons from place to place unless replicate belt transects are used. This is because adjacent quadrats are not strictly independent of each other whether they are at different or at the same shore levels. Personally, we feel that this approach should be avoided unless the shore being worked is particularly small in extent, as in the Mediterranean, or you are primarily interested in mapping distribution patterns and do not want to carry out any statistical analyses.

A less labour-intensive way of mapping an area of shore is to sample on a regular grid. This also allows contours of abundance to be constructed in relation to horizontal and vertical axes and greater coverage can be achieved than with a belt transect. The same statistical objections apply as for belt transects because each sampling point is not strictly independent of the others. Regular sampling is useful if a map of the numbers or the biomass of a species is needed, but comparisons with other shores, or of the same shore at different times, cannot easily be made statistically unless the gap between the sampling points is sufficiently large to be considered biologically independent. The problems that arise with non-independence of quadrats are discussed in detail by Legendre (1993) and Schneider and Gorewich (1994).

The best and most generally used method is **stratified random sampling**. This method uses previous knowledge, perhaps from qualitative observations or a quick semi-quantitative survey, to randomly sample within various strata (in the statistical, not biological sense). If we wish to study shore zonation, then the strata are different shore heights which can be established at set intervals, for example 0.5 m vertical intervals above low water mark. Alternatively, sampling can be within biologically defined strata, such as a particular zone like the mussel zone or a zone of a particular kelp or fucoid. Within each stratum, replicated randomly sited quadrats can be examined. If zonation patterns were being studied, this would be done perpendicularly either side of the vertical station, parallel with the water mark at distances determined by random number tables. If a quadrat placed in this way is located in a pool, crevice or some other microhabitat, it can be justifiably rejected as the main aim is to consider zonation patterns on freely draining open rock or sand. Alternatively, that quadrat can be counted but scored as being in a pool or crevice, allowing determination of the relative proportions of the different habitats at each site. If the latter approach is taken, many more samples will have to be taken. If different transects on the same shore or on different shores are to be compared, then the process is repeated at each site. For between-site comparisons, more than one survey must be made at each shore level. Similarly for regional comparisons, several shores must be sampled. Thus a 'nested' or 'hierarchical' approach needs to be adopted. Clearly this process will take far more time than using semi-quantitative abundance scales, but between-site comparisons can be made statistically and confidence intervals of estimates may be determined.

Deciding on the size and number of samples

Whether the survey is on a rocky or a sediment shore, involves destructive or non-destructive techniques, or is for one species or for many species, two basic decisions must be made: what is the appropriate size of sampling unit and how many are needed?

If all shore organisms were evenly distributed then the problems would be easily overcome. The sampling units would be about the same size as the spacing between individuals and few samples would be needed because the numbers in each sample would be similar. Unfortunately, few organisms are regularly spaced out, the exceptions being territorial species such as fiddler crabs or gardening limpets. Most species have an uneven pattern of distribution. Sometimes this is random, but more often it is clumped or patchy. The between-quadrat variability is an interesting property of the community or population as it reflects its spatial variation, but it also presents a sampling problem (Elliott, 1977; Holme and McIntyre, 1984; Littler and Littler, 1985;

Andrew and Mapstone, 1987; Morrisey *et al.*, 1992; Legendre, 1993). It is difficult to compare the densities of populations with large or different variances if only small numbers of samples are taken because the differences between populations may be less than the between-sample variation within a population. Therefore, it is important that sufficient samples be taken both to describe the spatial distribution of the species and to allow statistical comparisons between sites or sampling occasions. The smaller the quadrat or core is in relation to the patchiness, the greater the between-sample variation will be. Therefore choosing a sensible size of sampling unit will minimize the number of sampling units required.

The size of quadrat or corer is often determined on practical and pragmatic grounds. It should be as big as, or bigger than, the largest organism present, and it must be easy to use in the field. Quadrats of 1 m × 1 m are unwieldy, and larger ones are only likely to be used in special circumstances. The sampling unit must not be so big that very large numbers of the animals or plants have to be counted, nor so small that many zero counts occur. As a general guide, up to 100 or so individuals can easily be counted in a single quadrat on a rocky shore, but beyond a few hundred individuals, subsampling or a smaller quadrat are clearly necessary. When sampling in sediments, there are mechanical constraints. A volume of sand 25 × 25 × 25 cm can just about be sieved over a 1 mm mesh or smaller and it also weighs a lot! Larger volumes of sand can only be put through a coarser mesh which will retain only the larger animals. For instance, a volume of 50 × 50 × 25 cm must be dealt with using a 2 mm mesh. When dealing with sediment samples, large numbers of animals are not too much of a problem because after sorting they can be evenly dispersed by shaking or stirring in a jar and subsamples taken. Because between-sample variation is always greater than within-sample variation (you have just homogenized the sample, remember), it is probably better to take more samples and subsample from them. It is also better to count and identify organisms in fewer subsamples per sample and put more effort into taking more samples. The best approach is to do a quick pilot study with a variety of different-sized sampling units and choose those which are most convenient and most appropriate to the question being asked.

Also remember that what is a good sample size for a species of barnacle (5 × 5 cm quadrat) will not be for a limpet species (where a 25 × 25 cm, or a 50 × 50 cm quadrat would be more appropriate) and will be useless for a large seaweed where a 1 m × 1 m quadrat would be best. Clearly no single quadrat will do for most community studies on rocky shores. Here it is best to use a large quadrat and then subsample within using a smaller quadrat. On sediment shores this is less of a problem as most of the macrofauna are of a more similar size. Large worms or bivalves would be best sampled by a separate strategy, however, such as larger sampling areas sieved through a coarser mesh. Alternatively, the number encountered within a set digging time could be used for large and rare species. Special scaled-down methods are required for the much more numerous meiofauna (see below).

The number of quadrats required depends on the spatial dispersion pattern (even, random or contagious) of the organisms under study, their average density and the desired precision of the estimate (Elliott, 1977). For a given quadrat size, evenly distributed organisms have a ratio of the variance to the mean approaching zero. Randomly distributed organisms have a variance/mean ratio approximating to 1, and clumped or 'contagiously' distributed species have a variance/mean ratio much greater than 1. It will be more difficult to get a precise estimate of a clumped organism. The number of quadrats required to get the desired precision can be formally calculated (see Holmes and McIntyre, 1984; Krebs, 1989). To do this an estimation of the population mean and variance is required from a pilot study where you deliberately oversample. This then allows you to gauge how many quadrats will be adequate. At worst, this approach can be used as a check of your methods on completion of the study, but ideally you should have thought about this before starting. This method is only as good as the number of samples in the pilot study and it would have to be repeated for different shores. Finally, it should be remembered that if you are interested in spatial pattern *per se*, your impression of what this is will vary with quadrat size (Legendre, 1993).

A simpler and more pragmatic approach to estimating how many samples you need is to randomize the order of the quadrats or cores from your pilot study and see how the mean, and the confidence limit of that mean, changes with two, then three, then four samples, and so on. By inspection you should be able to see when the estimates damps down on a graph (Fig. 8.2) and thus whether taking more samples significantly improves your estimate. This can also be done for estimates of diversity: the diversity for the first sample is calculated, and then the data pooled with the second sample and so on. For species richness the cumulative number of species is plotted against sampling effort; where the curve flattens out you have the number of quadrats needed to estimate the total number of species in the community. In reality the curve rarely flattens because rare species are occasionally added (Hawkins and Hartnoll, 1980). However, you can calculate the number of quadrats needed to encounter, say, 90% of the species.

Because most species are unevenly spaced in nature, it is very unlikely that counts from quadrats or cores will ever be normally

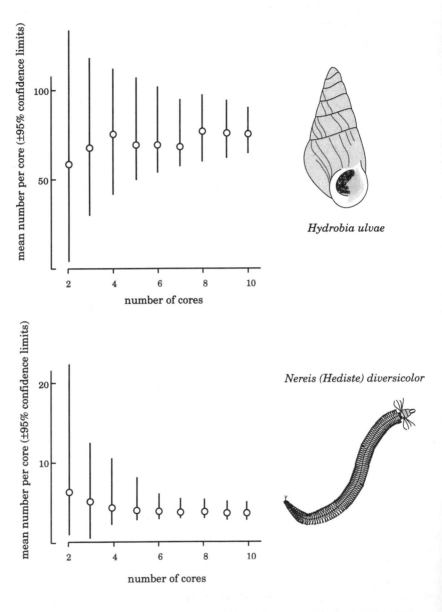

Fig. 8.2 Changes in the estimate of the mean and 95% confidence limits with increasing sample size for two estuarine invertebrates, (a) the mudsnail *Hydrobia ulvae* and (b) the ragworm *Nereis diversicolor*. For each species, six cores need to be taken to get the best estimate of the mean with the minimum of effort. Further sampling does not significantly improve the estimate. The unequal confidence intervals are because the data were transformed before calculating the mean and confidence interval and then back transformed for graphical presentation.

distributed. Similarly, when comparing two areas, variances are unlikely to be the same. Therefore care must be exercised when using standard parametric statistical techniques such as *t*-tests or analysis of variance. First the data must be appropriately transformed. If they are randomly distributed then square rooting is the best transformation. For clumped distributions it is usually safe to use a log transformation (Underwood, 1981). An alternative approach is to use non-parametric methods. These are based on fewer assumptions, but are less powerful than their parametric equivalents, and interaction terms cannot be calculated as with analysis of variance.

8.2 GENERAL METHODS FOR USE ON BOTH ROCKY AND SANDY SHORES

8.2.1 SURVEYING SHORE PROFILES

A variety of methods are available for surveying the profile of the shore or locating sampling stations at particular heights (Hawkins and Jones, 1992). These range from the simple and not very accurate, to those involving the most sophisticated levelling devices. Split-prism levels, accurate to the nearest cm, are relatively inexpensive and are recommended for most uses. Cheap levels using two vertical poles connected by either a taut piece of string or a third pole with a spirit level are very good for levelling over short distances on steep shores. Whatever method is used it has to allow levelling in relation to a set reference point or chart datum. Tide tables usually give heights above chart datum, and by taking high or low water on a set day an approximate level can be calculated. It helps if the sea is calm and the atmospheric pressure is neither too high (this depresses water level) nor too low (this raises water level). If a surveying benchmark is near by, then the shore can be levelled in relation to this reference point. This is labour intensive and for most purposes reference to the water level of a known tide is adequate. This can be improved by knowing the height in relation to chart datum of a tide mark in a sheltered harbour. If one observer notes the level on this marker at the time of low tide, this level can be related to that observed on the nearby shore by another person at the same time.

8.2.2 ESTIMATING EXPOSURE TO WAVE ACTION

The dramatic effects of exposure on distribution patterns have encouraged the development of objective methods for assessing how much wave action a shore experiences. Measurements of the forces generated by impacting waves have been made on rocky shores by fixing various gauges, meters or transducers to the rock surface (Field, 1968; Jones and Demetropoulos, 1968; Harger, 1970; Denny, 1982, 1983; Palumbi, 1984; Underwood and Jernakoff, 1984; Galley, 1991; Bell and Denny, 1994). Maximum wave action at a particular locality can be directly measured by simple devices, such as drogues attached to spring balances. However, if the exposure of the shore is to be characterized in a meaningful way, then measures need to be repeated over a long time – certainly over the same order of magnitude as the life of the organisms under study, because they may later be affected by rare catastrophic events. Conversely, recording maximum forces of waves tells us little about average water movement, which may be more important for nutrient uptake in seaweeds and filter-feeding in barnacles. The rate of dissolution of balls of plaster of Paris has been used to get an idea of water movement and breakdown of the boundary layer near to organisms (Muus, 1968). Recently, various force transducers have become more cheaply available and microcomputers can easily handle the vast amount of data generated (Denny, 1982), but these are beyond the budget of most students.

An alternative approach, applicable to both hard and soft shores, is to visually assess the wind or wave characteristics at a site, again over a reasonably long period. However, much of this routine work can be avoided by integrating standard weather data and information from hydrographic charts to derive an index of exposure (Thomas, 1986). This approach is fine for broad-scale comparisons but takes no account of local topography. Simple examination of detailed maps can often be used to rank shore exposures very effectively.

It is possible to estimate wave height and upward wave swash directly (e.g. Southward, 1953), but it is a wet and dangerous activity. One approach is to make observations against markings on a jetty or projecting rock or pole. Again the results are only as good as the period of observation but it would allow comparisons of different places on the same day and hence provide a rough ranking of shores in terms of the exposure they experience.

Many shore ecologists can only visit a site once to collect biological material, and cannot invest the time and effort required by the methods described above. Several workers have therefore derived scales for the rapid assessment of exposure, based on an extensive knowledge of the ways in which the distributions of species change with increasing and decreasing wave action. Ballantine's (1961) scheme (Fig. 8.3) is best known in the UK. In other words, exposure is defined by the overall biology of the shore itself. There are, however, pitfalls with this approach; if biological aspects are subsequently interpreted in terms of exposure, as is often done, then the reasoning becomes circular. These scales are most useful for relating physiological or morphological characteristics of organisms to exposure, for studies of physical effects on shores, such as erosion of cliffs, and for providing a readily appreciated thumbnail sketch of what the shore looks like. They can be used with caution when studying the abundance of a single species if it has not been used in compiling the index. If used for studies of organisms which form part of the index, then clearly this approach is fraught with problems. The major advantage of biological exposure scales is that they represent an integration of conditions over many years; limpets and barnacles can live for over 10 years and *Ascophyllum* for over 50 years.

With a little practice it is quite easy to use biological exposure scales to rank a series of rocky shores relative to one another within a restricted geographical area, but the larger the survey, the less reliable the rankings become, because of biogeographical factors. Thus, Ballantine's scheme works best for south-west Britain but modifications have to be made for more distant localities (e.g. Crothers, 1983). The scheme devised by Lewis (1964) is more generalized, but should really be restricted to British shores. An important assumption underlying the use of such scales is that the shore under investigation has reached some kind of biological equilibrium. However, when the succession is restarted by disturbance, species may temporarily attain a high abundance uncharacteristic of that shore (e.g. fucoid algae on exposed coasts, Chapters 4 and 7). Natural fluctuations on shores in the middle section of exposure scales such as Ballantine's can give rise to different rankings at different times (e.g. Hartnoll and Hawkins, 1985).

Rapid assessment of exposure using biological information is not practical on soft shores because of the extensive sampling required to find the fauna. McLachlan (1980) describes an exposure index for sandy beaches based on relatively easily quantified physical parameters of sediment and wave characteristics. The scheme is useful in that it can be applied to sandy shores throughout the world, but it cannot be used to discriminate between more sheltered flats with finer particle sizes.

8.3 METHODS APPLICABLE TO ROCKY SHORES

8.3.1 ESTIMATING PERCENTAGE COVER

On rocky shores there are various kinds of cover: canopy cover of large seaweeds, understorey cover by turfs of algae and sessile animals, and

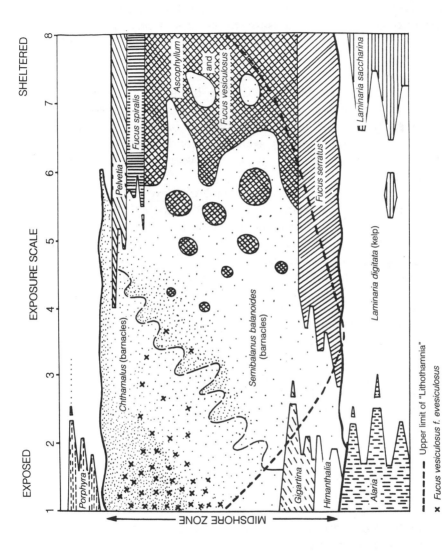

Fig. 8.3 Example of a biological exposure scale for rocky shores based on the Pembroke area, Wales, UK (modified with permission from Ballantine, 1961.)

encrustations on the rock surface itself. It is best to consider each of these separately. First, estimate the canopy cover just as it lies before pushing it aside to look at the understorey cover. Cover can be estimated using the subdivisions of a quadrat as a guide and with practice quite reliable estimates can be made. A better and more objective way of estimating cover is to use the percentage of 'hits' underneath the cross-wires of a quadrat, or the dots or holes on a transparent overlay sheet. A double layer of sighting points avoids parallax errors. Alternatively, a pin-frame can be used, which is more precise but cumbersome. Good estimates can be made with 25 or more sighting points: the larger the number, the more accurate the estimate (30–50 seem appropriate for most cases). These sighting points can be arranged regularly or randomly. For most purposes regular arrays are probably best. Random or regular dot overlays can also be used to measure percentage cover on photographs. PC-based digitizers or image analysis systems have become much cheaper and are excellent for estimating cover from photographs (M.S. Foster *et al.*, 1991; Meese and Tomich, 1992). Often the data can be entered directly into statistical packages for analysis. An assessment of various approaches is given in Meese and Tomich (1992).

8.3.2 SURFACE TOPOGRAPHY

The roughness of the surface will influence the number of refuges and the drainage of the rock surface. A good way of measuring this factor is to place a line taut across the surface and then run it along the contours of the rock. The actual length along the surface divided by the taut length provides an index of roughness (rugosity). Measurements can be made across the two diagonals of a quadrat to give a roughness figure for a particular quadrat. Alternatively a series of 1 m, 5 m or 10 m lengths of shore can be selected at random at a given shore level and the rugosity assessed using a line, as above (Trudgill, 1988).

8.3.3 OTHER FACTORS

The list of factors that can be measured is potentially very large (temperature, humidity, salinity, etc.) and the reader should consult Baker and Crothers (1987) for further information on these. Various probes and meters (e.g. salinity, relative humidity) are available and have become increasingly cheap in recent years.

8.4 METHODS APPLICABLE TO SEDIMENT SHORES

8.4.1 THE PHYSICAL AND CHEMICAL ENVIRONMENT

Particle size distribution

The main factor influencing organisms on sediment shores is the nature of the sediment. The sediment can be characterized by passing a sample through a graduated tower of sieves of decreasing mesh size and drying the sediments retained on the different sieves to work out the percentage of each grade. The Wentworth scale for grading sediments has unequal size units which can be converted to *phi* units of equal intervals for convenience. These data allow cumulative percentage curves to be plotted from which the median (average) particle size and the 25% and 75% quartiles (a measure of the spread about the average) can be found (Fig. 8.4). If the curve is very flat then the particles in a particular locality are poorly sorted by water movement; if steep, then only a very narrow range of sizes is present. Additionally, if the coarse and fine particles are equally sorted by water movement, the region between the 25% and 75% quartiles will be flat; if they are not equally well sorted, the graph will be skewed one way or the other. Various sediment parameters can therefore be derived from these curves: the slope of the interquartile range is a measure of sorting; the median particle size reflects the average grade and the skewness indicates whether fine or coarse particles are equally well sorted.

Organic matter

The amount of organic matter in the sediment will alter the density of deposit-feeding animals and the chemistry of the sediments (Chapter 1). Sediments rich in organic matter will be low in oxygen and rich in sulphides due to bacterial activity. The amount of organic matter present in the sediment can be determined by oxidizing the carbon in a sample of weighed sand and noting the loss in weight. This can be achieved by adding a very strong oxidizing agent, such as chromic acid, followed by back-titrating with an alkali to estimate how much acid was used and hence how much organic material was originally present in the sample. Whilst this method is accurate, it is also potentially dangerous and must be done under carefully controlled laboratory conditions, in a fume cupboard and with adequate safety precautions. An alternative, and usually just as satisfactory technique, is burning off the carbon in a muffle furnace at 450 °C and noting the loss in weight of the sample. Inevitably, some inorganic carbon is also oxidized but this error is usually small and presents no problems in most cases. However, it

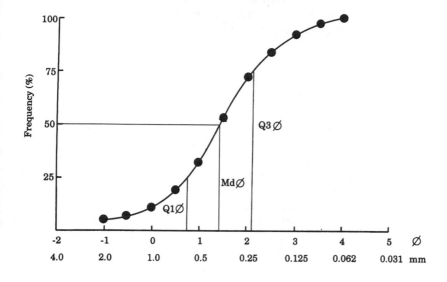

$$Md\emptyset = 1.36 \quad QD\emptyset = (2.1-0.80)/2 = 0.65$$

$$Q1\emptyset = 0.80$$

$$Q3\emptyset = 2.10$$

mm	ø	%wt. on sieve	Cumulative %	Grade
4.00	-2.0	0	0	4.00
2.83	-1.5	0	0	4.00-2.83
2.00	-1.0	5	5	2.83-2.00
1.41	-0.5	3	8	2.00-1.41
1.00	0.0	4	12	1.41-1.00
0.71	+0.5	8	20	1.00-0.71
0.50	+1.0	11	31	0.71-0.50
0.351	+1.5	23	54	0.50-0.351
0.250	+2.0	19	73	0.351-0.250
0.177	+2.5	9	82	0.250-0.177
0.125	+3.0	8	90	0.177-0.125
0.088	+3.5	6	96	0.125-0.088
0.062	+4.0	4	100	0.088-0.062

Fig. 8.4 Graphical method for estimating sediment granulometric properties. Sediment is dry fractionated by passing down a tower of sieves of decreasing mesh size. The cumulative percentage of the weight of sediment retained on each sieve is plotted against the mesh size of that sieve to produce the curve shown. The average or median particle size (Md), and the sorting coefficient (Qd), a measure of the spread of particles about the average, are then determined as shown.

would be unwise to use this technique for estimating the amount of organic material in a shell beach, where almost all of the inorganic particles will be of calcium carbonate.

Redox potential

The fauna of sediments responds markedly to the physico-chemical environment. The chemistry and microbiology of sediments are complex topics and their analytical techniques cannot be covered here. However, it is possible for the general student to obtain a broad picture of the physico-chemistry of the sediment by measuring its ability to reduce or oxidize compounds – its redox potential. This is measured as the electric potential set up between a reference and a platinum electrode placed in the sediment. By measuring this potential (often abbreviated to Eh) at different depths, the location of the redox potential discontinuity (RPD) layer is revealed (Pearson and Stanley, 1979). The RPD represents a switch from oxidizing to reducing conditions, broadly associated with aerobic and anaerobic environments, although free oxygen may not be present in much of the sediment above the RPD (Chapter 1).

On many beaches the RPD can be seen as a change in sediment colour from brown (oxidized iron) to grey or black (sulphide-reduced iron), and its depth can be measured with a ruler if a spade is used to lift and thereby fracture the sediment to reveal a clear profile. If done with care, this approach is useful for rapid surveys of beaches and for a relative ranking of beaches.

Sorting animals

The choice of mesh size when separating the fauna from the sediment using a sieve depends on the size of the animals being studied. For most bivalves, a 2 mm mesh works well, but 1 mm is the standard size used for most community studies of sandy beaches. A wide variety of animals are retained, yet using the sieve and sorting the animals is not too time consuming. A 0.5 mm mesh retains even more animals, especially juveniles, but on sandy beaches it will rapidly clog with larger particles, making subsequent sorting a nightmare. However, a 0.5 mm mesh is essential if the main subjects of interest are small worms or the early stages of infauna.

Meiofauna

The densities of meiofauna in beaches can be many times those of macrofauna (Chapter 6) and sampling devices need to be scaled down considerably. On sandy beaches, a plastic corer of only 2 cm or 3 cm in diameter will usually provide hundreds of nematodes, and on muddy flats one can often get away with a 1–2 cm diameter corer. A cheap and useful device is a plastic disposable syringe with the nozzle wall ground off. The injector piston can be pulled out gradually as the corer is inserted into the sediment, thereby preventing compression of the core and providing a good vacuum ensuring its successful removal, which is always a problem with small corers. The piston can then be used to push out discrete sections of sediment so that depth distributions of meiofauna can be investigated.

Meiofauna are usually smaller than the particles making up the sediment they inhabit and thus cannot be extracted by sieving. In coarse sand habitats, individuals can be removed by first fixing the sample with formalin (or alcohol) containing the stain Rose Bengal. This releases sticky species from the sediment and stains the meiofauna bright red, aiding subsequent sorting and counting. The sample is then shaken with water to suspend both the meiofauna and the sediment particles. After a few seconds, the heavier sand particles will sink to the bottom of the container, whilst most of the meiofauna will remain suspended in the overlying water. This can then be carefully poured through a 45 μm mesh which retains the meiofauna. If this process is repeated several times then up to 95% of the meiofauna can be extracted. The efficiency of this method should be checked by scanning the sediment afterwards. Any remaining red-stained meiofauna show up very clearly. This elutriation approach can be semi-automated by passing sea water from a tap supply through a sample at a sufficient rate to keep the material in suspension, but not so forcefully as to carry over sediment particles into the sieve. Of course, heavy organisms, such as ostracods, foraminiferans and bivalve spat, cannot be efficiently separated from sand particles using this technique.

Elutriation is not feasible for fine-grained or muddy sediments where particles have more or less the same settling velocities as the animals. The most useful way of dealing with such samples is to suspend them in a solution which has a specific gravity of about 1.15 (e.g. supersaturated magnesium sulphate), less than animal tissue, but more than that of inorganic particles, and leave the suspension to stand overnight. Sediment particles sink through the solution and the less dense meiofauna float to the surface where they can later be skimmed off.

There are many refinements to these methods appropriate for different types of samples and different meiofaunal taxa. Some taxa are in fact impossible to identify once they have gone through these processes and must be examined alive. Readers who are at all serious about working on meiofauna should consult the excellent texts by Hulings and Gray (1971) and Higgins and Thiel (1988).

8.5 LONG-TERM STUDIES

Long-term studies of whole communities can be made along transects, but in many cases this demands too much in time and resources. Many monitoring studies have therefore concentrated on fixed areas at selected shore levels of interest. During the 1970s there was considerable interest in monitoring rocky shores around the British coastline. The 'fixed site – key species' approach was advocated by Lewis and co-workers (Lewis, 1976). The restriction of study to a few important species has much to commend it. In any rocky shore community, there are usually a limited number of space-dominating species (mussels, barnacles and algae), and a few important sedentary predators and grazers (Chapter 4). Following events at a fixed site is a good way of understanding what is happening in that area, such as the relocation of elements in a species mosaic and the rate at which it occurs (Hawkins and Hartnoll, 1983b; Hartnoll and Hawkins, 1985). It can also give good indications of recruitment and mortality. Unfortunately, a fixed-site approach presents problems with subsequent statistical analysis as the repeated measures are clearly not independent of each other. Comparisons with other areas are virtually impossible.

An alternative approach is to define areas of shore and use sets of random quadrats on each sampling occasion. This will not tell you how the different elements of a species mosaic change place with time, but will allow comparisons of the overall abundance and distribution of species in an area, and allow statistical comparisons between times and places. This is the only option for sediment shores. Care must be taken that a large enough area is sampled to minimize sampling damage and so that areas previously sampled are not re-sampled. If resources are limited then one should concentrate on repeat random sampling. Again it is also important to use a nested approach and study at least two areas on a particular shore, and more than one shore in a particular region. Good photographs of the whole shore and the whole area under study are strongly recommended. Photographs are particularly useful for monitoring settlement or the population dynamics of barnacles in small areas.

The frequency of sampling depends on the aims of the study. If seasonal patterns are to be described, then monthly sampling is needed, with perhaps more frequent sampling during key periods such as barnacle settlement seasons. Longer-term studies demand much less frequent sampling. A frequency as low as six or four times a year on rocky shores will still give some idea of seasonal differences between years. One advantage of reducing the sampling frequency is that it lessens damage due to trampling. On sediment shores in temperate areas it has been suggested that an early-spring and a late-autumn sample will give sufficient information for long-term studies. The massive increases during the spring and summer settlement season will not be detected, but more stable longer-term patterns of numbers following the intense mortality that accompanies juvenile settlement will be recorded. However, if the aim of the study is to look at recruitment variation then more frequent sampling must be made.

In all long-term monitoring programmes it is important to be able to separate short-term variation from longer-term trends which might be due to human activities. The programme must therefore be designed so as to maximize the chances of detecting change and to ensure that the cause of any between-location differences can be separated from chance area-effects. Good advice on how to design such programmes is given by Morrisey *et al.* (1992) and Underwood (1992).

8.6 FIELD EXPERIMENTS

8.6.1 ROCKY SHORES

Rocky shores provide an ideal system for experimental ecology. Some of the earliest ecological experiments were done on rocky shores (Chapters 3 and 4) and the results of these have entered the general ecological literature, particularly those studies on biological interactions. An important difference between field and laboratory experiments is that in the latter, all factors are held constant except for the one under investigation. In contrast, for field experiments it is assumed that all factors will vary, but in a similar way in the treatments and controls, except for the factor under investigation.

Up until the mid 1970s, doing experimental ecology at all was an advance on the traditional descriptive approach. Unfortunately, a significant number of the rush of field experiments carried out in the 1970s and 1980s were not rigorously designed or analysed. Their results need to be viewed with caution, although the conclusions drawn from them may still be correct. These include some of the experiments now regarded as classics. Two main problem areas have been identified. The first is lack of true replication and the tendency for pseudoreplication. The second is lack of independence. Good and very readable accounts of the design requirements of field experiments and the subsequent analyses of the data are given in Hurlbert (1984), Underwood (1983, 1986, 1988, 1991), Hairston (1989), Krebs (1989), and Maxwell and Delaney (1990).

For the results of an experiment to be convincing there must be clear-cut differences between the treatment and control. The results will be more convincing and amenable to statistical analysis if the differences are consistently shown in more than one treatment area compared with more than one control area. Otherwise it might be argued that the effects are due to the different locations on the shore rather than the experimental treatment. Thus it is imperative that there be sufficient replicates of the experimental and control areas (Fig. 8.5).

In other words, each treatment (e.g. predator or grazer removal) must be dispersed several times throughout the community under investigation. Dividing the shoreline into two and removing predators from one half has in the past yielded dramatic changes which are convincing to all but the harshest critics. Good examples are the early grazer removal experiments of Jones (1948) and Burrows and Lodge (1950) described in Chapter 4. Strictly speaking, however, such experiments cannot be analysed using statistics because there is no replication of treatments. Few would doubt the findings of Jones or Burrows and Lodge, but it would have made more sense perhaps to have many smaller areas of shore cleared of limpets with controls sited alongside each cleared area. Nevertheless, for many questions, small-scale experiments may not deliver the right information. A good example is the problem of trying to predict the impact and consequences of a regional-scale disaster like the *Torrey Canyon* oil spill (Chapters 4 and 7) from the results of small-scale experiments on limpet exclusion. The importance of a correct and acceptable experimental design, which takes account of appropriate scale, is paramount where the significance and generality of particular shore processes is the subject of debate, as in the case of keystone predators (Chapter 4).

For effects to be convincing it is important, therefore, to maximize the number of experimental (and control) areas. Taking many replicate samples within each experimental and control area is *not* equivalent to treatment replication and has been termed **pseudoreplication** (Hurlbert, 1984; Fig. 8.5). Both of us have been guilty of this sin. There is no reason why the investigator should not take many samples within a treatment replicate, but it would be good practice to pool these so that there is only a single estimate (but one in which we have a lot of confidence) for each replicate. Often a great deal of labour and expense is involved in setting up field experiments, and ecologists cannot afford the luxury of enormous numbers of treatment replicates. On the other hand, they cannot afford to get the design wrong. There are procedures which

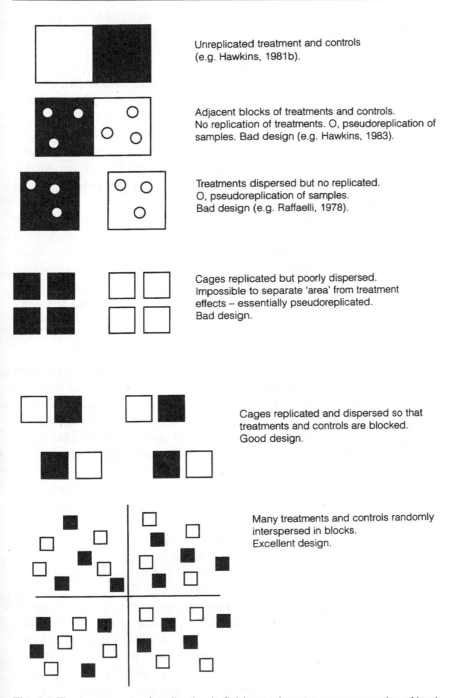

Fig. 8.5 The importance of replication in field experiments: some examples of bad and good designs, some perpetrated by the authors.

allow one to estimate the amount of replication required to demonstrate an effect of a given size, if the normal variability of the parameter being measured (e.g. prey density) is known; Cohen (1977) provides a good introduction to this **power analysis** (Fig. 8.6). Often, however, field experiments yield surprising and unexpected results – indeed that is one of their main attractions – and then it becomes impossible to predict what the number of replicates should be. Even if you do not carry out power analysis to plan the experiment, there is a lot to be said for performing a retrospective analysis so that you can decide how effective the experimental design was. This is particularly important for interpreting experiments in which no treatment effects were apparent (see, for example, Raffaelli and Hall, 1992).

Apart from the problem of appropriate scale discussed above, there are other circumstances where replication is just not possible. A good example would be a study comparing the influence on shore populations of cessation of human activities in a nature reserve compared with adjacent affected areas. It is difficult enough getting one nature reserve established, let alone replicates. In this case the only answer is to do several studies in the control areas either side of the reserve – at least results in the reserve can be put into the context of the between-site variation (Morrisey *et al.*, 1992; Underwood, 1992).

The problem of independence is well illustrated by experiments with contiguous arrays of treatments and controls. What goes on in one part of the experiment may influence what goes on next door. A good example would be a limpet exclusion experiment of the following design: a central treatment with a contiguous fenced control on one side and a contiguous unfenced control on the other side. Such matched-set experiments are often set up to ensure that the conditions in the treatment and the controls are as similar as possible, but most experimenters would secretly admit that it also saves on the number of holes needed to fix cages and on fencing materials! In the example used here, it could be argued that highly mobile herbivores, such as amphipods and isopods, might take advantage of the new habitat afforded by the seaweed springing up in the limpet exclosure (A.J. Underwood, pers. comm.). These could forage in the immediately adjacent control areas and prevent further colonization by weed, hence exaggerating any differences between treatments and controls. Thus there would remain a question mark over the results of this experiment. If the experimental treatments and controls had been located further apart (although still grouped in blocks), this problem would not have arisen (Thrush *et al.*, 1994). However, the experiment should not be so large in spatial extent that the blocks are sited at different tidal heights or experience different degrees of exposure. Furthermore, if the results are

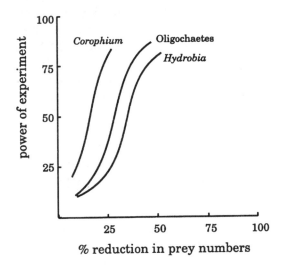

Fig. 8.6 Analysis of the power of a predator-exclosure experiment carried out on an intertidal mudflat. The graph shows how powerful the experimental design employed was for detecting different reductions in the densities of three prey species (modified with permission from Hall *et al.*, 1991). For instance, there was a 75% chance of detecting a 25% reduction in *Corophium* numbers.

to be of general significance, they should be consistent on several shores, which in turn means there should be within-shore replication giving a hierarchical experimental design. In practice this is not often possible.

The above example illustrates the problem of lack of independence in space, but similar problems occur with lack of independence in time. A good example is where the time course of an effect needs to be examined, for example an increase in size or a decrease in numbers with time. Statistical analysis cannot be carried out if the same replicate is examined several times: what you see in an area at time 2 depends on what was present in that area at time 1. One way around this is to have large numbers of replicates and only examine a few at random on each occasion, these never being sampled again. A less resource-intensive approach is to treat the whole time course as just one replicate and fit a line or a curve through the data. Treatment effects can then be assessed by comparing the different lines, using analysis of covariance or matched-set statistics, especially if several replicates are available per treatment. Although there are also statistical difficulties with this approach, at least the pattern with time is described, which in many cases is important information. Furthermore, testing for an interaction between treatment and time is sometimes a trivial analysis – you expect things to grow or multiply. What matters is the between-treatment difference in rates, intercepts or the asymptotes.

If the statistical methods to be employed are considered when designing the experiment, it often becomes clear that certain assumptions will be violated. The experiment can then be modified to allow the statistical analysis desired, or an alternative statistical method can be sought. If this is done at the start, then all sorts of problems can be avoided. Too often statistical analysis is an afterthought. It is particularly important to employ designs suitable for analysis of variance when exploring the possible interactions between two or more factors. Underwood (1981) gives a good introduction to analysis of variance in marine experimental ecology and Maxwell and Delaney (1990) provide more general advice on experimental design and statistical analysis.

A frequently encountered problem in designing field experiments is deciding what constitutes a sensible control. Often several kinds of control have to be built. For instance, if limpets are excluded by a fence, then they should be enclosed within a fence as one control and a further control should be a totally untouched area. If there are no differences between the two types of control, then anything that occurs in the treatment must be real. Controls for exclusion cages present all kinds of problems. Is a cage with the sides removed and having only a roof really a control? Does the behaviour of predators contained within a cage at background densities really represent what happens under natural conditions? A little common sense and ingenuity can often overcome these problems. One approach is to design the experiment so that any bias is against the hypothesized result. If a difference does occur between a treatment and a control, you can be confident that it is a real effect.

Constantly removing animals from an area as an alternative to exclusion cages is at first sight an attractive proposition (but not for the animals or for conservation). There are no cages to build or maintain and no complex controls for cage effects. However, if a grazer or a predator is consistently removed from an area, the density in the overall area will go down. This could influence density in the controls unless they are some distance away. If they are, the likelihood of their representativeness will be diminished. Therefore cages are preferred for these kinds of experiments. An alternative approach is to make use of isolated rocks or outcrops of large boulders with natural barriers to immigration, such as a stretch of sand or a gully. Removal experiments work well under these circumstances.

One further problem that must be considered is the size of treatment and control plots, exclusion cages, canopy removal areas, etc. If there are many very small plots, spatial variation can be exaggerated compared with that in larger plots. For example, early-settling algae may dominate an area, but it could be too small for a larger but rarer alga to establish, which would be more likely in a larger plot. Edge effects and cage effects will also be exaggerated in smaller treatment plots: the ratio of edge to enclosed area will be higher. Larger cages are, however, more difficult to maintain and are likely to sustain damage. With canopy removal experiments it is essential to have large plots with a bordering cordon sanitaire or buffer zone to prevent sweeping effects around the edge. Most intertidal experiments (rocky and sandy shore) involve treating areas of less than 1 m². Clearly this is not adequate for large seaweeds of more than 1 m length.

At the practical level the biggest problem encountered on rocky shores is marking areas and fixing structures. Drilling holes with a power tool of some kind is the most usual solution. In very soft rock a hand drill can be used. In slightly harder rock a cordless drill is fine, but lots of rechargeable batteries need to be taken into the field. Drills can be powered by compressed air from scuba cylinders and this is a cheap (but cumbersome) option if you have a ready supply of cylinders and access to a compressor (Hawkins and Hartnoll, 1979). For really hard granites or consolidated sandstones the most effective solution is a good petrol-powered drill. These can be bought for about £400–600 and are excellent. A less satisfactory alternative, from the safety perspective, is to use a portable generator and an ordinary hammer drill, remembering that great care must be taken with electricity in a salty, wet environment. Once a hole has been drilled, plastic plugs provide secure fastenings and good markers. Waterproof epoxy glues are now widely available and these are extremely useful for fixing transplanted rocks or boulders on the shore and can also be used to attach cages and other structures. Quick-set, do-it-yourself cement is also useful for temporary fixtures, but it does not last very long (a few weeks at most).

8.6.2 FIELD EXPERIMENTS IN SEDIMENT SYSTEMS

Field experiments on beaches and tidal flats are nowhere nearly as straightforward as on rocky shores. Exclosure and enclosure experiments are sometimes carried out with little appreciation of the potential difficulties involved and it is likely that many such experiments come to grief and are never reported in the literature. Those that do get published have served as a warning to others not to get too involved with this kind of work (e.g. Virnstein, 1978). It would be wrong, though, to think that the potential problems cannot be overcome, and often they evaporate altogether.

Of course, all of the experimental design considerations outlined above for rocky shores apply to experiments in sediment. In addition, the physical construction of the exclosure or enclosure becomes very important. Whenever a structure such as a cage is placed on a beach it is likely to change the physical and hence the biological characteristics of the caged area to some degree. For instance, not only will a mesh exclude or include predators, but it will also disrupt or slow down the flow of water passing over, through and around the cage. This water will carry fine particles in suspension and any decrease in velocity as the water passes through the mesh will result in finer particles settling out and the cage silting-up with fine sediment rich in organic matter. Living particles (e.g. larvae) may also be deposited passively within the cage by the same process, or larvae and juveniles may actively select the finer sediments. When such artefacts occur, it is difficult to interpret any biological differences between caged and uncaged areas. The differences could be due either to the experimental treatment (e.g. absence of the predators) or to the artefacts created by changes in local hydrography.

The scale of these unwanted artefacts will vary with the specifications of the mesh. Small-aperture mesh made of thick or heavy-gauge material will slow the flow of water much more than a large mesh made of fine wire. The investigator should opt for the largest mesh possible with the maximum 'open' area. Mesh problems can also be alleviated by regular cleaning and maintenance to prevent clogging by detached weed, which effectively decreases the mesh aperture.

Another potential problem with caging experiments in sediments is the scouring of the sediment around the cage supports and where the mesh makes contact with the sediment surface. Where this is thought likely, the cage should cover a sufficiently large area of sediment to ensure that those regions affected need not be sampled. This is a good policy anyway, even if the risk of scouring is not thought important, because edge effects are likely to occur in any caging experiment.

The very presence of a cage may locally increase the abundance of small epibenthic predators within and around the structure. Fish are attracted to underwater objects, and smaller species, such as shrimps and crabs, may gain protection inside cages, leading to overpredation within the enclosure. This has been a problem in some caging experiments in the shallow sublittoral (Pihl, 1985; Hall *et al.*, 1990b), but it may be less common in intertidal studies, where most epibenthic predators are not normally present on mudflats throughout the tidal cycle. Large intertidal cages can also provide a perch for birds, such as crows and gulls during the low tide period. These species may feed (and defecate) more than usual around the cages and their presence may scare away waders.

OVERVIEW

When studying the seashore, try to observe patterns over several shores and over time. Sometimes experiments can be established straight away on the basis of qualitative observations, but usually it is better to do some quantitative work to describe the observed patterns first. Monitoring will provide useful information about temporal changes in populations and may lead to the formulation of hypotheses for testing with a well-designed experiment. Keep the appropriate statistics in mind from the very beginning and erect alternative hypotheses to be tested. It is always better to set up experiments properly at the beginning and to be aware of potential problems; but sometimes these can appear so overwhelming to students that the experiment is never attempted. In our experience, many of the potential problems never actually materialize, but you will not find this out until you try. It is better to do an experiment with an element of uncertainty (you are unlikely to have covered every eventuality) than to attempt no experiment at all. The experience gained will lead to a better design next time.

Most importantly, get out on the shore. Get to know the fauna and flora and appreciate their natural history. Good ideas usually come from an intuitive appreciation of the system under study. These ideas can then be logically scrutinized and formally tested. Trying to shoehorn your system into some trendy theoretical concept is not always the best way to proceed, although an appreciation of general theoretical ideas will help put your work into a broader context. Whatever you do, it is bound to be interesting!

References

Al-Ogily, S.M. (1985) Further experiments on larval behaviour of the tubiculous polychaete *Spirorbis inornatus* L'Hardy and Quiévreux. *J. exp. mar. Biol. Ecol.*, **86**, 285–95.

Alfredson, G. (1984) The aboriginal use of St. Helena Island, Moreton Bay – the Archaeological evidence, in *Focus on Stadbroke* (eds R.J. Coleman, J. Covacevich and P. Davie), Boolarong Publications, Brisbane, pp. 1–15.

Allen, P.L. and Moore, J.J. (1987) Invertebrate macrofauna as potential indicators of sandy beach instability. *Est. coast. Shelf Sci.*, **24**, 109–25.

Alves-Coelho, P. and Ramos-Porto, M. (1980) Bentos litoraneo do nordeste oriental do Brasil, 1. Povoamento dos substratos duros. *Bol. Inst. Oceanogr. Sao Paulo*, **29**, 133–4.

Alzieu, C., Sanjuan, J., Deltreil, J.P. and Borel, M. (1986) Tin contamination in Arcachon Bay, effects on oyster shell anomalies. *Mar. Poll. Bull.*, **17**, 494–8.

Ambrose, W.G. (1991) Are infaunal predators important in structuring marine soft-bottom communities? *Am. Zool.*, **31**, 849–60.

Amoureux, L., Rullier, F. and Fishelson, L. (1978) Systematics and ecology of Polychaeta on the Sinai Peninsula. *Isr. J. Zool.*, **27**, 57–163.

Andrew, N.L. and Mapstone, B.D. (1987) Sampling and the description of spatial pattern in marine ecology. *Oceanogr. mar. Biol. ann. Rev.*, **25**, 39–90.

Appleton, R.D. and Palmer, A.R. (1988) Water-borne stimuli released by predatory crabs and damaged prey induce more predator–resistant shells in a marine gastropod. *Proc. Natl. Acad. Sci. USA*, **85**, 4387–91.

Ardré, F. (1970) Contribution à l'étude des algues marines du Portugal. I. La flore. *Portug. Acta Biol. Ser. B Biogeogr.*, **10**, 137–555.

Armisen, R. and Galatas, F. (1987) Properties and uses of agar, in *Production and Utilization of Products from Commercial Seaweeds.* (ed. D.J. McHugh) *FAO Fish. tech. Pap.* no. 288, pp. 1–57.

Asmus, H. and Asmus, R. (1985) The importance of the grazing food chain for energy flow and production in three intertidal sand bottom communities of the northern Wadden Sea. *Helgoländer Wiss. Meeresunters*, **39**, 273–301.

Audouin, J.V. and Milne-Edwards, H. (1832) *Recherches pour Servir a l'Histoire Naturelle au Littoral de la France*. Tome I. *Introduction.* Cruchard, Paris, 302 pp.

Augier, H. (1980) La cartographie biocenotique repetitive a grande echelle et a reperage metrique pour l'etude fine de l'evolution des peuplements marins benthiques. Application dans l'ile de Porquerolles (Mediterranee-France). *Trav. Sci. Parc. Natl. Port Cros.*, **8**, 11–18.

Baird, D. and Milne, H. (1981) Energy flow in the Ythan Estuary, Aberdeenshire, Scotland. *Est. coast. Shelf Sci.*, **12**, 217–32.

Baird, D., Evans, P.R., Milne, H. and Pienkowski, M.W. (1985) Utilisation by shorebirds of benthic invertebrate production in intertidal areas. *Oceanogr. mar. Biol. ann. Rev.*, **23**, 575–97.

Baker, J.M. (1979) Responses of salt marsh vegetation to oil spills and refinery effluents, in *Ecological Processes in Coastal Environments* (eds R.L. Jefferies and A.J. Davy), Blackwell Scientific, Oxford, pp. 529–42.

Baker, J.M. and Crothers, J.H. (1987) Intertidal rock (biological surveys/coastal zone), in *Biological Surveys of Estuaries and Coasts* (eds J.M. Baker and W.J. Wolff), Cambridge University Press, Cambridge, 449 pp.

Baker, J.M. and Wolff, W.J. (eds) (1987) *Biological Surveys of Estuaries and Coasts.* (Estuar. Brackish Wat. Sci. Assoc. Handb.), Cambridge University Press, Cambridge.

Baker, S.M. (1909) On the causes of the zoning of brown seaweeds on the seashore, II. *New Phytol.*, **9**, 54–67.

Ballantine, W.J. (1961) A biologically defined exposure scale for the comparative description of rocky shores. *Field Studies*, **1**, 1–19.

Bally, R. (1981) The ecology of three sandy beaches on the west coast of South Africa, PhD thesis, University of Cape Town, 404 pp.

Bally, R. and Griffiths, C.L. (1989) Effects of human trampling on an exposed rocky shore. *Int. J. env. Stud.*, **43**, 115–25.

Banus, M.D. (1983) The effects of thermal pollution on red mangrove seedlings, small trees, and on mangrove reforestation, in, *Coral Reefs, Seagrass Beds and Mangroves: Their Interactions in the Coastal Zones of the Caribbean.* Report of a workshop held at West Indies Laboratory, St. Croix, U.S. Virgin Islands (eds J.C. Ogden and E.H. Gladfelter), UNESCO Div. Mar. Sci., Paris, France, Vol. 23, pp. 114–27.

Barilotti, D.C. and Zertuche-Gonzales, J.A. (1990) Ecological effects of seaweed harvesting in the Gulf of California and Pacific Ocean of Baja California and California. *Hydrobiologia*, **204/205**, 35–40.

Barkai, A. and Branch, G.M. (1988) The influence of predation and substratal complexity on recruitment to settlement plates, a test of the theory of alternative states. *J. exp. mar. Biol. Ecol.*, **124**, 215–37.

Barkai, A. and McQuaid, C. (1988) Predator–prey role reversal in a marine benthic ecosystem. *Science*, **242**, 62–4.

Barnes, H. (1956) Surface roughness and the settlement of *Balanus balanoides* (L.). *Arch. Soc. Zool. Bot. Fenn. Vanamo*, **10**, 164–8.

Barnes, H. and Barnes, M. (1957) Resistance to desiccation in intertidal barnacles. *Science*, **126**, 358.

Barnes, H. and Barnes, M. (1977) The importance of being a "littoral" nauplius. In *Biology of Benthic Organisms* (eds B.F. Keegan, P. O'Ceidigh and P.J.S. Boaden), Pergamon Press, Oxford, pp 45–56.

Barnes, H. and Powell, H.T. (1950) The development, general morphology and subsequent elimination of barnacle populations, *Balanus crenatus* and *B. balanoides*, after heavy initial settlement. *J. Anim. Ecol.*, **19**, 175–9.

Barnes, R.S.K. (1980) *Coastal Lagoons*, Cambridge University Press, Cambridge, 106 pp.

Barnes, R.S.K. and Hughes, R.N. (1982) *An Introduction to Marine Ecology*, Blackwell Scientific, Oxford.

Barratt, J.H. and Yonge, C.M. (1958) *Collins Pocket Guide to the Sea-Shore*. Collins, London.

Bayne, B.L. (1964) Primary and secondary settlement in *Mytilus edulis* (Mollusca). *J. Anim. Ecol.*, **33**, 516–23.

Bebianno, M.J. and Langston, W.J. (1991) Metallothionein induction in *Mytilus edulis* exposed to cadmium. *Mar. Biol.*, **108**, 91–6.

Bebianno, M.J. and Langston, W.J. (1992a) Cadmium induction of metallothionein synthesis in *Mytilus galloprovincialis*. *Comp. Biochem. Physiol.*, **103C**, 79–85.

Bebianno, M.J. and Langston, W.J. (1992b) Metallothionein induction in *Littorina littorea* (Mollusca, Prosobranchia) on exposure to cadmium. *J. mar. biol. Ass. U.K.*, **72**, 329–42.

Beek, F.A., Rijnsdorp, A.D. and Clerck, R. (1989) Monitoring juvenile stocks of flatfish in the Wadden Sea and the coastal areas of the southeastern North Sea. *Helgoländer wiss. Meeresunters.*, **43**, 3–4.

Beer, S. and Kautsky, L. (1992) The recovery of net photosynthesis during rehydration of three *Fucus* species from the Swedish coast following exposure to air. *Bot. Mar.*, **35**, 487–91.

Begon, M., Harper, J.L. and Townsend, C.R. (1986) *Ecology: Individuals, Populations and Communities*, 1st edn, Blackwell Scientific, Oxford 945pp.

Behrens-Yamada, S. (1987) Geographic variation in the growth rates of *Littorina littorea* and *L. saxatilis*. *Mar. Biol.*, **96**, 529–34.

Behrens-Yamada, S. and Mansour, R.A. (1987) Growth inhibition of native *Littorina saxatilis* (Olivi) by introduced *L. littorea* (L.). *J. exp. mar. Biol. Ecol.*, **105**, 187–96.

Bell, E.C. and Denny, M.W. (1994) Quantifying "wave exposure": a simple device for recording maximum velocity and results of its use at several field sites. *J. exp. mar. Biol. Ecol.*, **181**, 9–29.

Bell, S.S. and Coull, B.C. (1980) Experimental evidence for a model of juvenile macrofauna–meiofauna interactions, in *Marine Benthic Dynamics* (eds K.C. Tenore and B.C. Coull), University of South Carolina Press, Columbia, SC, pp. 179–92.

Ben-Tuvia, A. (1973) Man-made changes in the Eastern Mediterranean Sea and their effect on the fishery resources. *Mar. Biol.*, **19**, 197–203.

Benedetti-Cecchi, L. and Cinelli, F. (1992a) Effects of canopy cover, herbivores and substratum type on patterns of *Cystoseira* spp. settlement and recruitment in littoral rockpools. *Mar. Ecol. Progr. Ser.*, **90**, 183–91.

Benedetti-Cecchi, L. and Cinelli, F. (1992b) Canopy removal experiments in *Cystoseira*-dominated rockpools from the western coast of the Mediterranean (Ligurian Sea). *J. exp. mar. Biol. Ecol.*, **155**, 69–83.

Bennett, I. and Pope, E.C. (1960) Intertidal zonation of the exposed rocky shores of Tasmania and its relationship with the rest of Australia. *Aust. J. mar. freshw. Res.*, **11**, 182–221.

Berrill, M. and Berrill, D. (1981) *A Sierra Club Naturalist's Guide to the North Atlantic Coast, Cape Cod to Newfoundland*, Sierra Club Books, San Francisco, 440 pp.

Bertness, M.D. (1985) Fiddler crab regulation of *Spartina alterniflora* production on a New England salt marsh. *Ecology*, **66**, 1042–55.

Bertness, M.D. (1992) The ecology of a New England saltmarsh community. *Am. Scient.*, **80**, 260–68.

Bertness, M.D., Gaines, S.D., Bermudez, D. and Sanford, E. (1991) Extreme spatial variation in the growth and reproductive output of the acorn barnacle *Semibalanus balanoides*, *J. exp. mar. biol. and ecol.*, **75**, 91–100.

Beukema, J.J., Wolff, W.J. and Brouns, J.J.W.M. (eds) (1990) *Expected Effects of Climate Change on Marine Coastal Ecosystems*, Kluwer, Dordrecht, 221 pp.

Biebl, R. (1952) Ecological and non-environmental constitutional resistance of the protoplasm of marine algae. *J. mar. biol. Ass. U.K.*, **31**, 307–15.

Bishop, M.W.H. (1951) Distribution of barnacles by ship. *Nature*, **167**, 531.

Boaden, P.J.S. (1989) Meiofauna and the origins of the metazoa. *Zool. J. Linn. Soc.*, **96**, 217–27.

Boaden, P.J.S. and Platt, H.M. (1971) Daily migration patterns in an intertidal meiobenthic community. *Thalass. Jugosl.*, **7**, 1–12.

Boates, J.S. and Smith, P.C. (1979) Length–weight relationships, energy content and the effects of predation on *Corophium volutator*. *Proc. Nova Scotia Inst. Sci.*, **29**, 489–99.

Bodin, P. and Klinger, T. (1986) Coastal uplift and mortality of intertidal organisms caused by the September 1985 Mexico earthquakes. *Science*, **233**, 1071–3.

Bokenham, N.A.M. (1938) The colonization of denuded rock surfaces in the intertidal region of the Cape Peninsula. *Ann. Natal Mus.*, **9**, 47–81.

Bolton, J.J.G. (1981) Community analysis of vertical zonation patterns on a Newfoundland rocky shore. *Aquat. Bot.*, **10**, 299–316.

Boorman, L.A., Goss-Custard, J.D. and McGrorty, S. (1989) *Climatic Change, Rising Sea Level and the British Coast*, HMSO, London, 24 pp.

Bowman, R.S. and Lewis, J.R. (1977) Annual fluctuations in the recruitment of *Patella vulgata*. *J. mar. biol. Ass. U.K.*, **57**, 793–815.

Brafield, A.E. (1978) *Life in Sandy Shores* (Institute of Biology Studies in Biology No. 89), Edward Arnold, London.

Branch, G.M. (1976) Interspecific competition experienced by South African *Patella* species. *J. Anim. Ecol.*, **45**, 507–29.

Branch, G.M. (1979) Aggression by limpets against invertebrate predators. *Anim. Behav.*, **27**, 408–10.

Branch, G.M. (1981) The biology of limpets: physical factors, energy flow and ecological interactions. *Oceanogr. mar. Biol. ann. Rev.*, **19**, 235–380.

Branch, G.M. (1985) Limpets: evolution and adaptation, in *The Mollusca*, Vol. 10, *Evolution* (eds E.R.Trueman and M.R. Clarke), Academic Press, New York, pp. 187–220.

Branch, M. and Branch, G.M. (1981) *The Living Shores of Southern Africa*, C. Struik, Cape Town, 272 pp.

Branch, G.M. and Griffiths, C.L. (1988) The Benguela ecosystem. Part 5. The coastal zone. *Oceanogr. mar. Biol. ann. Rev.*, **26**, 395–486.

Branch, G.M. and Marsh, A.C. (1978) Tenacity and shell shape in 6 *Patella* species, adaption features. *J. exp. mar. Biol. Ecol.*, **34**, 111–30.

Branch, G.M. and Newell, R.C. (1978) A comparative study of metabolic energy expenditure in the limpets *Patella cochlear*, *P. oculus* and *P. granularis*. *Mar. Biol.*, **49**, 351–61.

Branch, G.M., Harris, J.M., Parkins, C., Bustamante, R.H. and Eekhout, S. (1992) Algal 'gardening' by grazers, a comparison of the ecological effects of territorial fish and limpets, in *Plant–Animal Interactions in the Marine Benthos* (eds D. John, S.J. Hawkins and J. Price) (Systematics Association Special Volume No 46), Clarendon Press, Oxford, pp. 405–24.

Brattström, H. (1980) Rocky-shore zonation in the Santa Maria area, Columbia. *Sarsia*, **65**, 163–226.

Brock, J.C., McClain, C.R., Luther, M.E. and Hay, W.W. (1991) The phytoplankton bloom in the northwestern Arabian Sea during the southwest monsoon of 1979. *J. Geophys. Res. C Oceans*, **96**, 623–42.

Broekhuysen, G.J. (1941) A preliminary investigation of the importance of desiccation, temperature and salinity as factors controlling the vertical distribution of certain intertidal marine gastropods in False Bay, South Africa. *Trans. Roy. Soc. S. Afr.*, **28**, 255–92.

Brosnan, D.M. (1992) Ecology of tropical rocky shores, plant–animal interactions in tropical and temperate latitudes, in *Plant–Animal Interactions in the Marine Benthos* (eds D. John, S.J. Hawkins and J. Price) (Systematics Association Special Volume No 46), Clarendon Press, Oxford, pp. 101–32.

Brown, A.C. (1982) The biology of sandy beach whelks of the genus *Bullia* (Nassariidae). *Oceanogr. mar. Biol. ann. rev.*, **20**, 309–61.

Brown, A.C. and McLachlan, A. (1990) *Ecology of Sandy Shores*, Elsevier, Amsterdam, 328 pp.

Brown, B.E. (1990) Coral bleaching. *Coral Reefs*, **8**, 153–232.

Brown, D.S. (1971) Ecology of Gastropoda in a South African mangrove swamp. *Proc. malac. Soc. Lond.*, **39**, 263–79.

Brown, V.B., Davies, S.A. and Synnot, R.N. (1990) Long term monitoring of the effects of treated sewage effluent on the intertidal macroalgal community near Cape Schranck, Victoria, Australia. *Bot. Mar.*, **33**, 85–98.

Brusca, R.C. (1980) *Common Intertidal Invertebrates of the Gulf of California*, University of Arizona Press, Tucson, 513 pp.

Bryan, G.W. and Gibbs, P.E. (1991) Impact of low concentrations of tributyltin (TBT) on marine organisms, a review, in *Metal Ecotoxicology: Concepts and Applications* (eds M.C. Newman and A.W. McIntosh), Lewis Publishers, Boston, pp. 323–61.

Bryan, G.W. and Hummerstone, L.G. (1971) Adaptation of the polychaete *Nereis diversicolor* to estuarine sediments containing high concentrations of heavy metals. I. General observations and adaptations to heavy metals. *J. mar. biol. Ass. U.K.*, **51**, 845–63.

Bryan, G.W., Gibbs, P.E., Hummerstone, L.G. and Burt, G.R. (1986) The decline of the gastropod *Nucella lapillus* around S.W. England: evidence for the effect of tributyltin from antifouling paints. *J. mar. biol. Ass. U.K.*, **66**, 611–40.

Bryant, D.M. (1987) Wading birds and wildfowl of the estuary of the Firth of Forth, Scotland. *Proc. Roy. Soc. Edinb.*, **93B**, 509–20.

Bull, K.R., Every, W.J., Freestone, P., Hall, J.R. and Osborn, D. (1983) Alkyl lead pollution and bird mortalities on the Mersey Estuary 1979–1981. *Env. Poll.* (Ser. A), **31**, 239–59.

Bullock, T.H. (1953) Predator recognition and escape responses of some intertidal gastropods in the presence of starfish. *Behaviour*, **5**, 130–40.

Burns, K.A., Garrity, S.D. and Levings, S.C. (1993) How many years until mangrove ecosystems recover from catastrophic oil spills? *Mar. Poll. Bull.*, **26**, 239–48.

Burrows, E.M. and Lodge, S.M. (1950) Note on the inter-relationships of *Patella*, *Balanus* and *Fucus* on a semi-exposed coast. *Rep. mar. biol. Stn Port Erin*, **62**, 30–34.

Burrows, E.M., Conway, E., Lodge, S.M. and Powell, H.T. (1954) The raising of intertidal zones of algae on Fair Isle. *J. Ecol.*, **42**, 283–8.

Burrows, M.T. (1988) The comparative biology of *Chthamalus stellatus* (Poli) and *Chthamalus montagui* Southward, PhD thesis, University of Manchester.

Burrows, M.T., Hawkins, S.J. and Southward, A.J. (1992) A comparison of reproduction in co-occurring chthamalid barnacles, *Chthamalus stellatus* (Poli) and *Chthamalus montagui* Southward. *J. exp. mar. Biol. Ecol.*, **160**, 229–49.

Butman, C.A., Grassle, J.P. and Webb, C.M. (1988) Substrate choices made by marine larvae in still water and in a flume flow. *Nature*, **333**, 771–3.

Cadee, G.C. (1990) Feeding traces and bioturbation by birds on a tidal flat, Dutch Wadden Sea. *Ichnos*, **1**, 23–30.

Calman, W.T. (1927) Zoological results of the Cambridge Expedition to the Suez Canal, 1924. XIII. Report on the Crustacea Decapoda (Brachyura). (With Appendix by H. Munro Fox). *Trans. Zool. Soc. Lond.*, **22**, 211–19.

Carefoot T. (1977) *Pacific Seashores*, J.J. Douglas, Vancouver, 208 pp.

Carlton, J.T. (1982) The historical biogeography of *Littorina littorea* on the Atlantic coast of N. America and implications for the interpretation of the structure of New England intertidal communities. *Malacol. Rev.*, **15**, 146.

Carlton, J.T. (1989) The introduced marine and estuarine mollusks of North America. An end of the century perspective on four centuries of human mediated introductions. *J. Shellfish Res.*, **8**, 465.

Carlton, J.T., Cheney, D.P. and Vermeij, G.J. (conveners) (1982) A minisymposium and workshop. Ecological effects and biogeography of an introduced marine species: the periwinkle, *Littorina littorea*. (A zoological, botanical and ecological symposium). (Abstracts of papers). *Malacol. Rev.*, **5**, 143–50.

Carter, R.W.G. (1988) *Coastal Environments*, Academic Press, London, 617 pp.

Castilla, J.C. and Duran, L.R. (1985) Human exclusion from the rocky intertidal zone of central Chile: the effects on *Concholepas concholepas* (Gastropoda). *Oikos*, **45**, 391–9.

Castilla, J.C. and Olivera, D. (1990) Ecological consequences of coseismic uplift on the intertidal kelp belts of *Lessonia nigrescens* in central Chile. *Estuar. coast. Shelf Sci.*, **31**, 45–56.

Caswell, H. (1978) A general formula for the sensitivity of population growth rate to changes in life history parameters. *Theor. Pop. Biol.*, **14**, 215–30.

Chapgar, B.F. (1991) *Seashore Life of India*, Oxford University Press, Oxford, 78 pp.

Chapman, A.R.O. (1979) *Biology of Seaweeds*, University Park Press, Baltimore, 330 pp.

Chapman, A.R.O. (1990) Effects of grazing, canopy cover and substratum type on the abundances of common species of seaweeds inhabiting littoral fringe tide pools. *Bot. Mar.*, **33**, 319–26.

Chapman, A.R.O. and Johnson, C.R. (1990) Disturbance and organisation of macroalgal assemblages in the Northwest Atlantic. *Hydrobiologia*, **192**, 77–121.

Chapman, M.G. and Underwood, A.J. (1992) Foraging behaviour of marine benthic grazers, in *Plant–animal Interactions in the Marine Benthos* (eds D. John, S.J. Hawkins and J. Price) (Systematics Association Special Volume No. 46) Clarendon Press, Oxford, pp, 289–317.

Chapman, V.J. (1977) Introduction, in *Wet Coastal Ecosystems* (ed. V.J. Chapman), Elsevier, Amsterdam, pp. 1–29.

Chia, F.S. (1989) Differential settlement of benthic marine invertebrates, in *Reproduction, Genetics and Distributions of Marine Organisms*, (Proc. 23rd Eur. Mar. Biol. Symp. School of Biological Sciences, Univ. of Wales, Swansea, 5th–9th Sept. 1988), Olsen and Olsen, Denmark, pp. 3–12.

Chia, F.S. and Warwick, R.M. (1969) Assimilation of labelled glucose from seawater by marine nematodes. *Nature, Lond.*, **224**, 720–21.

Chia, F.S., Buckland-Nicks, J. and Young, C.M. (1984) Locomotion of marine invertebrate larvae: a review. *Can. J. Zool.*, **62**, 1205–22.

Clarke, R.B. (1992) *Marine Pollution*, 3rd edn, Clarendon Press, Oxford.

Cleary, J.J. and Stebbing, A.R.D. (1985) Organotin and total tin in coastal waters of southwest England. *Mar. Poll. Bull.*, **16**, 350–55.

Coghalan, A. (1987) Lethal paint makes for open sea. *New Scient.*, **128** (1746), 16.

Cohen, J. (1977) *Statistical Power Analysis for the Behavioural Sciences*, Rev. edn, Academic Press, New York, 474 pp.

Cole, H.A. (1942) The American whelk tingle, *Urosalpinx cinerea* (Say), on British oyster beds. *J. mar. biol. Ass. U.K.*, **25**, 477–508.

Colman, J.S. (1933) The nature of intertidal zonation of plants and animals. *J. mar. biol. Ass. U.K.*, **61**, 71–93.

Commito, J.A. and Ambrose, W.G.J. (1984) Predatory infauna and trophic complexity in soft-bottom communities, in *Proceedings Of The Nineteenth European Marine Biology Symposium, Plymouth, Devon, UK* (ed. P.E. Gibbs), Cambridge University Press, Cambridge, pp. 323–34.

Commito, J.A. and Shrader, P.B. (1985) Benthic community response to experimental additions of the polychaete *Nereis virens*. *Mar. Biol.*, **86**, 101–7.

Conan, G. (1982) The long-term effects of the Amoco Cadiz oil spill. *Phil. Trans. R. Soc. Lond.*, **297B**, 232–333.

Connell, J.H. (1961a) The influence of intra-specific competition and other factors on the distribution of the barnacle *Chthamalus stellatus*. *Ecology*, **42**, 710–23.

Connell, J.H. (1961b) Effects on competition, predation by *Thais lapillus*, and other factors on natural populations of the barnacle *Balanus balanoides*. *Ecol. Monogr.*, **31**, 61–104.

Connell, J.H. (1970) A predator–prey system in the marine intertidal region. I. *Balanus glandula* and several predatory species of *Thais*. *Ecol. Monogr.*, **40**, 49–78.

Connell, J.H. (1972) Community interactions on marine rocky intertidal shores. *Ann. Rev. Ecol. Syste.*, **3**, 169–92.

Connell, J.H. (1973) Population ecology of reef-building corals, in *Biology and Geology of Coral Reefs*, Vol. II, *Biology I*, (eds O.A. Jones and R. Endean), Academic Press, New York, pp. 271–324.

Connell, J.H. and Slatyer, R.O. (1977) Mechanisms of succession in natural communities and their role in community stability and organisation. *Am. Nat.*, **111**, 1119–44.

Connor, V.M. and Quinn, J.F. (1984) Stimulation of food species growth by limpet mucus. *Science*, **225**, 843–4.

Conway, N. and Capuzzo, J.M. (1990) The use of biochemical indicators in the study of trophic interactions in animal–bacteria symbioses, *Solemya velum*, a case study, in *Trophic Relationships in the Marine Environment* (Proc. 24th Eur. Mar. Biol. Symp.) (eds M. Barnes and R.N. Gibson), Aberdeen University Press, Aberdeen, pp. 553–64.

Cook, W. (1991) Studies on the effects of hydraulic dredging on cockle and other macroinvertebrate populations 1989–1990. North Western and North Wales Sea Fisheries Committee. 30 pp.

Creese, R.G. and Ballantine, W.J. (1986) Rocky shores of exposed islands in north–eastern New Zealand. *New Zealand Department of Lands and Survey Information Series*, **16**, 85–102.

Crisp, D.J. (1974) Factors influencing the settlement of marine invertebrate larvae, in *Chemoreception in Marine Organisms* (eds P.T. Grant and A.M. Mackie), Academic Press, London, pp. 177–265.

Crisp, D.J. (1976) Settlement responses in marine organisms, in *Adaptations to Environment, Essays on the Physiology of Marine Animals*, (ed. R.C. Newell), Butterworths, London, pp. 83–124.

Crisp, D.J. and Barnes, H. (1954) The orientation and distribution of barnacles at settlement with particular reference to surface contour. *J. Anim. Ecol.*, **23**, 143–62.

Crisp, D.J. and Chipperfield, P.N.J. (1948) Occurrence of *Elminius modestus* (Darwin) in British waters. *Nature*, **161**, 64.

Crisp, D.J. and Southward, A.J. (1958) The distribution of intertidal organisms along the coasts of the English Channel. *J. mar. biol. Ass. U.K.*, **37**, 1031–48.

Critchley, A.T., Farnham, W.F., Yoshida, T. and Norton, T.A. (1990) A bibliography of the invasive alga *Sargassum muticum* (Yendo) Fensholt. (Fucales, Sargassaceae). *Bot. Mar.*, **33**, 551–62.

Croker, R.A. and Hatfield, E.B. (1980) Space partitioning and interactions in an intertidal sand-burrowing amphipod guild. *Mar. Biol.*, **61**, 79–88.

Crosson, P.R. (1989) Climate change: problems of limits and policy responses, in *Greenhouse Warming: Abatement and Adaptation* (eds N.J. Rosenberg, W.E. Easterling, P. Crosson and J. Darmstadter), Resources for the Future, Washington, DC, pp. 69–82.

Crothers, J.H. (1983) Variation in dog-whelk shells in relation to wave action and crab predation. *Biol. J. Linn. Soc., Lond.,* **20,** 85–102.

Crothers, J.H. (1985) Dog-whelks, an introduction to the biology of *Nucella lapillus* (L.). *Field Studies,* **6,** 291–360.

Cunliffe, B. and Hawkins, S.J. (1988) The shell midden deposits, in *Mount Batten Plymouth – A Prehistoric and Roman Port* (Oxford University Committee for Archaeology, Monograph 26) (ed. B. Cunliffe), pp. 35–8.

Currey, J.D. and Hughes, R.N. (1982) Strength of the Dogwhelk *Nucella lapillus* and the Winkle *Littorina littorea* from different habitats. *J. Anim. Ecol.,* **51,** 47–56.

Currie, R.J. (1992) Circulation and upwelling off the coast of South-East Arabia. *Oceanol. Acta,* **15,** 43–60.

Dahl, E. (1952) Some aspects of the ecology and zonation of the fauna on sandy beaches. *Oikos,* **4,** 1–27.

Dakin, W.J. (1953) *Australian Seashores,* Angus & Robertson, London.

Davey, J.T. and George, C.L. (1986) Species interactions in soft sediments: factors in the distribution of *Nereis (Hediste) diversicolor* in the Tamar estuary. *Ophelia,* **26,** 151–64.

Davies, M.S., Hawkins, S.J. and Jones, H.D. (1990) Mucus production and physiological energetics in *Patella vulgata* L. *J. moll. Stud.,* **56,** 499–503.

Davies, M.S., Hawkins, S.J. and Jones, H.D. (1992a) Pedal mucus and its influence on the microbial food supply of two intertidal gastropods, *Patella vulgata* L. and *Littorina littorea* (L.). *J. exp. mar. Biol. Ecol.,* **161,** 57–77.

Davies, M.S., Jones, H. and Hawkins, S. (1992b) Physical factors affecting the fate of pedal mucus produced by the common limpet *Patella vulgata. J. mar. Biol. Ass. U.K.,* **72,** 633–43.

Davison, I.R., Dudgeon, S.R. and Ruan, H.M. (1989) Effect of freezing on seaweed photosynthesis. *Mar. Ecol. Progr. Ser.,* **58,** 1–2.

Dawes, C.J., McCoy, E.D. and Heck, K.L. (1991) The tropical western Atlantic including the Caribbean Sea, in *Intertidal and Littoral Ecosystems* (Ecosystems of the World 24) (eds A.C. Mathieson and P.H. Nienhuis), Elsevier, Amsterdam, pp. 215–34.

Day, J.W.J., Hall, C.A.S., Kemp, W.M. and Yanez, A.A. (1989) *Estuarine Ecology,* Wiley Interscience, New York.

Dayton, P.K. (1971) Composition disturbance and community organization. The provision and consequent utilisation of space. *Ecol. Monogr.,* **41,** 357–89.

Dayton, P.K. (1972) Towards an understanding of community resilience and the potential effects of enrichments to the benthos at McMurdo Sound, Antarctica, in *Colloquium on Conservation Problems in Antarctica* (ed. B.C. Parker), Allen Press, London, pp. 81—96.

Dayton, P.K. (1975) Experimental evaluation of ecological dominance in a rocky intertidal algal community. *Ecol. Monogr.,* **45,** 137–59.

Dayton, P.K. (1984) Patch dynamics and stability of some California kelp communities. *Ecol. Monogr.,* **54,** 253–89.

Dayton, P.K., Rosenthal, R.J., Mahen, L.C. and Antezana, T. (1977) Population structure and foraging biology of the predaceous Chilean asteroid *Meyenaster gelatinosus* and the escape biology of its prey. *Mar. Biol.*, **39**, 361–70.

Dayton, P.K., Tegner, M.J., Parnell, P.E. and Edwards, P.B. (1992) Temporal and spatial patterns of disturbance and recovery in a kelp forest community. *Ecol. Monogr.*, **62**, 421–45.

Della Santina, P. and Naylor, E. (1994) Endogenous rhythms in the homing behaviour of the limpet *Patella vulgata* Linnaeus. *J. moll. Stud.*, **59**, 87–91.

Den Hartog, C. (1959) The epilithic algal communities occurring along the east coast of the Netherlands. *Wentia*, **1**, 1–241.

Den Hartog, C. (1970) The Seagrasses of the World. *Verh. K. Ned. Akad. Wet. Afd. Natuurkd.*, **59**, 1–275.

Denley, E.J. and Underwood, A.J. (1979) Experiments on factors influencing settlement, survival and growth of two species of barnacles in New South Wales. *J. exp. mar. Biol. Ecol.*, **36**, 269–93.

Denny, M.W. (1982) Forces on intertidal organisms due to breaking ocean waves: design and application of a telemetry system. *Limnol. Oceanogr.*, **27**, 178–83.

Denny, M.W. (1983) A simple device for recording the maximum force exerted on intertidal organisms. *Limnol. Oceanogr.*, **28**, 1269–74.

Denny, M.W. (1988) *Biology and the Mechanics of the Wave-swept Environment*, Princeton University Press, Princeton, NJ.

Dicks, B. and Hartley, J.P. (1982) The effects of repeated small oil spillages and chronic discharges, in *The Long-Term Effects of Oil Pollution on Marine Populations, Communities and Ecosystems. Phil. Trans. R. Soc. Lond.*, Ser. B, **297**, 285–307.

Donze, M. (1968) The algal vegetation of the Ria de Arosa (N.W. Spain). *Blumea*, **16**, 159–92.

Doty, M.S. (1946) Critical tide factors that are correlated with the vertical distribution of marine algae and other organisms along the Pacific coast. *Ecology*, **27**, 315–28.

Dring, M.J. (1982) *The Biology of Marine Plants*, Edward Arnold, London, 199 pp.

Dring, M.J. and Brown, F.A. (1982) Photosynthesis of intertidal brown algae during and after periods of emersion: a renewed search for physiological causes of zonation. *Mar. Ecol. Progr. Ser.*, **8**, 301–8.

Druehl, L.D. (1973) Marine transplantations. *Science*, **179**, 12.

Dungan, M.L. (1986) Three-way interactions, barnacles, limpets and algae in a Sonora Desert rocky intertidal zone. *Am. Nat.*, **127**, 292–316.

Durante, K.M. (1991) Larval behaviour, settlement preference, and induction of metamorphosis in the temperate solitary ascidian *Molgula citrina* Alder and Hancock. *J. exp. mar. Biol. Ecol.*, **145**, 175–87.

Dutt, S. and Raju, K.V. (1983) A new record of *Sardinella aurita* Valenciennes, 1847 (Teleostei, Clupeidae) from the eastern part of the Arabian Sea: a possible immigrant through the Red Sea? in *Marine Science in the Red Sea. Proceedings of an International Conference of Marine Science in the Red Sea Celebrating the 50th Anniversary of Al Ghardaqa Marine Biological Station* (eds A.F.A. Latif, A.R. Bayoumi and M.F. Thompson) *Bull. Inst. Oceanogr. Fish., Cairo*, **9**, 264–70.

Dye, A.H. (1987) Aerial and aquatic oxygen consumption in two spihonariid limpets (Pulmonata: Siphonariidae). *Comp. Biochem. Physiol.*, **87A**, 695–8.

Dye, A.H. and Lasiak, T.A. (1986) Microbenthos, meiobenthos and fiddler crabs: trophic interactions in a tropical mangrove sediment. *Mar. Ecol. Progr. Ser.*, **32**, 2–3.

Eales, N.B. (1950) *The Littoral Fauna of the British Isles: a Handbook for Collectors*, Cambridge University Press, Cambridge.

Ebling, F.J., Sleigh, M.A., Sloane, J.F. and Kitching, J.A. (1960) The ecology of Lough Ine, VII. Distribution of some common plants and animals of the littoral and shallow sublittoral regions. *J. Ecol.*, **48**, 29–53.

Eckman, J.E. (1985) Flow disruption by an animal–type mimic affects sediment bacterial colonization. *J. mar. Res.*, **43**, 419–35.

Eckman, J.E. (1987) The role of hydrodynamics in recruitment, growth, and survival of *Argopecten irradians* (L.) and *Anomia simplex* (D'Orbigny) within eelgrass meadows. *J. exp. mar. Biol. Ecol.*, **106**, 165–91.

Edmondson, C.H. and Wilson, I.H. (1940) The shellfish resources of Hawaii. *Proc. 6th Pacif. Sci. Congr. 1939*, **3**, 241–3.

Eleftheriou, A. and McIntyre, A.D. (1976) The intertidal fauna of sandy beaches – a survey of the Scottish coast. *Scottish Fisheries Research Report*, **6**, 1–61.

Eleftheriou, A. and Nicholson, M. (1975) The effects of exposure on beach fauna. *Cah. Biol. Mar.*, **16**, 695–710.

Eleftheriou, A. and Robertson, M.R. (1988) The intertidal fauna of sandy beaches – a survey of the east Scottish coast. *Scottish Fisheries Research Report*, **38**, 1–52.

Elliott, A.J. and Savidge, G. (1990) Some features of the upwelling off Oman. *J. mar. Res.*, **48**, 319–33.

Elliott, J.M. (ed.) (1977) *Some Methods for the Statistical Analysis of Samples of Benthic Invertebrates*, 2nd edn, Freshwater Biological Association, Ambleside, Scientific Publication no. 25.

Elliott, M., O'Reilly, R.M.G. and Taylor, C.J.L. (1990) The Forth estuary: a nursery and overwintering area for North Sea fishes, in *North Sea Estuaries Interactions* (eds D.S. McLusky, V.N. de Jonge and J. Pomfret), Kluwer, Dordrecht, pp. 89–103.

Ellis, D.V. and Pattisina, L.A. (1990) Widespread neogastropod imposex, a biological indicator of global TBT pollution. *Mar. Poll. Bull.*, **21**, 248–53.

Elner, R.W. and Campbell, A. (1987) Natural diets of lobster *Homarus americanus* from barren ground and macroalgal habitats off southwestern Nova Scotia, Canada. *Mar. Ecol. Progr. Ser.*, **37**, 2–3.

Elner, R.W. and Vadas, R.L. (1990) Inference in ecology, the sea urchin phenomenon in the North-west Atlantic. *Am. Nat.*, **136**, 108–25.

Elton, C.S. (1958) *The Ecology of Invasions by Animals and Plants*, Methuen, London.

Eltringham, S.K. (1971) *Life in Mud and Sand*, English University Press, London, 218 pp.

Engelmann, T.W. (1884) Untersuchungen über die quantitataven Beziehungen zurischen Absorption des Lichtes und Assimilation Pfhanzenzellen. *Botan. z.*, **42**, 82–95.

Estes, J.A. and Duggins, D.O. (1995) Sea otters and kelp forests in Alaska: generality and variation in a community ecological paradigm. *Ecol. Monogr.*, **65**, 75–100.

Estes, J.A. and Palmisano, J.F. (1974) Sea otters, their role in structuring nearshore communities. *Science*, **185**, 1058–60.

Evans, R.G. (1947) The intertidal ecology of selected localities in the Plymouth neighbourhood. *J. mar. biol. Ass. U.K.*, **17**, 173–218.

Evans, R.G. (1957) The intertidal ecology of some localities on the Atlantic coast of France. *J. Ecol.*, **45**, 245–71.

Evans, S. and Tallmark, B. (1979) A modified drop–net method for sampling mobile epifauna on marine shallow sandy bottoms. *Holarct. Ecol.*, **2**, 58–64.

Fairweather, P.G. and Underwood, A.J. (1991) Experimental removals of a rocky intertidal predator: variations within two habitats in the effects on prey. *J. exp. mar. Biol. Ecol.*, **154**, 29–75.

Farnham, W.F., Fletcher, R.L. and Irvine, L.M. (1973) Attached *Sargassum* found in Britain. *Nature, Lond.*, **243**, 231–2.

Farrell T.M. (1991) Models and mechanisms of succession, an example from a rocky intertidal community. *Ecol. Monogr.*, **61**, 95–113.

Feare, C.J. and Summers, R.W. (1985) Birds as predators on rocky shores, in, *The Ecology of Rocky Coasts*, (eds P.G. Moore and R. Seed), Hodder and Stoughton, London, pp. 249–64.

Fenchel, T. (1969) The ecology of marine microbenthos. IV. Structure and function of the benthic ecosystem, its chemical and physical factors and the microfauna communities with special reference to the ciliated Protozoa. *Ophelia*, **6**, 1–182.

Fenchel, T. (1975a) Factors determining the distribution patterns of mud snails (Hydrobidae). *Oecologia*, **20**, 1–17.

Fenchel, T. (1975b) Character displacement and coexistence in mud snails (Hydrobidae). *Oecologia*, **20**, 19–32.

Fenchel, T.M. and Riedl, R.J. (1970) The sulphide system: a new biotic community underneath the oxidised layer of marine sand bottoms. *Mar. Biol.*, **7**, 255–68.

Ferguson, A. (ed.) (1984) *Intertidal Plants and Animals of the Landels–Hill Big Creek Reserve*, Centre for Marine Studies, University of California, Santa Cruz, 106 pp.

Field, J.G. (1968) The "turbulometer", an apparatus for measuring the relative exposure to wave action on shores. *Zool. Afr.*, **3**, 115–18.

Field, J.G. (1983) Flow patterns of energy and matter, in *Marine Ecology. A Comprehensive, Integrated Treatise on Life in Oceans and Coastal Waters*, Vol. 5, (ed. O. Kinne), Wiley Interscience, Chichester, pp. 758–94.

Field, J.G. and Griffiths, C.L. (1991) Littoral and sublittoral ecosystems of Southern Africa, in *Intertidal and Littoral Ecosystems* (Ecosystems of the World 24) (eds A.C. Mathieson and P.H. Nienhuis), Elsevier, Amsterdam, pp. 323–46.

Field, J.G., Griffiths, C.L., Griffiths, R.S., Jarman, N., Zoutendyk, P., Velimirov, B. and Bowes, A. (1980a) Variation in structure and biomasss of kelp communities along the south west Cape coast. *Trans. Roy. Soc. S. Afr.*, **44**, 113–50.

Field, J.G., Griffiths, C.L., Linley, E.A., Carter, R.A. and Zoutendyk, P. (1980b) Upwelling in a nearshore marine ecosystem and the biological implications. *Est. coast. mar. Sci.*, **11**, 133–50.

Filion-Mykelbust, C. and Norton, T.A. (1981) Epidermis shedding in the brown seaweed *Ascophyllum nodosum* (L.) Le Jolis and its ecological significance. *Mar. Biol. Lett.*, **2**, 45–51.

Fischer, E. (1929) Recherches di bionomie et d'oceanographie littorales sur la Rance et le littral de la Manche. *Ann Inst. Oceanogr. Monaco*, N.S., **5**.

Fischer-Piette, E. (1931) Reparition des principales especes fixees sur les rochers battus des cotes et des iles de la Manche de Lannion a Fecamp. *Ann. Inst. Oceanogr., Paris*, **12**, 105–213.

Flach, E.C. (1992) Disturbance of benthic infauna by sediment reworking activities of the lugworm *Arenicola marina*. *Neth. J. Sea Res.*, **30**, 81–9.

Flatteley, F.W. and Walton, C.L. (1926) *The Biology of the Sea-shore*. London.

Fletcher, A. (1980) Marine and maritime lichens of rocky shores, their ecology, physiology and biological interactions, in *The Shore Environment* (eds J.H. Price, D.E.G. Irvine and W.F. Farnham), Academic Press, London, pp. 789–842.

Fletcher, R.L. (1975) Heteroantaonism observed in mixed algal cultures. *Nature, Lond.*, **253**, 534–5.

Fletcher, R.L. and Callow, M.E. (1992) The settlement, attachment and establishment of marine algal spores. *Br. Phycol. J.*, **27**, 303–29.

Fletcher, R.L. and Manfredi, C. (1995) The occurrence of *Undaria pinnatifida* (Phaeophyceae, Laminariales) on the south coast of England. *Botanica Mar.*, (in press).

Foster, B.A. (1971a) Desiccation as a factor in the intertidal zonation of barnacles. *Mar. Biol.*, **8**, 12–29.

Foster, B.A. (1971b) On the determinants of the upper limit of intertidal distribution of barnacles (Crustacea, Cirripedia). *J. Anim. Ecol.*, **40**, 33–48.

Foster, M.J., De Vogelaere, A.P., Harrold, C., Pearse, J.S. and Thum, A.B. (1988) Causes of spatial and temporal patterns in rocky intertidal communities of central and northern California. *Mem. Calif. Acad. Sci.*, **9**, 1–45.

Foster, M.J., De Vogelaere, A.P., Oliver, J.S., Pearse, J.S. and Harrold, C. (1991) Open coast intertidal and shallow subtidal ecosystems of the northeast Pacific, in *Intertidal and Littoral Ecosystems* (eds A.C. Mathieson and P.H. Nienhuis) Elsevier, Amsterdam, pp. 235–72.

Foster, M.S. (1990) Organisation of macroalgal assemblages in the northeast Pacific, the assumption of homogeneity and the illusion of generality. *Hydrobiologia*, **192**, 21—33.

Foster, M.S. (1992) How important is grazing to seaweed evolution and assemblage structure in the north-east Pacific? in *Plant–Animal Interactions in the Marine Benthos* (eds D. John, S.J. Hawkins and J. Price) (Systematics Association Special Volume No. 46), Clarendon Press, Oxford, pp. 61–85.

Foster, M.S. and Barilotti, D.C. (1990) An approach to determining the ecological effects of seaweed harvesting, a summary, in *Proc. 13th Int. Seaweed Symp.* (eds. S.C. Lindstrom and P.W. Gabrielson) (Developments in Hydrobiology. *Hydrobiologia*, 58) Kluwer, Dordrecht, pp. 15–16.

Foster, M.S., Tarpley, J.A. and Dearn, S.L. (1990) To clean or not to clean: the rationale, methods and consequences of removing oil from temperate shores. *Northwest Environ. J.*, **6**, 105–20.

Foster, M.S., Harrold, C. and Hardin, D.D. (1991) Point vs. photo quadrat estimates of the cover of sessile marine organisms. *J. exp. mar. Biol. Ecol.*, **146**, 193–203.

Fralick, R.A., Turgeon, K.W. and Mathieson, A.C. (1974) Destruction of kelp populations by *Lacuna vincta* (Montagu). *Nautilus*, **88**, 112–14.

Gaines, S.D. and Bertness, M.D. (1992) Dispersal of juveniles and variable recruitment in sessile marine species. *Nature*, **360**, 579–80.

Gaines, S.D. and Roughgarden, J. (1985) Larval settlement rate, a leading determinant of structure in an ecological community of the intertidal zone. *Proc. Natn. Acad. Sci. USA*, **82**, 3707–11.

Gaines, S.D., Brown, S. and Roughgarden, J. (1985) Spatial variation in larval concentrations as a cause of spatial variation in settlement for the barnacle, *Balanus glandula*. *Oecologia*, **67**, 267–72.

Gallagher, E.D., Jumars, P.A. and Trueblood, D.D. (1983) Facilitation of soft-bottom benthic succession by tube builders. *Ecology*, **64**, 1200–1216.

Galley, D.J. (1991) The Ergopod, a simple device to measure on-shore wave action. *Field Stud.*, **7**, 719–29.

Gamble, J.C. (1991) Mesocosms, statistical and experimental design consideration, in, *Enclosed Experimental Marine Ecosystems, a Review and Recommendations* (ed. C.M. Lalli), Springer-Verlag, New York, pp. 188–96.

Gardner, L.R., Wolaver, T.G. and Mitchell, M. (1988) Spatial variations in the sulfur chemistry of salt marsh sediments at North Inlet, South Carolina. *J. mar. Res.*, **46**, 815–36.

Gee, J.M. (1989) An ecological and economic review of meiofauna as food for fish. *Zool. J. Linn. Soc.*, **96**, 243–61.

Gee, J.M., Warwick, R.M., Davey, J.T. and George, C.L. (1985) Field experiments on the role of epibenthic predators in determining prey densities in an estuarine mudflat. *Estuar. coast. Shelf Sci.*, **21**, 429–48.

Geller, J.B. (1990) Reproductive responses to shell damage by the gastropod *Nucella emarginata* (Deshayes). *J. exp.mar. Biol. Ecol.*, **136**, 77–87.

Gerlach, S.A. (1971) On the importance of marine meiofauna for benthos communities. *Oecologia*, **6**, 176–90.

Gerlach, S.A. (1978) Food-chain relationships in subtidal silty sand marine sediments and the role of meiofauna in stimulating bacterial productivity. *Oecologia*, **33**, 55–69.

Gibbons, M.J. and Griffiths, C.L. (1986) A comparison of macrofaunal and meiofaunal distribution and standing stock across a rocky shore, with an estimate of their productivities. *Mar. Biol.*, **93**, 181–8.

Gibbs, P.E. (1993) Phenotypic changes in the progeny of *Nucella lapillus* (Gastropoda) transplanted from an exposed shore to sheltered inlets. *J. moll. Stud.*, **59**, 187–94.

Gibbs, P.E. and Bryan, G.W. (1986) Reproductive failure of the dogwhelk, *Nucella lapillus*, caused by imposex induced by tributyltin from antifouling paints. *J. mar. biol. Ass. U.K.*, **66**, 767–77.

Gibbs, P.E., Spencer, B.E. and Pascoe, P.L. (1991) The American oyster drill, *Urosalpinx cinerea* (Gastropoda): evidence of decline in an imposex affected population (R. Blackwater, Essex). *J. mar. biol. Ass. U.K.*, **71**, 827–38.

Gieselman, J.A. and McConnell, O.J. (1981) Polyphenols in brown algae *Fucus vesiculosus* and *Ascophyllum nodosum*: chemical defenses against the marine herbivorous snail, *Littorina littorea*. *J. chem. Ecol.*, **7**, 1115–33.

Giller, P.S., Hildrew, A.G. and Raffaelli, D.G. (1994) *Aquatic Ecology: Scale, Pattern and Process*, Blackwell, Oxford, 649 pp.

Godeaux, J. (ed.) (1990) A propos des migrations lessepsiennes. *Bull. Inst. Oceanogr. Monaco.* 152 pp.

Goss-Custard, J.D. and dit Durrel, S.E.A. le V. (1990) Bird behaviour and environmental planning: approaches in the study of wader populations. *Ibis*, **132**, 273–82.

Goss-Custard, J.D. and Moser, M.E. (1988) Rates of change in the numbers of dunlin, *Calidris alpina*, wintering in British estuaries in relation to the spread of *Spartina anglica*. *J. appl. Ecol.*, **25**, 95–109.

Goss-Custard, J.D., Jones, R.E. and Newbery, P.E. (1977) The ecology of the Wash. I. Distribution and diet of wading birds (Charadrii). *J. appl. Ecol.*, **14**, 681–700.

Goss-Custard, J.D., McGrorty, S. and Kirby, R. (1990) Inshore birds of the soft coasts and sea-level rise, in *Expected Effects of Climate Change on Marine Coastal Ecosystems* (eds J.J. Beukema, W.J. Wolff and J.J.W.N. Brouns), Kluwer, Dordrecht, pp. 189–93.

Goss-Custard, J.D., dit Durrell, S.E.A., McGrorty, S. and Reading, C.J. (1982) Use of mussel *Mytilus edulis* beds by oystercatchers *Haematopus ostralegus* according to age and population size. *J. anim. Ecol.*, **51**, 543–54.

Gosse, P.H. (1856) *Tenby: a Sea-side Holiday*, London.

Gowanloch, J.N. and Hayes, F.R. (1926) Contributions to the study of marine gastropods. 1. The physical factors, behaviour and intertidal life of *Littorina*. *Contr. Can. Biol., N.S.*, **3**, 133.

Grant, J. (1981) Factors affecting the occurrence of intertidal amphipods in reducing sediments. *J. exp. mar. Biol. Ecol.*, **49**, 203–16.

Grassle, J.F. and Grassle, J.P. (1974) Opportunistic life histories and genetic systems in marine benthic polychaetes. *J. mar. Res.*, **32**, 253–84.

Grassle, J.F. and Sanders, H.L. (1973) Life histories and the role of disturbance. *Deep-Sea Res.*, **20**, 643–59.

Gray, A.J. and Benham, P.E.M. (eds) (1990) *Spartina anglica* – a research review. Institute of Terrestrial Ecology, special publication, HMSO, London, 79pp.

Grenon, J.F. and Walker, G. (1982) Further fine structure studies of the "space" layer which underlies the foot sole epithelium of the limpet, *Patella vulgata* L. *J. moll. Stud.*, **48**, 54–63.

Griffiths, C.L., Stenton-Dozey, J. and Koop, K. (1983) Kelp wrack and energy flow through a sandy beach, in *Sandy Beaches as Ecosystems* (eds A. McLachlan and T. Erasmus), Junk, The Hague, pp. 547–56.

Griffiths, R.J. (1981) Population dynamics and growth of the bivalve *Choromytilus merdionalis* (kr.) at different tidal levels. *Estuar. coast. Shelf Sci.*, **12**, 101–18.

Gruet, Y. (1986) Spatio-temporal changes of Sabellarian reefs built by the sedentary polychaete *Sabellaria alveolata* (Linné). *Mar. Ecol. (Naples)*, **7**, 289–372.

Gunter, G. (1967) Some relationships of estuaries to the fisheries of the Gulf of Mexico, in *Estuaries* (ed. G.H. Lauff), *Am. Assoc. Adv. Sci. Publ.*, **83**, 621–38.

Haedrich, R.L. (1983) Estuarine fishes, in, *Estuaries and Enclosed Seas*, (ed. B.H. Ketchum), Elsevier, Amsterdam, pp. 183–207.

Hairston, N.G. Sr. (1989) *Ecological Experiments, Purpose, Design and Execution*, Cambridge University Press, Cambridge, 370 pp.

Hall, S.J., Basford, D.J. and Robertson, M.R. (1990a) The impact of hydraulic dredging for razor clams *Ensis* sp. on an infaunal community. *Neth. J. Sea Res.*, **27**, 119–25.

Hall, S.J., Raffaelli, D.G. and Thrush, S. (1994) Patchiness and disturbance in shallow water benthic assemblages, in *Aquatic Ecology: scale, pattern and process* (eds P. Giller, A.G. Hildrew and D. Raffaelli), Blackwell Scientific Publications, Oxford, pp. 333–73.

Hall, S.J., Raffaelli, D.G. and Turrel, W.R. (1990b) Predator-caging experiments in marine systems: a reexamination of their value. *Am. Nat.*, **136**, 657–72.

Hall, S.J., Basford, D.J., Robertson, M.R., Raffaelli, D.G. and Tuck, I. (1991) Patterns of recolonisation and the importance of pit–digging by the crab *Cancer pagurus* in a subtidal sand habitat. *Mar. Ecol. Progr. Ser.*, **72**, 1–2.

Hall, S.J., Robertson, M.R., Basford, D.J. and Heaney, S.D. (1993) The possible effects of fishing disturbance in the northern North Sea: an analysis of spatial patterns in community structure around a wreck. *Neth. J. Sea Res.*, **31**, 201–8.

Hardy, A. (1956) *The Open Sea – the World of Plankton*, Collins, London.

Harger, J.R.E. (1970) The effect of wave impact on some aspects of the biology of sea mussels. *Veliger*, **12**, 401–14.

Harris, L.G., Ebeling, A.W., Laur, D.R. and Rowley, R.J. (1984) Community recovery after storm damage: a case of facilitation in primary succession. *Science*, **224**, 1336–8.

Hartnoll, R.G. (1976) The ecology of some rocky shores in tropical East Africa. *Est. coast. mar. Sci.*., **15**,365–71

Hartnoll, R.G. and Hawkins, S.J. (1982) The emersion curve in semidiurnal tidal regimes. *Estuar. coast. Shelf Sci.*, **15**, 365–71.

Hartnoll, R.G. and Hawkins, S.J. (1985) Patchiness and fluctuations on moderately exposed rocky shores. *Ophelia*, **24**, 53–63.

Hartnoll, R.G. and Wright, J.R. (1977) Foraging movements and homing in the limpet *Patella vulgata* L. *Anim. Behav.*, **25**, 806–10.

Hately, S.G., Grant, A. and Jones, N.V. (1989) Heavy metal tolerance in estuarine populations of *Nereis diversicolor*, in *Reproduction, Genetics and Distribution of Marine Organisms* (eds J.S. Ryland and P.A. Tyler) *Proc. 23rd Eur. Mar. Biol. Symp., Swansea, UK*, pp. 379–85.

Hatton, H. (1938) Essais de bionomie explicative sur quelques especes intercotidals d'alques et d'animaux. *Ann. Inst. Oceanogr. Monaco*, **17**, 241–348.

Hawkins, S.J. (1979) Field studies on Manx rocky shore communities, PhD thesis, University of Liverpool, 354 pp.

Hawkins, S.J. (1981a) The influence of *Patella* grazing on the fucoid–barnacle mosaic in moderately exposed rocky shores. *Kiel. Meeresf. Sond.*, **5**, 537–43.

Hawkins, S.J. (1981b) The influence of season and barnacles on algal colonisation of *Patella* exclusion zones. *J. mar. biol. Ass. U.K.*, **61**, 1–15.

Hawkins, S.J. (1983) Interactions of *Patella* and macroalgae with settling *Semibalanus balanoides* (L). *J. exp. mar. Biol. Ecol.*, **71**, 55–72.

Hawkins, S.J. and Harkin, E. (1985) Preliminary canopy removal experiments in algal dominated communities low on the shore and in the shallow subtidal on the Isle of Man. *Bot. Mar.*, **28**, 223–30.

Hawkins, S.J. and Hartnoll, R.G. (1979) A compressed air drill powered by SCUBA cylinders for use on rocky shores. *Est. coast. mar. Sci.*, **9**, 819–20.

Hawkins, S.J. and Hartnoll, R.G. (1980) Small scale relationships between species number and area on rocky shores. *Est. coast. mar. Sci.*, **10**, 201–14.

Hawkins, S.J. and Hartnoll, R.G. (1982a) Settlement patterns of *Semibalanus balanoides* (L.) in the Isle of Man (1977–1981). *J. exp. mar. Biol. Ecol.*, **62**, 271–83.

Hawkins, S.J. and Hartnoll, R.G. (1982b) The influence of barnacle cover on the growth and behaviour of *Patella vulgata* on a vertical pier. *J. exp. mar. Biol. Ecol.*, **62**, 855–67.

Hawkins, S.J. and Hartnoll, R.G. (1983a) Grazing of intertidal algae by marine invertebrates. *Oceanogr. mar. Biol. ann. Rev.*, **21**, 195–282.

Hawkins, S.J. and Hartnoll, R.G. (1983b) Changes in a rocky shore community: an evolution of monitoring. *J. mar. env. Res.*, **9**, 131–81.

Hawkins, S.J. and Hartnoll, R.G. (1985) Factors determining the upper limits of intertidal canopy-forming algae. *Mar. Ecol. Progr. Ser.*, **20**, 265–71.

Hawkins, S.J. and Jones, H.D. (1992) *Rocky Shores* (Marine Conservation Society, Marine Field Course Guide 1) Immel Publishing, 144 pp.

Hawkins, S.J. and Southward, A.J. (1992) Lessons from the Torrey Canyon oil spill, recovery and stability of rocky shore communities, in *Restoring the Nation's Marine Environment* (Symposium on marine habitat restoration, National Oceanic and Atmospheric Administration, USA) (ed G.D. Thayer), Maryland Sea Grant Publications, pp. 584–631.

Hawkins, S.J., Burnay, L.P., Neto, A.I., Tristão da Cunha, R. & Frias Martins, A.M. (1990) A description of the zonation patterns of molluscs and other biota on the south coast of São Miguel, Azores. *Azoreana*, Supplement, 21–38.

Hawkins, S.J., Hartnoll, R.G., Kain, J.M. and Norton, T.A. (1992a) Plant animal interactions on hard substrata in the North-east Atlantic, in, *Plant–Animal Interactions in the Marine Benthos* (eds D.M. John, S.J. Hawkins and J.H. Price) (Systematics Association Special Volume No. 46), Clarendon Press, Oxford, pp. 1–32.

Hawkins, S.J., Allen, J.R., White, K.N., Conlan, K. Hendry, K., Jones, H.D. and Russell, G. (1992b) Restoring and managing disused docks: marine habitat creation and nature conservation in inner city areas, in *Proc. Symp. on Habitat Restoration, National Oceanic and Atmospheric Administration, September 1990* (ed. G. Thayer), Maryland Seagrant Publications.

Hay, M.E. and Fenical, W. (1988) Marine plant–herbivore interactions: the ecology of chemical defense. *Ann. Rev. Ecol. Syst.*, **19**, 111–45.

Hay, M.E. and Fenical, W. (1992) Chemical mediation of seaweed–herbivore interactions, in *Plant–Animal Interactions in the Marine Benthos* (eds D.M. John, S.J. Hawkins and J.H. Price) (Systematics Association Special Volume No. 46), Clarendon Press, Oxford, pp. 319–38.

Hay, M.E., Duffy, J.E., Pfister, C.A. and Fenical, W. (1987) Chemical defenses against different marine herbivores, are amphipods insect equivalents? *Ecology*, **68**, 1567—80.

Hekstra, G.P. (1989) Sea-level rise: regional consequences and responses, in *Greenhouse Warming: Abatement and Adaptation* (eds N.J. Rosenberg, W.E. Easterling, P. Crosson and J. Darmstadter) Resources for the Future, Washington, DC, pp. 53—67.

Hekstra, G.P. (1990) Man's impact on atmosphere and climate: a global threat? Strategies to combat global warming, in *Expected Effects of Climate Change on Marine Coastal Ecosystems* (eds J.J. Beukema, W.J. Wolff and J.J.W.M. Brouns), Kluwer, Dordrecht, pp. 5–16.

Herdman, W.A. (ed.) (1886–1900) *Reports upon the Fauna of Liverpool Bay and the Neighbouring Seas*. Longmans, Green & Company (Vol. 1)/Liverpool Marine Biological Committee (Vols 2–5).

Heslinga, G.A. (1981) Larval development, settlement and metamorphosis of the tropical gastropod *Trochus niloticus*. *Malacologia*, **20**, 349–57.

Hewatt, W.G. (1935) Ecological succession in the *Mytilus califonianus* habitat as observed in Monterey Bay, California. *Ecology*, **16**, 244–51.

Hewatt, W.G. (1937) Ecological studies on selected marine intertidal communities of Monterey Bay, California. *Am. Midl. Nat.*, **18**, 161–206.

Heywood, R.B. and Whittaker, T.M. (1984) The Antarctic marine flora, in, *Antarctic Ecology*, Vol. 2 (ed. R.M. Laws), Academic Press, London, pp. 373–419.

Hicks, G.R.F. (1985) Meiofauna associated with rocky shore algae, in *The Ecology of Rocky Coasts, Essays Presented to J.R. Lewis*, D.Sc. (eds P.G. Moore and R. Seed), Hodder and Stoughton, London, pp. 36–56.

Higgins, R.P. and Thiel, H. (eds.) (1988) *Introduction to the Study of Meiofauna*, Smithsonian Institution Press, Washington, DC, 488 pp.

Hildebrand, S.F. (1939) The Panama Canal as a passageway for fishes, with lists and remarks on the fishes and invertebrates observed. *Zoologica, NY*, **24**, 15–45.

Hill, A.S. and Hawkins, S.J. (1990) Methods for the investigation of microbial films on rocky shores. *J. mar. biol. Ass. U.K.*, **70**, 77–88.

Hill, A.S. and Hawkins, S.J. (1991) Seasonal and spatial variation of epilithic microalgal distribution and abundance and its ingestion by *Patella vulgata* on a moderately exposed rocky shore. *J. mar. biol. Ass. U.K.*, **71**, 403–25.

Hill, T.O. (1993) Algal zonation in the sublittoral fringe: the importance of competition, PhD thesis, University of Liverpool, 293 pp.

Hillman, K., Walker, D.I., Larkum, A.W.D. and McComb, A.J. (1989) Productivity and nutrient limitation, in *Biology of Seagrasses*, Vol. 2 (eds A.W.D. Larkum, A.J. McComb and S.A. Shepherd), Elsevier, Amsterdam, pp. 635–68.

Hines, A.H. and Loughlin, T.R. (1980) Observations of sea otters digging for clams at Monterey Harbour, California. *Fish. Bull. U.S.*, **78**, 159–63.

Hiscock, K. (1983) Water movement, in *Sublittoral Ecology. The Ecology of the Shallow Sublittoral Benthos* (eds R. Earll and D.G. Erwin), Oxford University Press, Oxford, pp. 58–96.

Hiscock, K. (1985) Aspects of the ecology of rocky sublittoral areas, in *The Ecology of Rocky Coasts. Essays Presented to J.R. Lewis*, D.Sc. (eds P.G. Moore and R. Seed), Hodder & Stoughton, London, pp. 290–328.

Hixon, M.A. and Brostoff, W.N. (1983) Damselfish as keystone species in reverse: intermediate disturbance and diversity of reef algae. *Science*, **220**, 511–13.

Hockey, P.A.R. and Bosman, A.L. (1986) Man as an intertidal predator in Transkei: disturbance, community convergence and management of a natural food resource. *Oikos*, **46**, 3–14.

Hoff, R.Z. (1993) Bioremediation: an overview of its development and its use for oil spill cleanup. *Mar. Poll. Bull.*, **26**, 476–81.

Hoffman, J.A., Katz, J. and Bertness, M.D. (1984) Fiddler crab deposit–feeding and meiofaunal abundance in salt marsh habitats. *J. exp. mar. Biol. Ecol.*, **82**, 2–3.

Holme, N.A. and McIntyre, A.D. (eds) (1984) *Methods for the Study of Marine Benthos*. Blackwell Scientific, Oxford.

Hooper, R. (1981) Recovery of Newfoundland benthic marine communities from sea ice, in *Proc. Int. Seaweed Symp.*, Vol. 8 (eds G.E. Fogg and W.E. Jones), UCNW Mar. Sci. Lab., Menai Bridge, pp. 360–66.

Horn, M.H. (1989) Biology of marine herbivorous fishes. *Oceanogr. mar. Biol. ann. Rev.*, **27**, 167–272.

Horn, M.H. (1992) Herbivorous fishes, feeding and digestive mechanisms, in *Plant–Animal Interactions in the Marine Benthos* (eds D.M. John, S.J. Hawkins and J.H. Price) (Systematics Association Special Volume No. 46), Clarendon Press, Oxford, pp. 339–62.

Houghton, J.T., Jenkins, G.J. and Ephraums, J.J. (1990) *Climate Change, the IPCC Scientific Assessment*, Cambridge University Press, Cambridge, 365 pp.

Hovenkamp, F. (1991) Immigration of larval plaice (*Pleuronectes platessa* L.) into the western Wadden Sea: a question of timing. Texel Flatfish Symposium, *Proc. 1st Int. Symp. Flatfish Ecol.*, Part 1, pp. 287–96.

Hruby, T. (1976) Observation of algal zonation resulting from competition. *Estuar. coast. mar. Sci.*, **7**, 531–3.

Huang, R. and Boney, A.D. (1984) Growth interactions between littoral diatoms and juvenile marine algae. *J. exp. mar. biol. Ecol.*, **81**, 21–46.

Hughes, R.N. (1980a) Predation and community structure, in *The Shore Environment*, Vol. 2, *Ecosystems* (eds D.E.G. Irvine and W.F. Farnham) (Systematics Association Special Volume No. 17B), Academic Press, London, pp. 699–728.

Hughes, R.N. (1980b) Optimal foraging in the marine context. *Oceanogr. mar. biol. ann. Rev.*, **18**, 423–81.

Hughes, R.N. and Burrows, M.T. (1993) Predatory behaviour of the intertidal snail, *Nucella lapillus*, and its effect on community structure, in *Mutualism and Community Organisation. Behavioural, Theroretical, and Food Web Approaches* (eds H. Kawanabe, J. Cohen and K. Iwasaki), Oxford University Press, Oxford, pp. 64–83.

Hughes, R.N. and Elner, R.W. (1979) Tactics of a predator, *Carcinus maenas*, and morphological responses of the prey, *Nucella lapillus*. *J. Anim. Ecol.*, **48**, 65–78.

Hulings, N.C. and Gray, J.S. (1971) *A Manual for the Study of Meiofauna* (Smithsonian Contributions to Zoology No. 78), Smithsonian Institution Press, Washington, DC, 84 pp.

Hurd, C.L. and Dring, M.J. (1991) Desiccation and phosphate uptake by intertidal fucoid algae in relation to zonation. *Br. Phycol. J.*, **26**, 327–33.

Hurd, C.L., Galvin, R.S., Norton, T.A. and Dring, M.J. (1993) Production of hyaline hairs by intertidal species of *Fucus* (Fucales) and their role in phosphate uptake. *J. Phycol.*, **29**, 160–65.

Hurlbert, S.H. (1984) Pseudoreplication and the design of ecological field experiments. *Ecol. Monogr.*, **54**, 187–211.

Hutchings, P.A. and Saenger, P. (1987) *The Ecology of Mangroves*, Australian University of Queensland Press, Brisbane, Queensland.

Ince, M. (1990) *The Rising Seas*, Earth Scan, London, 152 pp.

Ingolfsson, A. (1975) Life on seashores, in *Rit. Landverndar*, Vol. 4, *Votlendi*, (ed. A. Gardarsson), Landvernd, Reykjavik, pp. 61–99.

Innes, A.J. (1984) The effects of aerial exposure and desiccation on the oxygen consumption of intertidal limpets. *Veliger*, **27**, 134–9.

Irvine, J.R. and Northcote, T.G. (1983) Selection by young rainbow trout (*Salmo gairdneri*) in simulated stream environments for live and dead prey of different sizes. *Can. J. Fish. aquat. Sci.*, **40**, 1745–9.

Janke, K. (1990) Biological interactions and their role in the community structure in the rocky intertidal of Helgoland (German Bight, North Sea). *Helgoländ. wiss. Meeresunters.*, **44**, 219–63.

Jaquet, N. and Raffaelli, D. (1989) The ecological importance of the sand goby *Pomatoschistus minutus* (Pallas). *J. exp. mar. Biol. Ecol.*, **128**, 147–56.

Jensen, P. (1987) Feeding ecology of free-living aquatic nematodes. *Mar. Ecol. Progr. Ser.*, **35**, 187–96.

Jensen, P. (1988) Nematode assemblages in the deep-sea benthos of the Norwegian Sea. *Deep Sea Res.*, **35**, 1173–84.

Johannesson, K. (1989) The bare zone of Swedish rocky shores, why is it there? *Oikos*, **54**, 77–86.

John, D.M. and Lawson, G.W. (1991) Littoral ecosystems of tropical western Africa, in *Intertidal and Littoral Ecosystems*, (eds A.C. Mathieson and P.H. Nienhuis), Elsevier, Amsterdam, pp. 297–323.

Johnson, L.E. and Strathmann, R.R. (1989) Settling barnacle larvae avoid substrata previously occupied by a mobile predator. *J. exp. mar. Biol. Ecol.*, **128**, 87–103.

Johnson, R.G. (1973) Conceptual models of benthic communities, in *Models in Palaeobiology* (ed. T.J.M. Schopf), Freeman, Cooper & Co., San Francisco, pp. 148–59.

Jokiel, P.L. and Coles, S.L. (1990) Response of Hawaiian and other Indo–Pacific related reef corals to elevated temperature. *Coral Reefs*, **8**, 155–62.

Jones, H.D. (1984) Shell cleaning behaviour of *Calliostoma zizyphinum*. *J. moll. Stud.*, **50**, 245–7.

Jones, N.S. (1948) Observations and experiments on the biology of *Patella vulgata* at Port St. Mary, Isle of Man. *Proc. Trans. Liverpool biol. Soc.*, **56**, 60–77.

Jones, S.E. and Jago, C.F. (1993) In situ assessment of modification of sediment properties by burrowing invertebrates. *Mar. Biol.*, **115**, 133–42.

Jones, W.E. and Demetropoulos, A. (1968) Exposure to wave action, measurements of an important ecological parameter on rocky shores in Anglesey. *J. exp. mar. Biol. Ecol.*, **2**, 46–63.

Jordan, T.E. and Valicla, I. (1982) A nitrogen budget of the ribbed mussel, *Geukensi demisa*, and its significance in nitrogen flow in a New England salt-marsh. *Limnol. Oceanogr.*, **27**, 75–90.

Jorde, I. (1966) Algal associations of a coastal area south of Bergen, Norway. *Sarsia*, **23**, 1–52.

Jørgensen, C.B. (1990) *Bivalve Filter Feeding, Hydrodynamics, Bioenergetics, Physiology and Ecology*, Olsen & Olsen, Fredensborg.

Kafaji, A.K. and Boney, A.D. (1979) Antibiotic effects of crustose germlings of the red alga *Chondrus crispus* on benthic diatoms. *Ann. Bot.*, **43**, 321–33.

Kain, J.M. (1975) Algal recolonization of some cleared sub-tidal areas. *J. Ecol.*, **63**, 739—65.

Kain, J.M. and Jones, N.S. (1967) Subtidal algal colonization following the removal of *Echinus. Helgoländ. wiss. Meeresunters.*, **15**, 460–66.

Katz, L.C. (1980) Notes and discussions. Effects of burrowing by the fiddler crabs, *Uca pugnax* (Smith). *Est. coast. mar. Sci.*, **11**, 233–7.

Kemp, P.F. (1986) Direct uptake of detrital carbon by the deposit-feeding polychaete *Euzonus mucronatia* (Treadwell). *J. exp. mar. Biol. Ecol.*, **29**, 49–61.

Kendall, M.A., Bowman, R.S., Williamson, P. and Lewis, J.R. (1985) Annual variation in the recruitment of *Semibalanus balanoides* on the North Yorkshire coast 1969–1981. *J. mar. biol. Ass. U.K.*, **65**, 1009–30.

Kennish, M.J. (1986) *Ecology of estuaries*, Vol. 1, *Physical and Chemical Aspects*, CRC Press, Boca Raton FL, 272 pp.

Kent, A.C. and Day, R.W. (1983) Population dynamics of an infaunal polychaete: the effect of predators and an adult–recruit interaction. *J. exp. mar. Biol. Ecol.*, **73**, 185—203.

Keough, M.J. and Butler, A.J. (1979) The role of asteroid predators in the organisation of a sessile community on pier pilings. *Mar. Biol.*, **51**, 167–77.

Kerstan, M. (1991) The importance of rivers as nursery grounds for 0– and 1–group flounder (*Platichthys flesus* L.) in comparison to the Wadden Sea. Texel Flatfish Symposium. *Proc. 1st Int. Symp. Flatfish Ecol.*, Part 1. pp. 353–66.

King, R.J., Hutchings, P.A., Larkum, A.W.D. and West, R.J. (1991) Southeastern Australia, in *Intertidal and Littoral Ecosystems* (eds A.C. Mathieson and P.H. Nienhuis), Elsevier, Amsterdam, pp. 429–60.

Kingsford, M.J., Underwood, A.J. and Kennelly, S.J. (1991) Humans as predators on rocky reefs in New South Wales, Australia. *Mar. Ecol. Progr. Ser.*, **72**, 2–14.

Kingsley, C. (1856) *Glaucus; or, the Wonders of the Shore*, Macmillan & Company, Cambridge.

Kirchman, D., Graham, S., Reish, D. and Mitchell, R. (1981) Bacteria induce settlement and metamorphosis of *Jauna* (*Dexiospora*) *brasiliensis* Grube (Polychaeta, Spirorbidae). *J. exp. mar. Biol. Ecol.*, **56**, 153–63.

Kitching, J.A. and Ebling, F.J. (1967) Ecological studies at Lough Ine. *Adv. ecol. Res.*, **4**, 198–292.

Kitching, J.A., Muntz, L. and Ebling, F.J. (1966) The ecology of Lough Ine. XV. The ecological significance of shell and body forms in *Nucella. J. Anim. Ecol.*, **35**, 113–26.

Kneib, R.T. (1991) Indirect effects in experimental studies of marine soft-sediment communities. *Am. Zool.*, **31**, 874–85.

Knight-Jones, E.W. (1951) Gregariousness and some other aspects of the settling behaviour of *Spirorbis. J. mar. biol. Ass. U.K.*, **30**, 201–22.

Knight-Jones, E.W. (1953) Laboratory experiments on gregariousness during settling in *Balanus balanoides* and other barnacles. *J. exp. Biol.*, **30**, 584–98.

Knight-Jones, E.W. (1955) The gregarious settling reaction of barnacles as a measure of systematic affinity. *Nature*, **174**, 266.

Knox, G.A. (1986) *Estuarine Ecosystems: A Systems Approach* (2 vols), CRC Press, Boca Raton, FL.

Koop, K.R. and Lucas, M.I. (1983) Carbon flow and nutrient regeneration from the decomposition of macrophyte debris in a sandy beach microcosm, in *Sandy Beaches as Ecosystems* (eds A. McLachlan and T. Erasmus), Junk, The Hague, pp. 249–62.

Krebs, C.J. (1989) *Ecological Methodology*, Harper and Row, New York, 654 pp.

Kronberg, I. (1988) Structure and adaptation of the fauna in the black zone (littoral fringe) along rocky shores in northern Europe. *Mar. Ecol. Progr. Ser.*, **49**, 95–106.

Kussakin, O.G. (1971) Intertidal ecosystems of the seas of the U.S.S.R. *Helgoländ wiss. Meeresunters.*, **30**, 243–62.

Langston, W.J., Bryan, G.W., Burt, G.R. and Gibbs, P.E. (1990) Assessing the impact of tin and TBT in estuaries and coastal regions. *Funct. Ecol.*, **4**, 433–43.

Lassuy, D.R. (1980) Effects of 'farming' behaviour by *Eupomacentrus lividus* and *Hemiglyphidodon plagiometopon* on algal community structure. *Bull. mar. Sci.*, **30**, 304–12.

Lauff, G.H. (ed.) (1967) *Estuaries.* American Association for the Advancement of Science, Washington, DC, Publ. No. 83, 755 pp.

Lawson, G.W. and Norton, T.A. (1971) Some observations on littoral and sublittoral zonation at Tenerife (Canary Isles). *Bot. Mar.*, **14**, 116–20.

Lazo, L., Markham, J.H. and Chapman, A.R.O. (1996) Herbivory and harvesting, effects of sexual recruitment and vegetative modules of *Ascophyllum nodosum. Ophelia* (in press).

Lebednik, P.A. (1973) Ecological effects of intertidal uplifting from nuclear testing. *Mar. Biol.*, **20**, 197–207.

Legendre, P.L. (1993) Spatial autocorrelation, trouble or new paradigm? *Ecology*, **74**, 1659–73.

Levin, S. (1981) The role of theoretical ecology in the description and understanding of populations in heterogeneous environments. *Am. Zool.*, **21**, 865–75.

Levington, J.S. (1987) The body size–prey hypothesis and *Hydrobia. Ecology*, **68**, 229–31.

Lewin, R. (1986) Supply-side ecology. *Science*, **234**, 25–7.

Lewis, J.R. (1954) Observations on a high level population of limpets. *J. Anim. Ecol.*, **23**, 85–100.

Lewis, J.R. (1964) *The Ecology of Rocky Shores*, English Universities Press, London, 323 pp.

Lewis, J.R. (1972) Problems and approaches to baseline studies in coastal communities, in *Marine Pollution and Sea Life* (ed. M. Ruivo), Fishing News (Books) Ltd, London, pp. 401–4.

Lewis, J.R. (1976) Long-term ecological surveillance: practical realities in the rocky littoral. *Oceanogr. mar. Biol. ann. Rev.*, **14**, 371–90.

Liddell, W.D. and Ohlhurst, S.L. (1986) Changes in benthic community composition following the mass mortality of *Diadema* at Jamaica. *J. exp. mar. Biol. Ecol.*, **95**, 271–8.

Limia, J.M. (1989) Bioturbation of intertidal sediments, an experimental approach involving the amphipod *Corophium volutator* (Pallas), PhD thesis, University of Aberdeen, 130 pp.

Limpenny, D.S., Rowlatt, S.M. and Manning, P.M. (1992) Environmental impact of marine colliery waste disposal operations on the sea bed off Seaham, County Durham. *Aquat. Environ. Monit. Rep. Dir. Fish. Res. G.B.*, **33**, 19 pp.

Lindahl, O. (1993) Hydrodynamical processes: a trigger and source for flagellate blooms along the Skagerrak coasts? in *Toxic Phytoplankton Blooms in the Sea* (eds T.J. Smayda and Y. Shimizu) (Developments in Marine Biology, 3) Elsevier, Amsterdam, pp. 775–82.

Lipkin, Y. (1991) Life in the littoral of the Red Sea (with remarks on the Gulf of Aden), in *Intertidal and Littoral Ecosystems* (Ecosystems of the World 24) (eds A.C. Mathieson and P.H. Nienhuis), Elsevier, Amsterdam, pp. 391–427.

Lipkin, Y. and Safriel, U. (1971) Intertidal zonation on rocky shores at Mikmoret (Mediterranean, Israel). *J. Ecol.*, **59**, 1–30.

Little, C. (1989) Factors governing patterns of foraging activity in littoral marine herbivorous molluscs. *J. moll. Stud.*, **55**, 273–84.

Littler, M.M. and Littler, D.S. (1980) The evolution of thallus form and survival strategies in benthic marine macroalgae, field and laboratory test of a functional form model. *Am. Nat.*, **116**, 25–44.

Littler, M.M. and Littler, D.S. (eds) (1985) *Handbook of Phycological Methods. Ecological Field Methods*, Cambridge University Press, Cambridge.

Littler, M.M., Littler, D.S., Murray, S.N. and Seapy, R.R. (1991) Southern California rocky intertidal ecosystems, in *Intertidal and Littoral Ecosystems* (Ecosystems of the World 24) (eds A.C. Mathieson and P.H. Nienhuis) Elsevier, Amsterdam pp. 273–96.

Lodge, S.M. (1948) Algal growth in the absence of *Patella* on an experimental strip of foreshore, Port St. Mary, Isle of Man. *Proc. Trans. Liverpool biol. Soc.*, **56**, 78–83.

Long, S.P. and Mason, C.F. (1983) *Saltmarsh Ecology*, Blackie, Glasgow.

Loosanoff, V.L. (1955) The European oyster in American waters. *Science*, **121**, 119–21.

Lopez, G., Riemann, F. and Schrage, M. (1979) Feeding biology of the brackish-water oncholaimid nematode *Adoncholaimus thalassophygas*. *Mar. Biol.*, **54**, 311–18.

Lubchenco, J. (1978) Plant species diversity in a marine intertidal community, importance of herbivore food preference and algal competitive abilities. *Am. Nat.*, **112**, 23–39.

Lubchenco, J. (1980) Algal zonation in a New England rocky intertidal community, an experimental analysis. *Ecology*, **61**, 333–44.

Lubchenco, J. (1983) *Littorina* and *Fucus*: effects of herbivores, substratum heterogeneity, and plant escapes during succession. *Ecology*, **64**, 1116–23.

Lubchenco, J. and Cubit, J. (1980) Heteromorphic life histories of certain marine algae as adaptations to variations in herbivory. *Ecology*, **61**, 676–87.

Lubchenco, J. and Gaines, S.D. (1981) A unified approach to marine plant–herbivore interactions. I. Populations and communities. *Ann. Rev. Ecol. Syst.*, **12**, 405–37.

Lubchenco, J. and Menge, B.A. (1978) Community development and persistence in a low rocky intertidal zone. *Ecol. Monogr.*, **48**, 67–94.

Lucas, M.I., Walker, G., Howard, D.L. and Crisp, D.J. (1979) An energy budget for metamorphosis of cypris larvae of *Balanus balanoides* (L.). *Mar. Biol.*, **55**, 211–30.

Luckens, P.A. (1976) Settlement and succession on rocky shores at Auckland, North Island, New Zealand. *Mem. N.Z. Oceanogr. Inst.*, **70**, 3–63.

Lüning, K. (1990) *Seaweeds. Their environment, biogeography and ecophysiology.* Wiley, New York, 528pp.

Lüning, K. and Asmus R. (1991) Physical characteristics of littoral ecosystems, with special reference to marine plants, in *Intertidal and Littoral Ecosystems* (Ecosystems of the World 24) (eds A.C. Mathieson and P.H. Nienhuis), Elsevier, Amsterdam, pp. 7–25.

Lutz, P.L. (1975) Adaptive and evolutionary aspects of the ionic content of fish. *Copeia*, **1975**, 369–73.

McCall, P. (1977) Community patterns and adaptive strategies of the infaunal benthos of Long Island Sound. *J. mar. Res.*, **35**, 221–66.

McIntyre, A.D. (1970) The range of biomass in intertidal sand, with special reference to *Tellina tenuis*. *J. mar. biol. Ass. U.K.*, **50**, 561–75.

McIntyre, A.D. and Murison, D.J. (1973) The meiofauna of a flatfish nursery ground. *J. mar. biol. Ass. U.K.*, **53**, 93–118.

McLachlan, A. (1980) The definition of sandy beaches in relation to exposure, a simple rating system. *S. Afr. J. Sci.*, **76**, 137–8.

McLachlan, A. (1990) Dissipative beaches and macrofauna communities on exposed intertidal sands. *J. coast. Res.*, **6**, 57–71.

McLachlan, A. (1992) Sand beach ecology, swash features relevant to the macrofauna. *J. coast. Res.*, **8**, 398–407.

McLachlan, A. and Lewin, J. (1981) Observations on surf phytoplankton blooms along the coasts of South Africa. *Bot. Mar.*, **24**, 553–7.

McLachlan, A. and Romer, G. (1990) Trophic relationships in a high energy beach and surf-zone ecosystem, in *Trophic Relationships in the Marine Environment* (Proc. 24th Eur. Mar. Biol. Symp.) (eds M. Barnes and R.N. Gibson) Aberdeen University Press, Aberdeen, pp. 356–71.

McLachlan, A., Jaramillo, E., Donn, T.E. and Wessels, F. (1993) Sandy beach macrofauna communities and their control by the physical environment, a geographical comparison. *J. coast. Res.*, **9**, 000–00.

MacLulich, J.H. (1986) Experimental evaluation of methods for sampling and assaying intertidal epilithic microalgae. *Mar. Ecol. Progr. Ser.*, **34**, 275–80.

McLusky, D.S. (1989) *The Estuarine Ecosystem*, 2nd edn (Tertiary level biology series), Blackie and Son Ltd, Glasgow, 215 pp.

McLusky, D.S., Bryant, D.M. and Elliott, M. (1991) The impact of reclamation on benthic production and prey availability in the Forth Estuary, eastern Scotland, in *The Changing Coastline* (ed. J. Pethick) Kluwer, Dordrecht, The Netherlands.

Mann, K.H. (1973) Seaweeds, Their productivity and strategy for growth. *Science*, **182**, 975–81.

Mann, K.H. (1982) *The ecology of coastal waters: a systems approach.* Blackwell Scientific Publishers, Oxford.

Mann, K.H. (1982) Kelp, sea urchins and predators, a review of strong interactions in rocky subtidal systems of eastern Canada 1970–1980. *Neth. J. Sea Res.*, **16**, 414–23.

Marsh, C.P. (1986) Impact of avian predators on high intertidal limpet populations. *J. exp. mar. Biol. Ecol.*, **104**, 185–201.

Mathieson, A.C., Penniman, C.A. and Harris, L.G. (1991) Northwest Atlantic rocky shore ecology, in *Intertidal and Littoral Ecosystems* (eds A.C. Mathieson and P.H. Nienhuis), Elsevier, Amsterdam, pp. 109–92.

Maxwell, S.E. and Delaney, H.D. (1990) *Designing Experiments and Analysing Data*, Wadsworth Publishing Company, Belmont, CA.

Meehan, B. (1982) *Shell Bed to Shell Midden*, Australian Institute of Aboriginal Studies, Globe Press, Melbourne.

Meese, R.J. and Tomich, P.A. (1992) Dots on the rocks, a comparison of percent cover estimation methods. *J. exp. mar. Biol. Ecol.*, **165**, 59–73.

Menge, B.A. (1976) Organization of the New England rocky intertidal community: role of predation, competition and environmental heterogeneity. *Ecol. Monogr.*, **46**, 355–93.

Menge, B.A. (1978a) Predation intensity in a rocky intertidal community: relationship between predator foraging activity and environmental harshness. *Oecologia*, **34**, 1–16.

Menge, B.A. (1978b) Predation intensity in a rocky intertidal community. Effect of an algal canopy, wave action and desiccation on predator feeding rates. *Oecologia*, **34**, 17–36.

Menge, B.A. (1991) Relative importance of recruitment and other causes of variation in rocky intertidal community structure. *J. exp. mar. Biol. Ecol.*, **145**, 69–100.

Menge, B.A. (1992) Community regulation: under what conditions are bottom-up factors important on rocky shores? *Ecology*, **73**, 755–65.

Menge, B.A. and Lubchenco, J. (1981) Community organisation in temperate and tropical rocky intertidal habitats. Prey refuges in relation to consumer pressure gradients. *Ecol. Monogr.*, **51**, 429–50.

Menge, B.A. and Olson, A.M. (1990) Role of scale and environmental factors in regulation of community structure. *Trends Ecol. Evol.*, **5**, 52–7.

Menge, B.A. and Sutherland, J.P. (1976) Species diversity gradients, synthesis of the roles of predation, competition and spatial heterogeneity. *Am. Nat.*, **110**, 351–69.

Menge, B.A. and Sutherland, J.P. (1987) Community regulation, variation in disturbance, competition and predation in relation to environmental stress and recruitment. *Am. Nat.*, **130**, 730–57.

Menge, B.A., Lubchenco, J. and Ashkenas, L.R. (1985) Diversity, heterogeneity and consumer pressure in a tropical intertidal community. *Oecologia*, **65**, 394–405.

Menge, B.A., Lubchenco, J., Ashkenas, L.R. and Ramsey, F. (1986) Experimental separation of the effects of consumers on sessile prey in the low zone of a rocky shore in the Bay of Panama, direct and indirect consequences of food web complexity. *J. exp. mar. Biol. Ecol.*, **100**, 225–70.

Meyers, M.B., Fossing, H. and Powell, E.N. (1987) Microdistribution of interstitial meiofauna, oxygen and sulfide gradients and the tubes of macrofauna. *Mar. Ecol. Progr. Ser.*, **33**, 223–41.

Mitchell, D. (1992) Effect of seed density on Manila clam growth and production at a British Columbia clam farm, in *Workshop on Water Quality and Alternate Species in the Canadian Mollusc Culture Industry* (eds N. Bourne and W. Heath). *Bull. Aquacult. Assoc. Can.*, **92**, 29–32.

Montgomery, W.L. (1980) Comparative feeding ecology of two herbivorous damsel-fish (Pomacentridae, Teleostei) from the Gulf of California, Mexico. *J. exp. mar. Biol. Ecol.*, **47**, 9–24.

Moore, H.B. (1934) The biology of *Balanus balanoides*. I. Growth rate and its relation to season and tide level. *J. mar. biol. Ass. U.K.*, **19**, 851–68.

Moore, H.B. and Kitching, S.A. (1939) The biology of *Chthamalus stellatus* (Poli). *J. mar. biol. Ass. U.K.*, **23**, 521–41.

Moore, P.G. and Seed, R. (eds.) (1985) *The Ecology of Rocky Coasts. Essays Presented to J.R. Lewis*, Hodder & Stoughton, London, 467 pp.

Mori, K. and Tanaka, M. (1989) Intertidal community structures and environmental conditions of exposed and sheltered rocky shores in Amakusa, Japan. *Publs Amakusa mar. biol. Lab.*, **10**, 41–64.

Morin, J.G., Kastendiek, J.E., Harrington, A. and Davis, N. (1985) Organisation and patterns of interactions in a subtidal sand community of an exposed coast. *Mar. Ecol. Progr. Ser.*, **27**, 163–85.

Morris, S. and Taylor, A.C. (1983) Diurnal and seasonal variations in physico-chemical conditions within intertidal rock pools. *Estuar. coast. Shelf Sci.*, **17**, 339–55.

Morrisey, D.J., Howitt, L. and Underwood, A.J. (1992) Spatial variation in soft-sediment benthos. *Mar. Ecol. Prog. Ser.*, **81**, 197–204.

Morse, A.N.C. (1992) Role of algae in the recruitment of marine invertebrate larvae, in *Plant–Animal Interactions in the Marine Benthos* (eds D.M. John, S.J. Hawkins and J.H. Price) (Systematics Association Special Volume No. 46), Clarendon Press, Oxford, pp. 385–404.

Morton, B. and Morton, J. (1983) *The Sea Shore Ecology of Hong Kong*, Hong Kong University Press, Hong Kong, 350 pp.

Morton, B.S., Hodgkiss, I.J. and Mak, M.S. (eds) (1993) *Proceedings of the First International Conference on the Biology of the South China Sea*, Hong Kong University Press, Hong Kong.

Morton, J. and Miller, M.C. (1973) *The New Zealand Sea Shore*, 2nd edn, Collins, London, 653 pp.

Munda, I.M. (1991) Shoreline ecology in Iceland, with special emphasis on the benthic algal vegetation, in *Intertidal and Littoral Ecosystems* (Ecosystems of the World 24) (eds A.C. Mathieson and P.H. Nienhuis) Elsevier, Amsterdam, pp. 67–81.

Murphy, D.J. (1979) A comparative study of the freezing tolerances of the marine snails *Littorina littorea* (L.) and *Nassarius obsoletus* (Say). *Physiol. Zool.*, **52**, 219–30.

Murphy, D.J. and Johnson, L.C. (1980) Physical and temporal factors influencing the freezing tolerance of the marine snail *Littorina littorea* (L.). *Biol. Bull.*, **158**, 220–32.

Muus, B.J. (1967) The fauna of Danish estuaries and lagoons. Distribution and ecology of dominating species in the shallow reaches of the mesohaline zone. *Meddelelser fra Danmarks Fiskeri-og Havundersogelser (NS)*, **5**, 3–316.

Muus, B.J. (1968) A field method for measuring "exposure" by means of plaster balls. A preliminary account. *Sarsia*, **34**, 61–8.

National Academy of Sciences (NAS) (1989) *Using Oil Spill Dispersants on the Sea*, National Academy Press, Washington, DC.

Naylor, E. (1965) Effects of heated effluents upon marine and estuarine organisms. *Adv. Mar. Biol.*, **3**, 63–103.

Naylor, E. (1985) Tidally rhythmic behaviour of marine animals. *Symp. Soc. Exp. Biol.*, **39**, 63–93.

Naylor, E. and Slinn, D.J. (1958) Observations on the ecology of some brackish water organisms in pools at Scarlett Point, Isle of Man. *J. Anim. Ecol.*, **27**, 15–25.

Neumann, D. and Heimbach, F. (1985) Circadian range of entrainment in the semilunar eclosion rhythm of the marine insect *Clunio marinus*. *J. Insect. Physiol.*, **31**, 549–57.

Newell, R.C. (1979) *Biology of Intertidal Animals*, Marine Ecological Surveys, Faversham.

Newton, L.C., Parkes, E.V.H. and Thompson, R.C. (1993) The effects of shell collection on the abundance of gastropods on Tanzanian shores. *Biol. Conserv.*, **63**, 241–5.

Niell, F.X. (1977) Rocky intertidal benthic systems in temperate seas, a synthesis of their functional performances. *Helgoländer wiss. Meeresunters*, **30**, 315–33.

Niemi, C.A. and Warheit, K.I. (1989) Variation in species diversity and the physical environment associated with primary treated sewage effluent. *Oceans '89*, **2**, 635–40.

North, W.J. (1968) Effects of canopy cutting on kelp growth, comparison of experimentation with theory, in, *Utilization of Kelp-bed Resources in Southern California* (eds. W.J. North and C.L. Hobbs). *Fish Bull. Calif. Dep. Fish Game*, **139**, 223–54.

Norton, T.A. (1981) Gamete expulsion and release in *Sargassum muticum*. *Bot. Mar.*, **24**, 465–70.

Norton, T.A. (1985) The zonation of seaweeds on rocky shores, in *The Ecology of Rocky Coasts* (eds P.G. Moore and R. Seed), Hodder & Stoughton, London, pp. 7–21.

Norton, T.A. (1991) Conflicting constraints on the form of intertidal algae. *Br. Phycol. J.*, **26**, 203–18.

Norton, T.A. (1992) *The Biology of Seaweed Propagules. Br. Phycol. J.*, **27**, 217.

Norton, T.A. and Benson, M.R. (1983) Ecological interactions between the brown seaweed *Sargassum muticum* and its associated fauna. *Mar. Biol.*, **75**, 169–77.

Norton, T.A. and Manley, N.L. (1990) The characteristics of algae in relation to their vulnerability to grazing snails, in, *Behavioural Mechanisms of Food Selection* (ed. R.N. Hughes) (NATO Advanced Study Institute Series, G20), Springer-Verlag, Berlin, pp. 461–78.

Norton, T.A., Hawkins, S.J., Manley, N.L., Williams, G.A. and Watson, D.C. (1990) Scraping a living, a review of littorinid grazing. *Hydrobiologia*, **193**, 117–38.

Nott, J.A. (1973) Settlement of the larvae of *Spirorbis spirorbis* L. *J. mar. biol. Ass. U.K.*, **53**, 437–53.

Nott, J.A. and Langston, W.J. (1989) Cadmium and the phosphate granules in *Littorina littorea. J. mar. biol. Ass. U.K.*, **69**, 219–27.

Odum, E.P. (1971) *Fundamentals of Ecology*, Saunders, Philadelphia.

Odum, W.E. and Heald, E.J. (1975) The detritus-based food web of an estuarine mangrove community, in *Estuarine Research*, Vol. 1. *Chemistry, Biology, and the Estuarine System* (ed. L.E. Cronin) (2nd Int. Estuar. Res. Conf., Myrtle Beach, SC), Academic Press, New York.

Odum, W.E., McIvor, C.C. and Smith, T.S. (1982) The ecology of mangroves of South Florida: a community profile. U.S. Fish Wildl. Serv., Office of Biological Services, Washington, DC. FWS/OBS–81/24.

Oerlemans, J. (1991) Possible changes in the mass balance of Greenland and Antarctic ice sheets and their effects on sea level, in *United Nations Environment Programme/CEP/USEPA International Workshop on Climate Change, Sea Level, Severe Tropical Storms and Associated Impacts* (eds T.M.L. Wigley and R.A. Warrick), Washington, DC.

Ojeda, F.F. and Santelices, B. (1984) Ecological dominance of *Lessonia nigrescens* (Phaeophyta) in central Chile. *Mar. Ecol. Progr. Ser.*, **19**, 83–91.

Olafsson, E.B. (1988) Inhibition of larval settlement to a soft bottom benthic community by drifting algal mats: an experimental test. *Mar. Biol.*, **97**, 571–4.

Olafsson, E.B. and Elmgren, R. (1991) Effects of biological disturbance by benthic amphipods *Monoporeia affinis* on meiobenthic community structure: a laboratory approach. *Mar. Ecol. Progr. Ser.*, **74**, 99–107.

Olafsson, E.B. and Persson, L. (1986) The interaction between *Nereis diversicolor* O.F. Muller and *Corophium volutator* Pallas as a structuring force in a shallow brackish environment. *J. exp. mar. Biol. Ecol.*, **103**, 103–117.

Oliva, D. and Castilla, J.C. (1986) The effect of human exclusion on the population structure of key-hole limpets *Fissurella crassa* and *F. limbata* on the coast of central Chile. *Mar. Ecol.*, **3**, 201–17.

Oliveira, E.C. de, jun. and Fletcher, A. (1977) Comparative observations on some physiological aspects of rocky shore and saltmarsh populations of *Pelvetia canaliculata* (Phaeophyta). *Bol. Bot. Univ. Sao Paulo*, **5**, 1–12.

Oliver, J.S., Kvitek, R.G. and Slattery, P.N. (1985) Walrus feeding disturbance: scavenging habits and recolonization of the Bering Sea benthos. *J. exp. mar. Biol. Ecol.*, **91**, 233–46.

Oltmann, F. (1892) Uber die Kultur und Lebensbedingungen der Meeresalgen. *Jb. wiss. Bot.*, **23**, 349–440.

Open University Course Team (OUCT) (1989) *Waves, Tides and Shallow Water Processes*, Pergamon Press, Oxford, 187 pp.

Orth, R.J., Heck, K.L. and Diaz, R.J. (1991) Littoral and intertidal systems in the mid-Atlantic coast of the United States, in *Intertidal and Littoral Ecosystems* (eds A.C. Mathieson and P.H. Nienhuis), Elsevier, Amsterdam, pp. 193–214.

Orton, J.H. (1929) Observations on *Patella vulgata* Part III. Habits and habitats. *J. mar. biol. Ass. U.K.*, **16**, 277–88.

Oswald, R.C. (1986) The epifaunal community of *Fucus serratus* (L.), ecology and physiology of association, PhD thesis, University of Wales, 282 pp.

Oswald, R.C., Telford, N. and Happey-Wood, C.M. (1984) The effect of encrusting bryozoans on the photosynthetic activity of *Fucus serratus* L. *Estuar. coast. Shelf Sci.*, **19**, 697–702.

Paine, R.T. (1966) Food web complexity and species diversity. *Am. Nat.*, **100**, 65–75.

Paine, R.T. (1969) The *Pisaster–Tegula* interaction: prey patches, predator food preferences and intertidal community structure. *Ecology*, **50**, 950–61.

Paine, R.T. (1971) A short-term investigation of resource partitioning in a New Zealand rocky intertidal habitat. *Ecology*, **52**, 1096–106.

Paine, R.T. (1974) Intertidal community structure, experimental studies on the relationship between a dominant competitor and its principal predator. *Oecologia*, **15**, 93–120.

Paine, R.T. (1977) Controlled manipulations in the marine intertidal zone and their contributions to ecological theory, in *The Changing Scenes in Natural Sciences 1776–1976*, Academy of Natural Sciences, New York, pp. 245–70.

Paine, R.T. (1980) Food webs: linkage, interaction strength and community infrastructure. *J. Anim. Ecol.*, **49**, 667–85.

Paine, R.T. (1994) *Marine Rocky Shores and Community Ecology: An Experimentalist's Perspective*, Ecology Institute, Nordbruite, 159 pp.

Paine, R.T. and Levin, S.A. (1981) Intertidal landscapes, disturbance and the dynamics of pattern. *Ecol. Monogr.*, **51**, 145–78.

Paine, R.T. and Suchanek, T.H. (1983) Convergence of ecological processes between independently evolved competitive dominants, a tunicate–mussel comparison. *Evolution*, **37**, 821–31.

Paine, R.T. and Vadas, R.L. (1969) The effects of grazing by sea urchins *Strongylocentrotus* spp. on benthic algal populations. *Limnol. Oceanogr.*, **14**, 710–19.

Paine, R.T., Castillo, J.C. and Cancino, J. (1985) Perturbation and recovery patterns of starfish-dominated assemblages in Chile, New Zealand and Washington State. *Am. Nat.*, **125**, 679–91.

Palmer, A.R. (1991) Effect of crab effluent and scent of damaged conspecifics on feeding, growth and shell morphology of the Atlantic dogwhelk *Nucella lapillus. Hydrobiologia*, **193**, 155–82.

Palmer, J.D. and Round, F.E. (1965) Persistent vertical migration rhythms in benthic microflora. I. The effect of light and temperature on the rhythmic behaviour of *Euglena obtusa. J. mar. biol. Ass. U.K.*, **95**, 567–82.

Palumbi, S.R. (1984) Measuring intertidal wave forces. *J. exp. mar. Biol. Ecol.*, **81**, 171–9.

Parks, R.J. and Buckingham, W.J. (1986) The flow of organic carbon through aerobic respiration and sulphate reduction in inshore marine sediments. *Proc. 4th Int. Symp. Microbial Ecology*, pp. 617–24.

Paterson, D.M., Crawford, R.M. and Little, C. (1986) The structure of benthic diatom assemblages: a preliminary account and evaluation of low tempertaure scanning electron microscopy. *J. exp. mar. Biol. Ecol.*, **96**, 279–89.

Pawlick, J.R. (1992) Chemical ecology of the settlement of benthic marine invertebrates. *Oceanograph. mar. Biol. ann. Rev.*, **30**, 273–335.

Pawlick, J.R. and Hadfield, M.G. (1990) A symposium on the chemical factors that influence the settlement and metamorphosis of marine invertebrate larvae: introduction and perspective. *Bull. mar. Sci.*, **46**, 450–54.

Pearse, J.S. (1980) Intertidal animals, in *The Natural History of Año Nuevo* (eds B.J. LeBouef and S. Kaza), Boxwood Press, Pacific Grove, CA, pp. 205–36.

Pearson, T.H. and Rosenberg, R. (1978) Macrobenthic succession in relation to organic enrichment and pollution of the marine environment. *Oceanogr. mar. Biol. ann. Rev.*, **16**, 229–311.

Pearson, T.H. and Stanley, S.O. (1979) Comparative measurements of the redox potential of marine sediments as a rapid means of assessing the effect of marine pollution. *Mar. Biol.*, **53**, 371–9.

Peres, J.M. (1967) The Mediterranean benthos. *Oceanogr. mar. Biol. ann. Rev.*, **5**, 449–533.

Perkins, E.J. (1977) The quality and biology of the environment adjacent to the Workington works of the British Steel Corporation. A report to the Cumbria Sea Fisheries Committee by the Marine Laboratory, University of Strathclyde. CSFC Scientific Report 77/1. 88 pp.

Peterson, B.J. and Fry, B. (1987) Stable isotopes in ecosystem studies. *Ann. Rev. Ecol. Syst.*, **18**, 293–320.

Peterson, C.H. (1977) Competitive organization of the soft-bottom macrobenthic communities of Southern California lagoons. *Mar. Biol.*, **43**, 343–59.

Peterson, C.H. (1979) The importance of predation and competition in organizing the intertidal epifaunal communities of Barnegat Inlet, New Jersey. *Oecologia*, **39**, 1–24.

Peterson, C.H. and Black, R. (1987) Resource depletion by active suspension feeders on tidal flats, influence of local density and tidal elevation. *Limnol. Oceanogr.*, **32**, 143–66.

Peterson, C.H. and Black, R. (1988) Density-dependent mortality caused by physical stress interacting with biotic history. *Am. Nat.*, **131**, 257–70.

Pethick, J. (1991) Marshes, mangroves and sea level rise. *Geography*, **76**, 330.

Petpiroon, S. and Morgan, S. (1983) Observations of the tidal activity rhythm of the periwinkle *Littorina nigrolineata* (Gray). *Mar. Behav. Physiol.*, **9**, 171–92.

Petraitis, P.S. (1984) Laboratory experiments on the effects of a gastropod (*Hydrobia totteni*) on survival of an infaunal deposit–feeding polychaete (*Capitella capitata*). *Mar. Ecol. Progr. Ser.*, **18**, 263–8.

Pickett, S.T.A. and White, P.S. (eds) (1985) *The Ecology of Natural Disturbance and Patch Dynamics*, Academic Press, New York.

Pihl, L. (1985) Food selection and consumption of mobile epibenthic fauna in shallow marine areas. *Mar. Ecol. Progr. Ser.*, **22**, 169–79.

Por, F.D. (1978) *Lessepsian Migration. The Influx of Red Sea Biota into the Mediterranean by Way of Suez Canal* (Ecological Studies, Analysis and Synthesis, Vol. 23), Springer-Verlag, Berlin, 228 pp.

Posey, M.H. (1986) Changes in a benthic community associated with dense beds of a burrowing deposit feeder *Callianassa californiensis*. *Mar. Ecol. Progr. Ser.*, **31**, 15–22.

Povey, A. and Keough, M.J. (1991) Effects of trampling on plant and animal populations on rocky shores. *Oikos*, **61**, 355–68.

Prater, A.J. (1981) *Estuary Birds of Britain and Ireland*, Poyser, Berkhampstead. 440 pp.

Pullen, J.S.H. and Rainbow, P.S. (1991) The composition of pyrophosphate heavy metal detoxification granules in barnacles. *J. exp. mar. Biol. Ecol.*, **150**, 249–66.

Puttick, G. (1979) Foraging behaviour and activity budgets of Curlew sandpipers. *Ardea*, **67**, 111.

Quammen, M.L. (1984) Predation by shorebirds, fish and crabs on invertebrates in intertidal mudflats: an experimental test. *Ecology*, **65**, 529–37.

Raffaelli, D. (1977) Observations of the copulatory behaviour of *Littorina rudis* (Maton) and *Littorina nigrolineata* (Gray) (Gastropoda: Prosobranchia). *Veliger*, **20**, 75–7.

Raffaelli, D. (1978) The relationship between shell injuries, shell thickness and habitat characteristics of the intertidal snail *Littorina rudis* (Maton). *J. moll. Stud.*, **44**, 166–70.

Raffaelli, D. (1979) The grazer–algae interaction in the intertidal zone on New Zealand rocky shores. *J. exp. mar. Biol. Ecol.*, **38**, 81–100.

Raffaelli, D.G. (1982) Recent ecological research on some European species of *Littorina*. *J. moll. stud.*, **48**, 342–54.

Raffaelli, D. (1992) Conservation of Scottish Estuaries. *Proc. Roy. Soc. Edinb.*, **100B**, 55–76.

Raffaelli, D. and Hall, S.J. (1992) Compartments and predation in an estuarine food web. *J. Anim. Ecol.*, **61**, 551–60.

Raffaelli, D.G. and Hughes, R.N. (1978) The effect of crevice size and availability on populations of *Littorina rudis* and *Littorina neritoides*. *J. Anim. Ecol.*, **47**, 71–83.

Raffaelli, D. and Milne, H. (1987) An experimental investigation of the effects of shorebird and flatfish predation on estuarine invertebrates. *Estuar. coast. Shelf Sci.*, **24**, 1–13.

Raffaelli, D., Conacher, A., McLachlan, H. and Emes, C. (1989a) The role of epibenthic crustacean predators in an estuarine food web. *Estuar. coast. Shelf Sci.*, **28**, 149–60.

Raffaelli, D., Hull, S. and Milne, H. (1989b) Long-term changes in nutrients, weed mats and shorebirds in an estuarine system. *Cah. Biol. Mar.*, **30**, 259–70.

Raffaelli, D., Falcy, V. and Galbraith, C. (1990a) Eider predation and the dynamics of mussel-bed communities, in *Trophic Relationships in the Marine Environment* (Proc. 24th Eur. Mar. Biol. Symp.) (eds M. Barnes and R.N. Gibson), Aberdeen University Press, Aberdeen, pp. 157–69.

Raffaelli, D., Richner, H., Summers, R. and Northcott, S. (1990b) Tidal migrations in the flounder (*Platichthys flesus*). *Mar. Behav. Physiol.*, **16**, 249–60.

Raffaelli, D., Karakassis, I. and Galloway, A. (1991) Zonation schemes on sandy shores: a multivariate approach. *J. exp. mar. Biol. Ecol.*, **148**, 241–53.

Ramus, J., Beale, S.I. and Mauzerall, D. (1976) Correlation of changes in pigment content with photosynthetic capacity of seaweeds as a function of water depth. *Mar. Biol.*, **37**, 231–8.

Rankin, J.C. and Davenport, J. (1981) *Animal Osmoregulation*, Blackie, Glasgow and London, 202 pp.

Rees, C.B. and Cattley, J.G. (1949) *Processa aequimana* Paulson in the North Sea. *Nature*, **164**, 367.

Reid, D.G. (1990) A cladistic study of the genus *Littorina* (Gastropoda): implications for evolution of reproductive strategies and for classification. *Hydrobiologia*, **193**, 1–20.

Reid, W.V. and Trexler, M.C. (1992) Responding to potential impacts of climatic change on U.S. coastal biodiversity. *Coast. Manage.*, **20**, 117–42.

Reidenauer, J.A. and Thistle, D. (1981) Response of a soft-bottom harpacticoid community to stingray (*Dasyatis sabina*) disturbance. *Mar. Biol.*, **65**, 261–7.

Reise, K. (1985) *Tidal Flat Ecology. An Experimental Approach to Species Interactions* (Ecological Studies 54), Springer-Verlag, Berlin.

Reise, K. (1992) The Wadden Sea as a pristine nature reserve, in *Present and Future Conservation of the Wadden Sea: Proc. 7th Int. Wadden Sea Symp., Ameland, 1990* (eds N. Dankers, C.J. Smit and M. Scholl). *Publ. Ser. Neth. Inst. Sea Res.*, **20**, 49–53.

Reise, K. and Ax, P. (1979) A meiofaunal "thiobios" limited to the anaerobic sulfide system of marine sand does not exist. *Mar. Biol.*, **54**, 225–37.

Renoux-Meunier, A., 1965. Etude de la végétation algale du cap Saint-Martin (Biarritz). *Bull. Cent. Etudes Rech. Sci. Biarritz*, **5**, 378–564.

Rhoads, D.C. (1974) Organism–sediment relations on the muddy sea floor. *Oceanogr. mar. Biol. ann. Rev.*, **12**, 263–300.

Rhoads, D.C. and Boyer, L.F. (1982) The effects of marine benthos on physical properties of sediments. A successional perspective, in *Animal Sediment Relations. The Biogenic Alteration Of Sediments* (eds P.L. McCall and M.J.S. Tevesz), Plenum Press, New York, pp. 3–52.

Rhoads, D.C., Yingst, J.Y. and Ullman, W. (1978) Seafloor stability in central Long Island Sound. Part I. Temporal changes in erodibility of fine-grained sediment, in *Estuarine Interactions* (ed. M.L. Wiley), Academic Press, New York, pp. 221–4.

Ricketts, E.F., Calvin, J., Hedgpeth, J.W. and Phillips, D.W. (1985) *Between Pacific Tides*, 5th edn Stanford University Press, Stanford, CA.

Riedl, R. (1971) Water movement, general introduction, in *Marine Ecology. A Comprehensive Integrated Treatise on Life in Oceans and Coastal Waters* (ed. O. Kinne), Vol. 1, pt. 2. Wiley Interscience, Chichester.

Riemann, F. and Schrage, M. (1978) The mucus-trap hypothesis on feeding of aquatic nematodes and implications for biodegradation and sediment texture. *Oecologia*, **34**, 75–88.

Rittschof, D., Branscomb, E.S. and Costlow, J. (1984) Settlement and behaviour in relation to flow and surface in larval barnacles, *Balanus amphitrite* Darwin. *J. exp. mar. Biol. Ecol.*, **82**, 131–46.

Roe, P. (1975) Aspects of the history and territorial behaviour in young individuals of *Platynereis bicanaliculata* and *Nereis vexillosa*. *Pac. Sci.*, **29**, 341–8.

Round, F.E. (1981) *The Ecology of the Algae*, Cambridge University Press, Cambridge.

Russell, E.S. (1907) Environmental studies on the limpet. *Proc. zool. Soc. Lond.*, (1907), 856–70.

Russell, G. (1972) Phytosociological studies on a two-zonal shore. I. Basic pattern. *J. Ecol.*, **60**, 539–45.

Russell, G. (1991) Vertical distribution, in *Intertidal and Littoral Ecosystems* (eds A.C. Mathieson and P.H. Nienhuis), Elsevier, Amsterdam, pp. 43–66.

Russell, G. and Veltkamp, C.J. (1982) Epiphytes and antifouling characteristics of *Himanthalia* (brown algae). *Br. Phycol. J.*, **17**, 239.

Russell, G., Hawkins, S.J., Evans, L.C., Jones, H.D. and Holmes, G.D. (1983) Restoration of a disused dock basin as a habitat for marine benthos and fish. *J. appl. Ecol.*, **20**, 43–58.

Ryland, J.S. (1959) Experiments on the selection of algal substrates by polyzoan larvae. *J. exp. Biol.*, **36**, 613–31.

Ryland, J.S. (1960) Experiments on the influence of light on the behaviour of polyzoan larvae. *J. exp. mar. Biol. Ecol.*, **37**, 783–800.

Ryland, J.S. (1962) The association between Polyzoa and algal substrata. *J. Anim. Ecol.*, **31**, 331–8.

Safriel, U. and Lipkin, Y. (1964) On the intertidal zonation of the rocky shores at Eilat (Red Sea, Israel). *Isr. J. Zool.*, **18**, 205–31.

Saldanha, L. (1974) Estudo do povoamento dos horizontes superiores da rocha litoral da Costa da Arr bida (Portugal). *Arquivos do museu Bocage*, Ser. 2, V, no. 1.

Salvat, B. (1964) Les conditions hydrodynamiques interstitielles des sediments meubles intertidaux et la repartition verticale de la faune endognee. *C.R. Acad. Sci. Paris*, **259**, 1576–9.

Salvat, B. (1967) La macrofauna carcinologique endogenee des sediments meubles intertidaux (tanaiduces, isopodes et amphipodes), ethologie, bionomic et cycle biologique. *Mem. Mus. natn. Hist. Nat., Paris, Ser. A.*, **45**, 1–275.

Sanders, H.L. (1977) The West Falmouth spill – Florida 1969. *Oceanus*, **20**, 15–24.

Sandison, E.E. (1950) Appearance of *Elminius modestus* in South Africa. *Nature*, **165**, 79–80.

Santelices, B. (1990) Patterns of organisations of intertidal and shallow subtidal vegetation in wave exposed habitats of central Chile. *Hydrobiologia*, **192**, 35–57.

Santelices, B. (1991) Littoral and sublittoral communities of continental Chile, in *Intertidal and Littoral Ecosystems* (eds A.C. Mathieson and P.H. Nienhuis), Elsevier, Amsterdam, pp. 347–70.

Santelices, B. (1992) Digestion survival in seaweeds, an overview, in *Plant–Animal Interactions in the Marine Benthos*, (eds D.M. John, S.J. Hawkins and J.H. Price) (Systematics Association Special Volume No. 46), Clarendon Press, Oxford, pp. 363–84.

Santos, S.L. and Bloom, S.A. (1983) Evaluation of succession in an estuarine macrobenthic soft-bottom community near Tampa, Florida. *Int. Rev. ges. Hydrobiol.*, **68**, 617–32.

Savidge, G., Lennon, J. and Matthews, A.J. (1990) A shore-based survey of upwelling along the coast of Dhofar region, southern Oman. *Cont. Shelf Res.*, **10**, 259–75.

Savidge, G., Elliott, A.E. and Hubbard, L. (1992) The monsoon-induced upwelling system of the southern coast of Oman. *Mar. Geol.*, **104**, 290–91.

Savidge, W.B. and Taghon, G.L. (1988) Passive and active components of colonization following two types of disturbance on an intertidal sandflat. *J. exp. mar. Biol. Ecol.*, **115**, 137–55.

Scheer, B.T. (1945) The development of marine fouling communities. *Biol. Bull.,* **89**, 103–21.

Scheibling, R. (1986) Increased macroalgal abundance following mass mortalities of sea urchins (*Strongylocentrotus droebachieusis*) along the Atlantic coast of Nova Scotia. *Oecologia,* **48**, 184–98.

Schiel, D.R. and Foster, M. (1986) The structure of subtidal algal stands in temperate waters. *Oceanogr. mar. Biol. ann. Rev.,* **24**, 265–307.

Schiel, D.R. and Nelson, W.A. (1990) The harvesting of macroalgae in New Zealand, in *Proc. 13th Int. Seaweed Symp.* (eds S.C. Lindstrom and P.W. Gabrielson). *Hydrobiologia,* **58**, 25–34.

Schneider, D.C. (1978) Equalisation of prey numbers by migratory shorebirds. *Nature,* **271**, 371–2.

Schneider, D. (1985) Migratory shorebirds: resource depletion in the tropics? in *Neotropical Ornithology* (eds P.A. Buckley, M.S. Foster, E.S. Morton, R.S. Ridgely, and F.G. Buckley), Am. Ornithologist's Union, Washington, DC, pp. 520–58.

Schneider, D. and Gorewich, S. (1994) *Design and Analysis of Ecological Experiments,* Chapman & Hall, New York, 445 pp.

Schneider, S.H. and Rosenberg, N.J. (1989) The greenhouse effect: its causes, possible impacts and associated uncertainties, in *Greenhouse Warming: Abatement and Adaptation,* (eds N.J. Rosenberg, W.E. Easterling, P. Crosson and J. Darmstadter) Resources for the Future, Washington, DC, pp. 7–34.

Schonbeck, M.W. and Norton, T.A. (1978) Factors controlling the upper limits of fucoid algae. *J. exp. mar. Biol. Ecol.,* **31**, 303–13.

Schonbeck, M.W. and Norton, T.A. (1979a) Drought-hardening in the upper shore seaweeds *Fucus spiralis* and *Pelvetia canaliculata. J. Ecol.,* **67**, 687–99.

Schonbeck, M.W. and Norton, T.A. (1979b) An investigation of drought avoidance in intertidal fucoid algae. *Bot. Mar.,* **22**, 133–44.

Schonbeck, M.W. and Norton, T.A. (1979c) The effects of brief periodic submergence on intertidal fucoid algae. *J. exp. mar. Biol. Ecol.,* **8**, 205–11.

Schonbeck, M.W. and Norton, T.A. (1980) Factors controlling the lower limits of fucoid algae on the shore. *J. exp. mar. Biol. Ecol.,* **43**, 131–50.

Schramm, W. (1991) Chemical characteristics of marine littoral ecosystems, in, *Intertidal and Littoral Ecosystems* (eds. A.C. Mathieson and P.H. Nienhuis), Elsevier, Amsterdam, pp. 27–38.

Schwinghamer, P. (1981) Characteristic size distributions of integral benthic communities. *Can. J. Fish. aquat. Sci.,* **38**, 1255–63.

Scofield, N.B. and Bryant, H.C. (1926) The striped bass in California. *Calif. Fish Game,* **12**, 55–74.

Scott, A. (1960) The fauna of the sandy beach, Village Bay, St. Kilda. A dynamical relationship. *Oikos,* **11**, 153–60.

Sebens, K.P. and Lewis, J.R. (1985) Rare events and population structure of the barnacle *Semibalanus cariosus* (Pallas, 1788). *J. exp. mar. Biol. Ecol.,* **87**, 55–65.

Seed, R. (1978) Observations on the adaptive significance of shell shape and body form in dogwhelks (*Nucella lapillus* (L.)) from N. Wales. *Nature Wales,* **16**, 111–22.

Shanks, A.L. (1986) Tidal periodicity in the daily settlement of intertidal barnacle larvae and an hypothesised mechanism for the cross shelf transport of cyprids. *Biol. Bull.*, **170**, 429–40.

Shanks, A.L. and Wright, W.G. (1986) Adding teeth to wave action: the destructive effects of wave-borne rocks on intertidal organisms. *Oecologia*, **69**, 420–28.

Sharp, G. (1987) *Ascophyllum nodosum* and its harvesting in Eastern Canada, in *Case Studies of Seven Commercial Seaweed Resources* (eds M.S. Doty, J.F. Caddy and B. Santelices), FAO, Rome, *FAO Fish. tech. Pap. no. 281*, pp. 3–48.

Sharp, G.J. and Pringle, J.D. (1990) Ecological impact of marine plant harvesting in the northwest Atlantic, a review, in *Proc. 13th Int. Seaweed Symp.* (eds. S.C. Lindstrom and P.W. Gabrielson). *Hydrobiologia*, **58**, 17–24.

Sheenan, I. (1989) Holocene crustal movements and sea-level changes in Great Britain. *J. Quaternary Sci.*, **4**, 77–89.

Siefert, W. (1990) Sea-level changes and tidal flat characteristics, in *Expected Effects of Climate Change on Marine Coastal Ecosystems*, (eds J.J. Beukema, W.J. Wolff and J.J.W.M. Brouns), Kluwer, Dordrecht, pp. 105–12.

Slocum, C.J. (1980) Differential susceptibility to grazers in two phases of an intertidal alga: advantages of heteromorphic generations. *J. exp. mar. Biol. Ecol.*, **46**, 99–110.

Smayda, T.S. (1989) Global epidemic of noxious phytoplankton blooms and food chain consequences. 1990 AAAS Annual Meeting Abstracts (ed. M.D. Games) p. 71.

Smayda, T.S. (1990) Novel and nuisance phytoplankton blooms in the sea: evidence for a global epidemic. in *Toxic Marine Phytoplankton* (eds E. Graneli, B. Sundstroem, L. Edler and D.M. Anderson) pp. 29–40.

Smith, A.M. (1991) The role of suction in the adhesion of limpets. *J. exp. Biol.*, **161**, 151–69.

Smith, J.E. (ed.) (1968) 'Torrey Canyon' Pollution and Marine Life, Cambridge University Press, London, 196 pp.

Smith, R.F., Swartz, A.H. and Massman, W.H. (1966) A symposium on estuarine fisheries. *Trans. Am. Fish. Soc. Suppl.* **95**, 1–154.

Smith, T.J. III, Boto, K.G., Frusher, S.D. and Giddins, R.L. (1991) Keystone species and mangrove forest dynamics: the influence of burrowing crabs on soil nutrient status and forest productivity. *Estuar. coast. Shelf Sci.*, **33**, 419–32.

Snedaker, S.C. (1989) Overview of ecology of mangroves and information needs for Florida Bay. *Bull. mar. Sci.*, **44**, 341–7.

Snelgrove, P.V.R. (1994) Hydrodynamic enhancement of invertebrate larval settlement in microdepositional environments, Colonization tray experiments in a muddy habitat. *J. exp. mar. Biol. Ecol.*, **176**, 149–66.

Sousa, W.P. (1979) Disturbance in marine intertidal boulder fields: the non-equilibrium maintenance of species diversity. *Ecology*, **60**, 1225–39.

Sousa, W.P. (1991) Can models of soft-sediment community structure be complete without parasites? *Am. Zool.*, **31**, 821–30.

Sousa, W.P. and Connell, J.H. (1992) Grazing and succession in marine algae, in *Plant—Animal Interactions in the Marine Benthos* (eds D.M. John, S.J. Hawkins and J.H. Price), Clarendon Press, Oxford, pp. 425–41.

Southgate, T., Wilson, K., Cross, T.F. and Myers, A.A. (1984) Recolonization of a rocky shore in S.W. Ireland following a toxic bloom of the dinoflagellate *Gyrodinium aureolum*. *J. mar. biol. Ass. U.K.*, **64**, 485–92.

Southward, A.J. (1953) The ecology of some rocky shores in the south of the Isle of Man. *Proc. Trans. Liverpool Biol. Soc.*, **59**, 1–50.

Southward, A.J. (1956) The population balance between limpets and seaweeds on wave-beaten rocky shores. *Rep. mar. biol. Stn Port Erin*, **68**, 20–29.

Southward, A.J. (1959) The zonation of plants and animals on rocky sea shores. *Biol. Rev.*, **33**, 137–77.

Southward, A.J. (1964) . impet grazing and the control of vegetation on rocky shores, in *Grazing in Terrestrial and Marine Environments* (ed. D.J. Crisp), Blackwell Scientific, Oxford, pp. 265–73.

Southward, A.J. (1965) *Life on the seashore*, 153 pp. Heinemann, London

Southward, A.J. (1967) Recent changes in the abundance of intertidal barnacles in south-west England, a possible effect of climatic deterioration. *J. mar. biol. Ass. U.K.*, **47**, 81–95.

Southward, A.J. (1976) On the taxonomic status and distribution of *Chthamalus stellatus* (Cirripedia) in the north-east Atlantic region with a key to the common intertidal barnacles of Britain. *J. mar. biol. Ass. UK.*, **56**, 1007–28.

Southward, A.J. (1980) The Western English Channel – an inconstant ecosystem. *Nature*, **285**, 361–6.

Southward, A.J. (1991) Forty years of changes in species composition and population density of barnacles on a rocky shore near Plymouth. *J. mar. biol. Ass. U.K.*, **71**, 495–513.

Southward, A.J. and Crisp, D.J. (1954) Recent changes in the distribution of the intertidal barnacles *Chthamalus stellatus* (Poli) and *Balanus balanoides* (L.) in the British Isles. *J. Anim. Ecol.*, **23**, 163–77.

Southward, A.J. and Orton, J.H. (1954) The effects of wave action on the distribution and numbers of the commoner plants and animals living on the Plymouth breakwater. *J. mar. biol. Ass. U.K.*, **33**, 1–19.

Southward, A.J. and Southward, E.C. (1978) Recolonization of rocky shores in Cornwall after the use of toxic dispersants to clean up the *Torrey Canyon* oil spill. *J. Fish Res. Bd Canada*, **35**, 682–706.

Southward, A.J., Hawkins, S.J. and Burrows, M.T. (1995) Seventy years' observations of changes in distribution and abundance of zooplankton and intertidal organisms in the western English channel in relation to rising sea temperature. *J. therm. Biol.*, **20**, 127–55.

Spencer-Davies, P. (1992) Endosymbiosis in marine cnidarians, in *Plant–Animal Interactions in the Marine Benthos* (eds D. John, S.J. Hawkins and J. Price) (Systematics Association Special Volume No. 46), Clarendon Press, Oxford, pp. 511–40.

Stanley, N. (1987) Products and uses of carrageenans, in *Production and Utilization of Products from Commercial Seaweeds* (ed. D.J. McHugh), FAO, Rome, *FAO Fish. tech. Pap., no. 288*, pp. 116–46.

Steele, J.H. and Baird, I.E. (1968) Production ecology of a sandy beach. *Limnol. Oceanogr.*, **13**, 14–25.

Steinbeck, J. and Ricketts, E. (1941) *The Log from the Sea of Cortez*, Viking Press, New York.

Steneck, R.S. (1992) Plant–herbivore coevolution: a reappraisal from the marine realm and its fossil record, in *Plant–Animal Interactions in the Marine Benthos* (eds D.M. John, S.J. Hawkins and J.H. Price) (Systematics Association Special Volume No. 46), Clarendon Press, Oxford, pp. 477–91.

Steneck, R.S. and Warling, L. (1982) Feeding capabilities and limitation of herbivorous molluscs: a functional group approach. *Mar. Biol.*, **68**, 299–319.

Stenton-Dozey, J.M.E. and Griffiths, C.L. (1983) The fauna associated with kelp stranded on a sandy beach, in *Sandy Beaches as Ecosystems* (eds A. McLachlan and T. Erasmus), Dr W. Junk, The Hague, pp. 557–68.

Stephenson, T.A. (1936) The marine ecology of the South African coast, with special reference to the habits of limpets. *Proc. Linn. Soc. Lond.*, **148**, 74–9.

Stephenson, T.A. and Stephenson, A. (1949) The universal features of zonation between tidemarks on rocky coasts. *J. Ecol.*, **38**, 289–305.

Stephenson, T.A and Stephenson, A. (1972) *Life between Tidemarks on Rocky Shores*, W.H. Freeman, San Francisco, 425 pp.

Stevenson, J.P. (1992) A possible modification of the distribution of the intertidal seastar *Patiriella exigua* (Lamarck) (Echinodermata, Asteroidea) by *Patiriella calcar* (Lamarck). *J. exp. mar. Biol. Ecol.*, **155**, 41–54.

Strathmann, R.R., Branscomb, E.S. and Vedder, K. (1981) Fatal errors in set as a cost of dispersal and the influence of intertidal flora on set of barnacles. *Oecologia*, **48**, 13–18.

Suchanek, A. (1985) Mussels and their role in structuring rocky shore communities, in, *The Ecology of Rocky Coasts. Essays Presented to J.R. Lewis* (eds P.G. Moore and R. Seed), Hodder & Stoughton, London, pp. 70–96.

Suchanek, T.H. (1992) Extreme biodiversity in the marine environment: mussel bed communities of *Mytilus californianus*. *Northwest Environ. J.*, **8**, 150–52.

Sutherland, J.P. and Karlson, R.H. (1977) Development and stability of the fouling community at Beaufort, North Carolina. *Ecol. Monogr.*, **47**, 425–46.

Svendsen, P. (1959) The algal vegetation of Spitsbergen. *Norsk Polarinst. Skr.*, **116**, 1–47.

Swinbanks, D.D. (1982) Intertidal exposure zones, a way to subdivide the shore. *J. exp. mar. Biol. Ecol.*, **62**, 69–86.

Takada, Y. and Kukuchi, T. (1990) Mobile molluscan communities in boulder shores and the comparison with other intertidal habitats in Amakusa. *Publ. Amakusa mar. biol. Lab.*, **10**, 145–68.

Takeda, S. and Kurihara, Y. (1987) The effects of burrowing of *Helice tridens* (De Haan) on the soil of a salt-marsh habitat. *J. exp. mar. Biol. Ecol.*, **113**, 79–89.

Talbot, M.M.B., Bate, G.C. and Campbell, E. (1990) A review of the ecology of surf-zone diatoms, with special reference to *Anaulus australis*. *Oceanogr. mar. Biol. ann. Rev.*, **28**, 155–75.

Tamaki, A. (1987) Comparison of resistivity to transport by wave action in several polychaete species on an intertidal sand flat. *Mar. Ecol. Progr. Ser.*, **37**, 181–9.

Tamaki, A. and Suzukawa, K. (1991) Co-occurrence of the cirolanid isopod *Eurydice nipponica* (Bruce and Jones) and the ghost shrimp *Callianassa japonica* (Ortmann) on an intertidal sand flat. *Ecol. Res.*, **6**, 87–100.

Taylor, J.D. (1971) Intertidal zonation at Aldabra Atoll. *Phil. Trans. R. Soc. Lond.*, **260B**, 173–213.

Teal, J.M. (1962) Energy flow in the saltmarsh ecosystem of Georgia. *Ecology*, **43**, 614–24.

Tegner, M.J. (1989) The California abalone fishery: production, ecological interactions, and prospects for the future, in, *Marine Invertebrate Fisheries, Their Assessment And Management* (ed. J.F. Caddy), Wiley, New York, pp. 401–20.

Thayer, G.W. (ed.) (1992) *Restoring the Nation's Marine Environment*, Maryland Sea Grant, College Park, MD, 716 pp.

Thayer, G.W., Adams, S.M. and LaCroix, M.W. (1975) Structural and functional aspects of a recently established *Zostera marina* community. *Estuar. Res.*, **1**, 518–40.

Thayer, G.W., Bjorndal, K.A., Ogden, J.C., Williams, S.L. and Zieman, J.C. (1984) Role of larger herbivores in seagrass communities. *Estuaries*, **7**, 351–76.

Thomas, M.L.H. (1985) Littoral community structure and zonation on the rocky shores of Bermuda. *Bull. mar. Sci.*, **37**, 857–70.

Thomas, M.L.H. (1986) A physically derived exposure index for marine shorelines. *Ophelia*, **25**, 1–13.

Thomas, M.L.H. and Page, F.H. (1983) Grazing by the gastropod, *Lacuna vincta*, in the lower intertidal area at Musquash Head, New Brunswick, Canada. *J. mar. biol. Ass. U.K.*, **63**, 725–36.

Thompson, J.M. (1952) The acclimatization and growth of the Pacific oyster (*Gryphaea gigas*) in Australia. *Aust. J. mar. freshw. Res.*, **3**, 64–73.

Thorhaug, A., Roessler, M.A., Bach, S.D., Hixon, R., Brook, I.M. and Josselyn, M.N. (1979) Biological effects of power-plant thermal effluents in Card Sound, Florida. *Env. Conserv.*, **6**, 127–37.

Thrush, S.F. (1986) Spatial heterogeneity in subtidal gravel generated by the pit-digging activities of *Cancer pagurus*. *Mar. Ecol. Progr. Ser.*, **30**, 221–7.

Thrush, S.F., Pridmore, R.D., Hewitt, J.E. and Cummings, V.J. (1991) Impact of ray feeding disturbances on sandflat macrobenthos: do communities dominated by polychaetes or shellfish respond differently? *Mar. Ecol. Progr. Ser.*, **69**, 245–52.

Thrush, S.F., Pridmore, R.D. and Hewitt, J.E. (1994) Impacts on soft-sediment macrofauna: the effects of spatial variation on temporal trends. *Ecol. Appl.*, **4**, 31–41.

Todd, C.D. and Havenhand, J.N. (1989) Nudibranch–bryozoan associations, the quantification of ingestion and some observations on partial predation among Doridoidea. *J. moll. Stud.*, **55**, 245–59.

Todd, C.D. and Lewis, J.R. (1984) Effects of low air temperature on *Laminaria digitata* L. in Southwestern Scotland. *Mar. Ecol. Progr. Ser.*, **16**, 199–201.

Trudgill, S. (1988) Integrated geomorphological and ecological studies on rocky shores in Southern Britain. *Field Studies*, **7**, 239–79.

Trueman, E.R. (1975) *The Locomotion of Soft-bodied Animals*, Arnold, London, 200 pp.

Trueman, E.R. (1983) Locomotion in molluscs, in *The Mollusca*, Vol. 4 (eds A.S.M. Saleuddin and K.M. Wilbur), Academic Press, New York, pp. 155–98.

Tseng, C.K. (1987) *Laminaria* mariculture in China, in, *Case Studies of Seven Commercial Seaweed Resources*, (eds M.S. Doty, J.F. Caddy and B. Santelices), FAO, Rome, *FAO Fish. tech. Paper*, no. 281, pp. 239–64.

Turner, S.J. and Todd, C.D. (1993) The early development of epifaunal assemblages on artificial substrata at two intertidal sites on an exposed rocky shore in St. Andrews Bay, N.E. Scotland. *J. exp. mar. Biol. Ecol.*, **166**, 251–72.

Turner, T.A. (1983) Facilitation as a successional mechanism in a rocky intertidal community. *Am. Nat.*, **121**, 729–38.

Turner, T.A. (1985) Stability of rocky intertidal surfgrass beds: persistence, preemption and recovery. *Ecology*, **66**, 83–92.

Ugolini, A., Scapini, F. and Pardi, L. (1986) Interaction between solar orientation and landscape visibility in *Talitrus saltator* (Crustacea, Amphipoda). *Mar. Biol.*, **90**, 449—60.

Underwood, A.J. (1975) Comparative studies on the biology of *Nerita atramentosa* Reeve, *Bembicium nancum* (Lamarck) and *Cellana tramoserica* (Sowerby) (Gastropoda: Prosobranchia) in S.E. Australia. *J. exp. mar. Biol. Ecol.*, **18**, 153–72.

Underwood, A.J. (1976a) Food competition between age-classes in the intertidal neritacean *Nerita atramentosa* Reeve (Gastropoda: Prosobranchia). *J. exp. mar. Biol. Ecol.*, **23**, 145–54.

Underwood, A.J. (1976b) Nearest neighbour analysis of spatial dispersion of intertidal prosobranch gastropods within two substrata. *Oecologia*, **26**, 257–66.

Underwood, A.J. (1978) A refutation of critical tidal levels as determinants of the structure of intertidal communities on British shores. *J. exp. mar. Biol. Ecol.*, **33**, 261–76.

Underwood, A.J. (1979) The ecology of intertidal gastropods. *Oceanogr. mar. Biol. ann. Rev.*, **16**, 111–210.

Underwood, A.J. (1980) The effects of grazing by gastropods and physical factors on the upper limits of distribution of intertidal macroalgae. *Oecologia*, **46**, 201–13.

Underwood, A.J. (1981) Techniques of analysis of variance in experimental marine biology and ecology. *Oceanogr. mar. Biol. ann. Rev.*, **19**, 513–605.

Underwood, A.J. (1983) Spatial and temporal problems in the design of experiments with marine grazers, in *Proceedings of the Inaugural Great Barrier Reef Conference* (eds J.T. Baker, R.M. Carter, P.W. Sammarco and K.P. Stark), James Cook University Press, Townsville, pp. 251–6.

Underwood, A.J. (1986) The analysis of competition by field experiments, in *Community Ecology, Patterns and Process* (eds J. Kikkawa and D.J. Anderson), Blackwell Scientific, Oxford, pp 240–68.

Underwood, A.J. (1988) Design and analysis of field experiments on competitive interactions affecting behaviour of intertidal animals, in *Behavioural Adaptation to Intertidal Life* (eds G. Chelazzi and M. Vannini), Plenum Press, New York, pp. 333–58.

Underwood, A.J. (1992) Competition and marine plant–animal interactions, in *Plant–Animal Interactions in the Marine Benthos* (eds D. John, S.J. Hawkins and J. Price) (Systematics Association Special Volume No. 46), Clarendon Press, Oxford, pp. 443–75.

Underwood, A.J. and Denley, E.J. (1984) Paradigms, explanations and generalizations in models for the structure of intertidal communities on rocky shores, in *Ecological Communities, Conceptual Issues and the Evidence* (eds

D.R. Strong, D. Simberloff, L.G. Abele and A.B. Thistle), Princeton University Press, Princeton, NH, pp. 151–80.

Underwood, A.J. and Fairweather, P.G. (1986) Intertidal communities: do they have different ecologies or different ecologists? *Proc. Ecol. Soc. Aust.*, **14**, 7–16.

Underwood, A.J. and Fairweather, P.G. (1989) Supply-side ecology and benthic marine assemblages. *Trends Ecol. Evol.*, **4**, 16–20.

Underwood, A.J. and Jernakoff, P. (1981) Effects of interactions between algae and grazing gastropods on the structure of a low intertidal algal community. *Oecologia*, **48**, 221–33.

Underwood, A.J. and Jernakoff, P. (1984) Effects of tidal height, wave exposure, seasonality and rock-pools on grazing and the distribution of intertidal macroalgae in New South Wales. *J. exp. mar. Biol. Ecol.*, **75**, 71–96.

Underwood, A.J., Denley, E.J. and Moran, M.J. (1983) Experimental analyses of the structure and dynamics of mid-shore rocky intertidal communities in New South Wales. *Oecologia*, **56**, 202–19.

Underwood, R.L. (1991) The logic of ecological experiments: a case history from studies of the distribution of macro-algae on rocky intertidal shores. *J. mar. biol. Ass. U.K.*, **71**, 841–66.

Vadas, R.L. (1977) Preferential feeding: an optimization strategy in sea urchins. *Ecol. Monogr.*, **47**, 377–81.

Vadas, R.L., Sr (1990) Littorinid grazing and algal patch dynamics, in *Proc. Third Int. Symp. Littorinid Biology, Dale Fort Field Centre, Wales* (eds J. Grahame, P.J. Mill and D.G. Reid), Malacological Society of London, London, pp. 197–210.

Vadas, R.L. and Elner, R.W. (1992) Plant–animal interactions in the north-west Atlantic, in *Plant–animal interactions in the Marine Benthos* (eds D.M. John, S.J. Hawkins and J.H. Price), Clarendon Press, Oxford, pp. 33–60.

Vadas, R.L. and Steneck, R.S. (1988) Zonation of deep water benthic algae in the Gulf of Maine. *J. Phycol.*, **24**, 338–46.

Vadas, R.L., Keser, M., Rusanowski, P.C. (1976) Influence of thermal loading on the ecology of intertidal algae, in *Thermal Ecology II* (eds G.W. Esch and R.W. McFarlane) ERDA Symposium Series (Conf–750425, NTIS), Augusta, GA, pp. 202–12.

Vadas, R.L., Wright, W.A. and Miller, St. L. (1990) Recruitment of *Ascophyllum nodosum*: wave action as a source of mortality. *Mar. Ecol Progr. Ser.*, **61**, 263–72.

Vaillant, L. (1891) Zones littorales. *Ann. Sci. Nat. Zool.*, **7**, 39–50.

Van Blaricom G.R. (1978) Disturbance, predation, and resource allocation in a high-energy sublittoral sand-bottom ecosystem: experimental analysis of critical structuring processes for the infaunal community, PhD thesis, University of California, San Diego, CA, 328 pp.

Van Blaricom G.R. (1982) Experimental analyses of structural regulation in a marine sand community exposed to ocean swell. *Ecol. Monogr.*, **52**, 283–305.

Van Impe, J. (1985) Estuarine pollution as a probable cause of increase in estuarine birds. *Mar. Poll. Bull.*, **16**, 271–6.

Velimirov, B., Field, J.G., Griffiths, C.L. and Zoutendyk, P. (1977) The ecology of kelp bed communities in the Benguela upwelling system. *Helgöländ wiss. Meeresunters*, **30**, 495–518.

Vermeij, G.J. (1978) *Biogeography and Adaptation: Patterns of Marine Life*, Harvard University Press, Cambridge, MA, 332 pp.

Virnstein, R.W. (1978) Predator caging experiments in soft sediments: caution advised, in *Estuarine Interactions* (ed M.L. Wiley), Academic Press, New York, pp. 261–73.

Wafar, M.V.M. (1990) Global warming and coral reefs, in, *Sea Level Variation and its Impact on Coastal Environment* (ed. C.V. Rajamanickam), pp. 411–37.

Wahl, M. (1989) Marine epibiosis. I. Fouling and antifouling, some basic aspects. *Mar. Ecol. Progr. Ser.*, **58**, 175–89.

Walker, M.I., Burrows, E.M. and Lodge, S.W. (1954) Occurrence of *Falkenbergia rufolanosa* in the Isle of Man. *Nature*, **174**, 315.

Wallentinus, I. (1991) The Baltic Sea Gradient, in *Intertidal and Littoral Ecosystems* (eds A.C. Mathieson and P.H. Nienhuis), Elsevier, Amsterdam, pp. 83–108.

Wallin, M. and Haakenson, L. (1991) Nutrient loading models for estimating the environmental effects of marine fish farms, in *Marine Aquaculture and Environment* (ed. T. Maekinen) pp. 39–55.

Walne, P.R. (1956) The biology and distribution of the slipper limpet *Crepidula fornicata* in Essex rivers with notes on the distribution of the larger epibenthic invertebrates. Ministry of Agriculture, Fisheries and Food, *Fishery Investigations Ser. II*, **260**(6), 1–50.

Walton, C.L. (1915) The distribution of some littoral Trochidae and Littorinidae in Cardigan Bay. *J. mar. biol. Ass. U.K.*, **10**, 114–22.

Warren, J.H. and Underwood, A.J. (1986) Effects of burrowing crabs on the topography of mangrove swamps in New South Wales. *J. exp. mar. Biol. Ecol.*, **102**, 2–3.

Warwick, R.M. (1984) Species size distributions in marine benthic communities. *Oecologia*, **61**, 32–41.

Warwick, R.M. (1989) The role of meiofauna in the marine ecosystem: evolutionary considerations. *Zool. J. Linn. Soc.*, **96**, 229–41.

Warwick, R.M., Joint, I.R. and Radford, P.J. (1975) Secondary production of the benthos in an estuarine mud-flat, in *Ecological Processes in Coastal Environments* (eds R.L. Jeffries and A.J. Davy), Blackwell Scientific Publications, Oxford, UK, pp. 429–50.

Warwick, R.M. and Price, J.R. (1979) Ecological and metabolic studies on free–living nematodes from an estuarine mud-flat. *Estuar. coast. mar. Sci.*, **9**, 257–71.

Warwick, R.M., George, C.L. and Davies, J.R. (1978) Annual macrofauna production in a *Venus* community (Bristol Channel/Severn Estuary). *Estuar. coast. mar. Sci.*, **7**, 215–42.

Warwick, R.M., Clarke, K.R. and Gee, J.M. (1990a) The effect of disturbance by soldier crabs *Mictyris platycheles* H. Milne Edwards on meiobenthic community structure. *J. exp. mar. Biol. Ecol.*, **135**, 19–33.

Warwick, R.M., Platt, H.M., Clarke, K.R., Agard, J. and Gobin, J. (1990b) Analysis of macrobenthic and meiobenthic community structure in relation to pollution and disturbance in Hamilton Harbour, Bermuda. *J. exp. mar. Biol. Ecol.*, **138**, 1–2.

Watling, L. (1991) The sedimentary milieu and its consequences for resident organisms. *Am. Zool.*, **31**, 789–96.

Watson, D.C. and Norton, T.A. (1985) The physical characteristics of seaweed thalli as deterrents to littorine grazers. *Bot. Mar.*, **28**, 383–7.

Watzin, M.C. (1983) The effects of meiofauna on settling macrofauna: meiofauna may structure macrofaunal communities. *Oecologia*, **59**, 163–6.

Webb, J.E. (1991) Hydrodynamics, organisms and pollution of coastal sands. *Ocean Shoreline Manage.*, **6**, 23–51.

Wells, S.M. (1981) International trade in Ornamental shells. *Proc. 4th Int. Coral Reef Symp., Manila*, **I**, 323–30.

Wells, S.M. and Alcala, A.C. (1987) Collecting corals and shells, In *Human Impacts on Coral Reefs, Facts, Recommendations* (ed. B. Salvat). *Antenne Museum EPHE, French Polynesia*, pp. 13–27.

Wells, S.M. and Edwards, A. (1989) Gone with the waves. *New Scient.*, **11**, 51.

Whittaker, R.H. (1974) *Communities and Ecosystems*, Macmillan, New York.

Whitton, B.A. (1975) *River Ecology*, Blackwell Scientific, Oxford, 725 pp.

Williams, G.A. (1990) *Littorina mariae* – a factor structuring low shore communities? *Hydrobiologia*, **193**, 139–46.

Williams, G.A. (1993) Seasonal variation in algal species richness and abundance in the presence of molluscan herbivores on a tropical rocky shore. *J. exp. mar. Biol. Ecol.*, **167**, 261–75.

Williams, G.A. and Seed, R. (1992) Interactions between macrofaunal epiphytes and their host algae, in *Plant–Animal Interactions in the Marine Benthos* (eds D. John, S.J. Hawkins and J. Price) (Systematics Association Special Volume 46), Clarendon Press, Oxford, pp.189–212.

Williams, J.A. (1983) The endogenous locomotor activity rhythm of four supralittoral peracarid crustaceans. *J. mar. biol. Ass. U.K.*, **36**, 481–92.

Wilson, D.P. (1937) The influence of the substratum on the metamorphosis of *Notomastus* larvae. *J. mar. biol. Ass. U.K.*, **22**, 227–43.

Wilson, D.P. (1954) The attractive factor in the settlement of *Ophelia bicornis* Savigny. *J. mar. biol. Ass., U.K.*, **33**, 361–80.

Wilson, D.P. (1971) *Sabellaria* colonies at Duckpool, North Cornwall, 1961–1970. *J. mar. biol. Ass. U.K.*, **51**, 509–80.

Wilson, W.H. (1983) The role of density dependence in a marine infaunal community. *Ecology*, **64**, 295–306.

Wilson W.H. (1986) Importance of predatory infauna in marine soft-sediment communities. *Mar. Ecol. Prog. Ser.*, **32**, 35–40.

Wilson W.H. (1991) Competition and predation in marine soft sediment communities. *A. Rev. Ecol. Syst.*, **21**, 221–41.

Withers, P.G., Farnham, W.F., Lewey, S., Jephson, N.A., Haythorn, J.M. and Gray, P.W.G. (1975) The epibionts of *Sargassum muticum* in British Waters. *Mar. Biol.*, **31**, 79–86.

Witman, J.D. (1987) Subtidal co-existence, storms, grazing, mutualism and the zonation of kelps and mussels. *Ecol. Monogr.*, **57**, 167–87.

Wolcott, F.G. (1973) Physiological ecology and intertidal zonation in limpets: a critical look at 'limiting factors'. *Biol. Bull.*, **145**, 389–422.

Wolff, W.J. (1977) A benthic food budget for the Grevelingen Estuary, the Netherlands, and a consideration of the mechanisms causing high benthic secondary production in estuaries, in *Ecology of Marine Benthos* (ed. B.C. Coull), University of South Carolina Press, Columbia, SC, pp. 267–80.

Wolff, W.J., Sandee, A.J.J. and De Wolf, L. (1977) The development of a benthic ecosystem. *Hydrobiologia*, **52**, 107–15.

Wood, E. and Wells, S.M. (1988) *The Marine Curio Trade, Conservation Issues.* A report for the Marine Conservation Society. Ross-on-Wye, Herefordshire, UK.

Woodin, S.A. (1974) Polychaete abundance patterns in a marine soft-sediment environment. The importance of biological interactions. *Ecol. Monogr.*, **44**, 171–87.

Woodin, S.A. (1985) Effects of defecation by arenicolid polychaete adults on spionid polychaete juveniles in field experiments. Selective settlement or differential mortality? *J. exp. mar. Biol. Ecol.*, **87**, 119–32.

Woodin, S.A. (1991) Recruitment of infauna: positive or negative cues? *Am. Zool.*, **31**, 797–807.

Woodroffe, C.D. (1985) Studies of a mangrove basin, Tuff Crater, New Zealand. III. The flux of organic and inorganic particulate matter. *Estuar. coast. Shelf Sci.*, **20**, 447–62.

Workman, C. (1983) Comparisons of energy partitioning in contrasting age-structured populations of the limpet *Patella vulgata* L. *J. exp. mar. Biol. Ecol.*, **68**, 81–103.

Wright, J.R. and Hartnoll, R.G. (1981) An energy budget for a population of the limpet *Patella vulgata. J. mar. biol. Ass. U.K.*, **61**, 627–46.

Xue, Q. (1992) Study on population dynamics on rocky shore in Qingdao, China. *Oceanol. Limnol. Sin. Haiyang Yu Huzhao*, **23**, 438–44.

Yamada, I. (1980) Benthic marine algal vegetation along the coasts of Hokkaido, with special reference to the vertical distribution. *J. Fac. Sci. Hokkaido Univ. Ser. 5 (Botany)*, **12**, 11–98.

Yonge, C.M. (1949) *The Sea Shore*, The New Naturalist, Collins, London.

Zajac, R.N. and Whitlatch, R.B. (1985) A hierarchical approach to modelling soft-bottom successional dynamics, in *Proc. 19th European Marine Biology Symp.* (ed. P.E. Gibbs), Cambridge University Press, Cambridge, pp. 265–76.

Species index

Page numbers in **bold** refer to figures, those in *italics* refer to tables.

SCIENTIFIC NAMES

Subject index

Page numbers in **bold** refer to figures, those in *italics* refer to tables.